In the Variorum Collected Studies Series

DAVID OLDROYD
Science of the Earth: Studies in the History of Mineralogy and Geology

DAVID M. KNIGHT
Science in the Romantic Era

ANDREW WEAR
Health and Healing in Early Modern England: Studies in Social and Intellectual History

DANIELLE JACQUART
La science médicale occidentale entre deux renaissances (XIIe s.–XVe s.)

FRANZ ROSENTHAL
Science and Medicine in Islam: A Collection of Essays

JOHN HEDLEY BROOKE
Thinking About Matter: Studies in the History of Chemical Philosophy

A. RUPERT HALL
Science and Society: Historical Essays on the Relations of Science, Technology and Medicine

VARIORUM COLLECTED STUDIES SERIES

Pristina Medicamenta

Professor Jerry Stannard

Jerry Stannard

Pristina Medicamenta

Ancient and Medieval Medical Botany

edited by
Katherine E. Stannard
and Richard Kay

Routledge
Taylor & Francis Group
LONDON AND NEW YORK

First published 1999 in the Variorum Collected Studies Series by Ashgate Publishing

Published 2017 by Routledge
2 Park Square, Milton Park, Abingdon, Oxon OX14 4RN
52 Vanderbilt Avenue, New York, NY 10017

Routledge is an imprint of the Taylor & Francis Group, an informa business

British Library CIP Data
Stannard, Jerry
Pristina Medicamenta: Ancient and Medieval Botany
(Variorum Collected Studies Series: CS646)
1. Botany, Medical – History. 2. Materia Medica, Vegetable – History.
3. Medicine – Early Works to 1800.
I. Title
580.9

US Library of Congress CIP Data
The Library of Congress Catalog Card Number has been preassigned as:
98–74833

ISBN 13: 978-0-86078-773-0 (hbk)

VARIORUM COLLECTED STUDIES SERIES CS646

CONTENTS

This volume contains xxii + 324 pages

Publisher's Note: The articles in this volume, as in all others in the Variorum Collected Studies Series, have not been given a new, continuous pagination. In order to avoid confusion, and to facilitate their use where these same studies have been referred to elsewhere, the original pagination has been maintained wherever possible.

Each article has been given a Roman number in order of appearance, as listed in the Contents. This number is repeated on each page and quoted in the index entries.

ACKNOWLEDGEMENTS

Grateful acknowledgement is made to the following persons, journals, institutions, and publishers for their permission to reproduce the articles included in this volume: the Johns Hopkins University Press, Baltimore (I and XII); the University of Chicago Press (II and XVIII); Leo S. Olschki, Publisher, Florence (III); Gregory J. Higby, the American Institute of the History of Pharmacy, Madison (VI); Ossolineum, Warsaw (VII); Biblioteca del Seminario Vescovile, Padua (illustrations in VIII); the Dumbarton Oaks Center for Byzantine Studies, Washington, DC (IX); the Oxford University Press, Oxford (X), the Pontifical Institute of Mediaeval Studies, Toronto (XIII), Matthiàs-Grünewald-Verlag, Mainz, and P. Manuel Merten OP, on behalf of the Dominikaner-Provinz Teutonia, Köln (XIV); Dr. Cecilia Prete on behalf of *Res Publica Litterarum* (XV); Librairie Scientifique et Technique Albert Blanchard, Paris (XVI).

INTRODUCTION

In over thirty-four years of research and publications, Jerry Stannard pioneered a new approach to the study of the past. Before Stannard, the herbal and medical lore found in ancient, medieval and early documents was regarded as 'fact literature' beyond the scholar's obligation to study. Moreover, the specialized technical vocabularies in Greek, Latin, and numerous medieval vernaculars made the enterprise so formidable a task that most modern scholars were disinclined to invest their time. How could quaint but useless, superstitious information about medical and pharmaceutical nonsense inform us when our understanding of chemistry, pharmacokinetics, physiology, molecular modeling, and receptor sites is so superior to what the ancients knew or could have known? Instead, historians followed the past in terms of great people causing great events, or, in medical history, overarching theory, anatomy, and physiology.

Stannard showed us why and how to study pre-modern texts about plants. In doing so, he caused us to reflect upon our boasts of our scientific knowledge. The ancient Greek was as capable of good thinking about medicine as about philosophy or religion, and the ancient Roman equally capable. Stannard taught us to understand that the underestimated medieval person had knowledge about plants and medicine that we, who know mostly our garden vegetables and flowers, no longer learn. After all, ancient and medieval peoples' lives depended on their knowing the plants and what each was capable of doing for humans. This is a point explicitly made and driven home by the articles in this volume and its companion *Herbs and Herbalism in the Middle Ages and Renaissance*.

A few biographical facts shed light upon how Stannard's scholarship was unique. First, there was the love of languages and rigorous training in classical history and philosophy. His earliest publications were in the history of classical philosophy, in which he was as knowledgeable about Socrates as Hippocrates. Second, there was Stannard's military service as a pharmacist's mate in the US Navy. Third, he enjoyed plants. His backyard in Lawrence, Kansas, would have made most medieval monastic gardeners envious and even incredulous as to how he coaxed Mediterranean herbs to grow in America's heartland. Fourth, Stannard cultivated and enjoyed a close professional relationship with librarians. Almost all his articles – and I suspect his speeches – express appreciation to those librarians who helped him track down the rare articles and texts seldom found in the best research libraries. Once he found a lead or reference, he was indefatigable in finding that information and learning from it, regardless of what was the language of its text. Then, he placed it in a footnote very carefully so that future scholars could be certain of a correct citation. Finally, his soul was filled with a combined joy of discovery and a high regard for precise detail. Little wonder, then, that he developed scholarship in a field where others feared to tread.

Because of the diversity of the subject matter, Stannard's writings were published in journals devoted to widely diverse fields, ranging from medical to philological to philosophical. His publications number more than one hundred. Together with his companion, namely his wife Katherine, Stannard habitually attended international congresses where he delivered his findings. Unfortunately, the congresses' published proceedings did not find their way into many libraries. No one today can keep apprised of the myriad fields with which Stannard's articles dealt. Consequently few people are aware of the breadth and depth in his scholarly sleuthing about matters passed over by others. For these reasons, this volume and the one to follow form a union, or in the classical word, a Συνουσίωσις whereby the reader can read articles never seen even by those who now follow the field that Stannard developed. The representative articles on *pristina medicamenta* are organized topically and chronologically. Volume One covers the scope of ancient and medieval pharmacy. Volume Two, which will appear soon after the present volume, will comprise the special areas of medieval herbals and late medieval recipe literature. This second volume will conclude with three delightful studies devoted to three medicinal herbs: the plant called *moly* squill, and dill.

Hippocratic pharmacology introduces this volume not only chronologically by beginning with Homer's knowledge of drugs but also by setting the context of ancient lore within the context of modern science. Stannard observes that the early Hippocratic physicians who followed Homer did not know the modern divisions of pharmacognosy, pharmacodynamics, and pharmacotherapy as such – even though the names are Greek-derived. Exploring ancient Greek medical writings, however, enabled Stannard to demonstrate that the substance of each of these divisions was available for astute observations by ancient physicians. Often medical theory determined the choice of therapy. Medicine was more than an empirical knowledge of things that had a beneficial effect and others that were inert or harmful.

There follows a series of articles on individual medical writers of the Roman period: Pliny, Lucian, Aretaeus, Marcellus of Bordeaux, and the much-maligned Pliny – 'gullible and uncritical' – as reevaluated by showing him to have critical abilities previously unappreciated. Here is where Stannard's knowledge of plants provides the means to reexamine Pliny's botanical descriptions, in which he went beyond the sources he was accused of following uncritically. Rightly, Stannard regards Theophrastus as the Father of Botany, but he produces a case for Pliny as the Father of Botanical History.

The second study on Pliny discloses how he made intelligent use of folk observations, perhaps pejoratively called 'folklore'. Any negative connotation to 'folklore' is modern, because ancient medieval peoples regarded 'folklore' as part of knowledge. A key to understanding how Pliny's distinctly Roman view derived from Roman folk wisdom is the concept that each plant has a function 'useful' to humans. The intelligent person discovers that use, regardless of whether that discovery derives from the philosopher or from the country bumpkin. Accordingly, Pliny noted how more than one thousand plants, which he specifically discussed, functioned as foods, medicines, and condiments.

The same philosophical guide that Stannard saw in Pliny is culled from the writings of a literary figure, Lucian. The poet's use of natural-history metaphors and proverbs indicates a thorough knowledge of nature in such expressions as 'mute as a fish' 'harder to grasp than sea urchins', and 'senseless as grasshoppers'. Lucian was alert to any information that had potential value – just as was Pliny – from wherever it came regardless of its degree of credibility. Moving from a *literatus* to a *medicus*, Stannard turns to one of the more enigmatic of the ancient physicians, Aretaeus, whose extensive writings left him admired but virtually unknown biographically. Stannard accepts that he had a kinship with the pneumatist school, although even this is uncertain. The extensive writings done in the early Roman Empire disclose systematic therapeutic addresses: bleeding, purging, dietary regimen, and drugs, each receiving a full measure of examination. Stannard reveals Aretaeus' insights into both dietary and pharmaceutical therapy. Importantly, Stannard sees him administering specific drugs for specific diseases – an uncommon phenomenon in a largely symptomatic therapy – a characteristic Stannard associates with the pneumatic school of medicine.

The last four studies in this section concern the transition from ancient to medieval pharmacy and to Byzantine medicine. Marcellus of Bordeaux was a late Roman writer on medicines, although he was, in Stannard's delicate phrase, 'innocent of physiological theory'. Marcellus combined a learned knowledge of the pharmacy literature with a respect for and dedication to folk knowledge of drugs. In accord with the spirit of his times, he placed more emphasis on what we call the magical elements, such as *incantamenta*, and polypharmacy (compound drugs). Looking back over antiquity and forward to the Middle Ages, Stannard studied the lost botanical writings of antiquity, such as Aristotle's lost (?) treatise on plants, Mithridates' works, and numerous other writers. Among the anonymous, lost works were the numerous dictionaries that were then, as now, so vital in the study of plants.

In two studies on Byzantine medicine, Stannard examines a millennium of medicine that has, surprisingly, received little scholarly attention. In a broad introduction to Byzantine *materia medica* he shows that, even though the Byzantines based their pharmaceutical knowledge on ancient Greek texts, they deceptively (from our perspective) either concealed their contributions in commentaries on classical writers or even published under false names such as Hippocrates and Galen. The centuries of medical experience, however, brought new words and new substances into the Greek vocabulary and a particular fascination with *lexica* or dictionaries to guide medical authors and individual practitioners. These guides contain errors that could have been disastrous to patients, but Stannard concludes that, given our removal in time, it would 'be uncharitable to catalogue' those errors. Instead, he accentuates the positive by pointing to an on-going corpus of medical knowledge adapting to changes. The 'Byzantine physician had at his disposal one or more means of treating nearly every complaint with which he was faced', Stannard concludes.

The second study on Byzantine pharmacy concentrates on the lexicography literature. These guides to medicinal herbs (including the entire plant kingdom) had practical purposes: 1) listing plants with ailments that they relieved; 2) providing descriptive data to identify the plant and prepare it as a medicine. Stannard employs various examples in an in-depth search for the insights provided by this genre of medical writings. As always, he worked from the primary sources in their original language. Few scholars before him have had the skill or energy to undertake this type of search because of the daunting abundance of texts, obscurely printed. Not limited to classical and Byzantine Greek, Stannard studied the same genre in Latin literature in the concluding essays in this volume.

Benedictus Crispus wrote a medical poem in the early eighth century that 'deserves' recognition in challenging the old view of a Dark Age in medical and science history. Significantly, Benedictus' knowledge was devoid of Greek sources except as to how their knowledge filtered through the Latin authors whom he consulted. Acting contrary to his sources and, more importantly, contrary to modern generalizations about his time period, Benedictus excluded amulets, charms, and other magical elements. Stannard provides a translation that adheres to medical sense rather than merely duplicating meter and recapturing 'the artless didacticism of the original'.

Italian and German medical botany receive Stannard's scrutiny with some expected and unexpected observations. Medieval Italian botany was similar to other medieval medical literature in its respect for authority, but with notable innovations. By Platearius' time in the thirteenth century, medical botany no longer discussed the identification of the plants themselves – that knowledge was assumed – but instead concerned nomenclature and medicinal uses. The rich genre of regimens of health reflects an intensively enlarged corpus of *scientia* regarding alimentary uses of plants, roughly but not the same as our nutrition. Finally, cumulative knowledge about plants caused medieval authors, many of whom are anonymous, to introduce personal observations, some made by apothecaries who knew drugs well, not to mention the tricks of the trade. German medical botany took some distinctive turns. First, the classical corpus was based largely on Mediterranean flora; indigenous plants were different in colder, wetter climes. Practical collections of recipes reflect climatological and other adaptations, and they became one of the most popular medieval genres. Earlier German guides were practical to the extent that neither pharmacological nor physiological theory was necessary for knowing that such-and-such was good for this-and-that. The next stage occurred when the impact of Salernitan and Arabic medical theory reached the north. Still the *Arzneibücher* (pharmaceutical guides) and *Kräuterbücher* (herbals) became some of the most enduring medical writings still being available down to the modern day in German, English, French and Spanish bookstores. The German guides influenced the other languages during the early modern period.

The next three articles move – but not as much as it first appears – from practical knowledge to the theoretical approach of one of medieval Europe's greatest minds, Albertus Magnus, *doctor universalis*. Stannard sees Albertus' *De vegetabilibus* in seven books as a traditional herbal written by one whom had a 'profound knowledge' of plants. Surprisingly, Albertus synthesizes traditional lore with scholarship by bringing to bear the great writers of the classical, Islamic, and medieval periods but integrating authorities with his own astute, personal, and practical observations. The complexity of Albert is seen in how he could, in his theoretical works, denounce abortion as wrong and, in this herbal, convey abortion recipes.

The second study establishes why Albertus' *De vegetabilibus* deserved but did not receive recognition as the greatest contribution to botany since Theophrastus' works in the third century BC. Albertus' botanical theory was delivered with a distinctive Aristotelian structure, and, at the same time, he continued practical botany in a mixed, complex, but brilliant assemblage. Stannard's final Albertian study concentrates on Book Six of *De vegetabilibus*, where the individual plants are discussed. Stannard's scholarship is witnessed clearly and impressively in the prodigious notes whereby he identifies the exact plant according to modern scientific nomenclature and the sources that Albertus employed in producing his account.

Three studies conclude the volume and each study embodies a different methodological approach. The medieval encyclopedia was another genre for relating information about plants. Bartholomaeus Anglicus wrote a massive encyclopedia around 1250 that contains a section on plants. Bartholomaeus followed Isidore of Seville's premise that much can be learned about a *res*, or 'thing', through exploring the *origines* (to use Isidore's word) of its name. An article on vegetable gums and resins examines how medieval recipe literature employs this form of plant substance. First, gums and resins were often *exotica*, that is to say, derived from aromatic desert plants imported into

Europe. They were used as fumigants, pomanders (perfumed aromatic 'apples' or balls to alter 'corrupted air'), and unguents, as well as administered internally. The paramedical uses, Stannard believes, were readily exploited by medieval apothecaries who carried such expensive wares on their shelves ready for sale. The concluding study concerns a fifteenth-century botanical glossary that previously had been passed over by scholars who simply listed its title. Stannard shows us the rich and accumulative knowledge captured by the anonymous author who attempted to identify useful plants through their names, synonyms, and medical and non-medical uses. Far from being a 'glossary' in our sense, this genre of medieval literature proves valuable when we seek to understand how medieval peoples sought to control their lives through specific knowledge of the world around them.

Future scholars of the classical and medieval periods should pay attention to Stannard's foot- and endnotes. People of our day, with its massive data bases and computer-generated bibliographies, should appreciate the scholarship of one whose sources were difficult to find, be they in manuscripts in numerous European libraries or in obscure eighteenth- and nineteenth-century scientific academy proceedings. Ably assisted by his friends the librarians, Stannard pioneered a new world of old world history: the relationships between plants and people.

JOHN M. RIDDLE

North Carolina State University
October 1998

BIBLIOGRAPHY OF PUBLICATIONS BY JERRY STANNARD

In his early years, Jerry Stannard contributed many signed abstracts to *Clio*, and in the 1960's he often reviewed books in professional journals; but he kept no record of these activities, and consequently neither one is included in the following list.

1. "What the Citizen Needs to Know About Philosophy." *Progressive Education* 30 (1953): 172-174.
2. "Jainism: A Systematic Philosophy." *The Voice of Ahinsa* (India) 3 (1953): 162-166.
3. "The Psychology of the Passions in the Old Stoa." Ph.D. diss., University of Illinois, 1958. [Cf. *Dissertation Abstracts* 18 (1958): 1743.]
4. "Philosophy and Liberal Education." *Journal of Education* (India) 1 (1959): 24-42.
5. "The Role of Categories in Historical Explanation." *Journal of Philosophy* 56 (1959): 429-447.
6. "Socratic Eros and Platonic Dialectic." *Phronesis* 4 (1959): 120-134.
7. "Plato, *Epistle* II, 312A." *Phronesis* 5 (1960): 53-55.
8. "Parmenidean Logic." *Philosophical Review* 69 (1960): 526-533.
9. "Platonism and Art." *Giornale di metafisica* 15 (1960): 576-593.
10. "Hippocratic Pharmacology." *Bulletin of the History of Medicine* 35 (1961): 497-518.
11. "The Plant Called Moly." *Osiris* 14 (1962): 254-307.
12. "Materia Medica in the Locke-Clarke Correspondence." *Bulletin of the History of Medicine* 37 (1963): 201-225.
13. Edwin Clarke and Jerry Stannard. "Aristotle on the Anatomy of the Brain." *Journal of the History of Medicine and Allied Sciences* 18 (1963): 130-148.
14. "A Fifteenth-Century Botanical Glossary (Huntington Library MS HM 64)." *Isis* 55 (1964): 353-367.
15. "Materia Medica and Philosophic Theory in Aretaeus." *Sudhoffs Archiv* 48 (1964): 27-53.
16. "Method and Logic in Presocratic Methodology." In *Contributions to Logic and Methodology in Honor of J. M. Bocheński*, ed. A.-T. Tymieniecka, 107-131. Amsterdam: North-Holland, 1965.
17. "Pliny and Roman Botany." *Isis* 56 (1965): 420-425.
18. "The Presocratic Origin of Explanatory Method." *Philosophical Quarterly* 15 (1965): 3-16.
19. "Historical Section of the Botanical Society of America." *Plant Science Bulletin* 11 (1965): 5-6.
20. Introduction to *Canadensium plantarum aliarúmque nondum editarum historia*, by Jacques Cornut. 2d ed. The Sources of Science, no. 37. New York: Johnson Reprint Corp., 1966; first published Paris, 1635.

21. "Dioscorides and Renaissance Materia Medica." In *Materia Medica in the XVIth Century: Proceedings of a Symposium of the Internationl Academy of the History of Medicine Held at the University of Basel, 7th September 1964*, ed. M. Florkin, 1-21. Analecta Medico-historica, vol. 1. Oxford: Pergamon Press, 1966.
22. "Early American Botany and Its Sources." In *Bibliography and Natural History: Essays Presented at a Conference Convened in June 1964*, ed. Thomas R. Buckman, 73-102. Lawrence: University of Kansas Libraries, 1966.
23. "P. A. Mattioli and Some Renaissance Editions of Dioscorides." *Books and Libraries at the University of Kansas* 4 (1966): 1-5.
24. "Pharmacy--From Library to Laboratory." *Colorado Journal of Pharmacy* 9 (1966): 23-29.
25. "Ludwig Edelstein (1902-1965)." *Isis* 57 (1966): 379-384.
26. "Benedictus Crispus, an Eighth Century Medical Poet." *Journal of the History of Medicine and Allied Sciences* 21 (1966): 24-26.
27. "L'Antiquité classique et nos jours." *Clio Medica* 2 (1967): 66.
28. "The Skirmishes of Medieval Science." *Colorado Engineer* 63 (1967): 9-15.
29. "The Graeco-Roman Background of the Renaissance Herbal." *Organon* 4 (1967): 141-145.
30. "Note on the Use of Opium in Antiquity." *Scientific American* 216, no. 4 (1967): 8.
31. "Lost Botanical Writings of Antiquity." In *Actes [du] XIe Congrès international d'histoire des sciences, Varsovie-Toruń-Kielce-Cracovie 24-31 août 1965*, 5:319-322. Warsaw: Ossolineum, 1968.
32. "Medieval Reception of Classical Plant Names." In *Actes du XIIe Congrès international des sciences, 25-31 août 1968, Paris* = *Revue de synthèse* 89, ser. 3, nos. 49-52 (1968): 153-162.
33. "Lucianic Natural History." In *Classical Studies Presented to Ben Edwin Perry by his Students and Colleagues at the University of Illinois, 1924-1960*, ed. John L. Heller, 15-26. Illinois Studies in Language and Literature, 58. Urbana: University of Illinois Press, 1969.
34. "Medical Botany." In *A Short History of Botany in the United States*, ed. Joseph Ewan, 146-152. New York: Hafner, 1969.
35. "The Historical Background of J. E. Purkyně's *De cellulis antherarum fibrosis* (Breslau, 1830)." *Abstracts of the Papers Presented at the XI International Botanical Congress, August 24--September 2, 1969 ... Seattle, Washington*, 79. n. p.: n. p., n. d.
36. "P. A. Mattioli: Sixteenth Century Commentator on Dioscorides." *Bibliographical Contributions* 1 (1969): 59-81.
37. "The Herbal as a Medical Document." *Bulletin of the History of Medicine* 43 (1969): 212-220.
38. "Old and New in the History of Biology." *The Historian* 31 (1969): 436-444.
39. "Aristotelian Influences and References in Harvey's *De motu locali animalium*." In *Studies in Philosophy and in the History of Science: Essays in Honor of Max Fisch*, ed. Richard Tursman, 122-131. Lawrence, Kansas: Coronado Press, 1970).
40. "Medieval Italian Medical Botany." In *Atti del XII Congresso internazionale di storia della medicina, Siena 22-28 Settembre 1968*, 2:1554-1565. Rome: Arti Grafiche e Cossidente, 1970.

41. "Alexander of Myndos." In *Dictionary of Scientific Biography*, 1:120-121. New York: Scribner's, 1970.
42. "Prospero Alpini." In *Dictionary of Scientific Biography*, 1:124-125. New York: Scribner's, 1970.
43. "Luigi Anguillara." In *Dictionary of Scientific Biography*, 1:167. New York: Scribner's, 1970.
44. "Archigenes." In *Dictionary of Scientific Biography*, 1:212-213. New York: Scribner's, 1970.
45. "Asclepiades." In *Dictionary of Scientific Biography*, 1:314-315. New York: Scribner's, 1970.
46. "Jerome Bock." In *Dictionary of Scientific Biography*, 2:218-220. New York: Scribner's, 1970.
47. "Bolos of Mendes." In *Dictionary of Scientific Biography*, 2:256-257 New York: Scribner's, 1970.
48. "Otto Brunfels." In *Dictionary of Scientific Biography*, 2:535-538. New York: Scribner's, 1970.
49. "Crateuas." In *The Oxford Classical Dictionary*, 2d ed. by N. G. L. Hammond and H. H. Scullard, 297. Oxford: Clarendon, 1970.
50. "New Studies on Hieronymus Bock and Leonhard Rauwolf." *Archives internationales d'histoire des sciences* 23 (1970): 228-232.
51. "Purkyně's Contributions to Plant Physiology and Botany." In *Jan Evangelista Purkyně, 1787-1869: Centenary Symposium Held at the Carolinum, Prague, 8.-10. September 1969*, ed. Vladislav Kruta, 213-226. Brno: Universita Jana Evangelisty Purkyně, 1971.
52. "Bartholomaeus Anglicus and Thirteenth Century Botanical Nomenclature." *Actes [du] XIIe Congrès international des sciences, Paris 1968, Paris*, 8:191-194. Paris: Albert Blanchard, 1971.
53. "Byzantine Botanical Lexicography." *Episteme* (Italy) 5 (1971): 168-187.
54. "Hans von Gersdorff and some Anonymous Strassburg Apothecaries." *Pharmacy in History* 13 (1971): 55-65.
55. "Botanical Nomenclature in Gersdorff's *Feldbüch der Wundartzney*." In *Science, Medicine and Society in the Renaissance: Essays to Honor Walter Pagel*, ed. Allen G. Debus, 1:87-103. New York: Science History Publications, 1972.
56. "The Botanico-Medical Background of Baptista Fiera's *Coena de herbarum virtutibus*." In *Civilità dell'Umanesimo: Atti del VI, VII, VIII Convegno del Centro di studi umanistici "A. Poliziano," Palazzo Tarugi, Montepulciano*, 327-344. Florence: Olschki, 1972.
57. "Greco-Roman Materia Medica in Medieval Germany." *Bulletin of the History of Medicine* 46 (1972): 455-468.
58. "Marcellus of Bordeaux and the Beginnings of Medieval Materia Medica." *Pharmacy in History* 15 (1973): 47-53.
59. "Classici and Rustici in Clusius' *Stirp. Pannon. Hist.* (1583)." In *Festschrift anlässlich der 400jährigen Wiederkehr der wissenschaftlichen Tätigkeit von Carolus Clusius (Charles de l'Escluse) im pannonischen Raum = Burgenländische Forschungen* (Eisenstadt), Sonderheft 5 (1973): 253-269.
60. "Squill in Ancient and Medieval Materia Medica, with Special Reference to its Employment for Dropsy." *Bulletin of the New York Academy of Medicine* ser. 2, 50 (1974): 684-713.

61. "Medieval Herbals and their Development." *Clio Medica* 9 (1974): 23-33.
62. "Eastern Plants and Plant Products in Medieval Germany." In *Proceedings of the XIIIth International Congress of the History of Science, Moscow, August 18-24, 1971*, 5:220-225. Moscow: Editions "Naouka", 1974.
63. "Hooker Family." In *Encyclopaedia Britannica*, 15th ed., 8:1066-1067. Chicago: Encyclopaedia Britannica, Inc., 1974.
64. "Pliny the Elder." In *Encyclopaedia Britannica*, 15th ed., 14:572-573. Chicago: Encyclopaedia Britannica, Inc., 1974.
65. Wolfgang-Hagen Hein and Jerry Stannard. "Themen und Quellen der mittelalterlichen Botanik." *Deutsche Apotheker-Zeitung* 114 (1974): 735-736.
66. "The Medicinal Uses of Vegetable Gums in the Eighteenth Century." In *Acta Congressus Internationalis XXIV Historiae Artis Medicinae, 25-31 Augusti 1974, Budapestini*, ed. Jòzsef Antall et al., 2:1069-1074. Budapest: Museum, Bibliotheca, et Archivum Historiae Artis Medicinae, 1976.
67. "Botanical Data in Medieval Medical Recipes." *Studies in the History of Medicine* 1 (1977): 80-87.
68. "Gummen und Harze in der mittelalterlichen Rezeptliteratur." In *Internationaler Kongress für Geschichte der Pharmazie in Bremen 29 Sept.--5 Oktober 1975*, ed. Herbert Hügel, 63. Stuttgart: Deutscher Apotheker-Verlag, 1977.
69. "Magiferous Plants and Magic in Medieval Medical Botany." *The Maryland Historian* 8 (1977): 33-46.
70. *Carl Linnaeus: A Bibliographical Measure of his Scientific Achievements*. Lawrence, Kansas: Friends of the Library at the University of Kansas, 1978.
71. Jerry Stannard and Peter Dilg. "Camerarius' Contributions to Medicine and Pharmacy: Observations on his *De theriacis et mithridateis commentariolus*." In *Joachim Camerarius (1500-1574): Beiträge zur Geschichte des Humanismus im Zeitalter der Reformation*, ed. Frank Baron, 152-186. Munich: Wilhelm Fink Verlag, 1978. [Part 1, "Pestilence and Theriaca," pp. 152-168, by Jerry Stannard.]
72. "Vegetable Gums and Resins in Medieval Recipe Literature." In *Acta Congressus Internationalis Historiae Pharmaciae, Bremae MCMLXXV ... Bremen, 29 September--5 Oktober 1975*, ed. Wolfgang-Hagen Hein et al. = *Veröffentlichungen der internationalen Gesellschaft für Geschichte der Pharmazie*, n.s. 45: 41-48. Stuttgart: Wissenschaftliche Verlagsgesellschaft, 1978).
73. "Natural History." In *Science in the Middle Ages*, ed. David C. Lindberg, 429-460. Chicago: University of Chicago Press, 1978).
74. "Bibliographical References in Medieval Pharmaceutical Writings." In *Die Vorträge des internationalen pharmaziehistorichen Kongresses, Innsbruck, Juni 7-11, 1977*, ed. K. Ganzinger = *Veröffentlichungen der Internationalen Gesellschaft für Geschichte der Pharmazie*, n.s. 47: 39-46. Stuttgart: Wissenschaftliche Verlagsgesellschaft, 1979.
75. "Identification of the Plants Described by Albertus Magnus' *De vegetabilibus* lib. VI." *Res Publica Litterarum* 2 (1979): 281-318.
76. "Albertus Magnus and Medieval Herbalism." In *Albertus Magnus and the Sciences: Commemorative Essays, 1980*, ed. James A. Weisheipl, 355-377. Toronto: Pontifical Institute of Mediaeval Studies, 1980.

77. "The Botany of St. Albert the Great." In *Albertus Magnus, Doctor Universalis, 1280/1980*, ed. Gerbert Meyer and Albert Zimmermann, 345-372. Mainz: Matthias-Grünewald-Verlag, 1980.
78. "Folk Medicine, Philosophy and Medieval Herbalism." *Res Publica Litterarum* 3 (1980): 229-236.
79. "Pharmaceutical Practices and Folk Beliefs in Albertus Magnus' *De vegetabilibus*." *Farmaceutisch Tijdschrift voor België* 58 (1981): 157-161.
80. "Pharmaceutical Practices and Folk Beliefs in Albertus Magnus' *De vegetabilibus*." In *Recente Bijdragen tot de Geschiedenis van de Farmacie: Liber Amicorum L. J. Vandewiele*, ed. P. Nijs and C. De Backer, 157-161. Brussels: Algemene pharmaceutische Bond, 1981. [Reprint of no. 78.]
81. "Rezeptliteratur as Fachliteratur." *Studies on Medieval Fachliteratur*, ed. William Eamon, 59-73. Scripta: Mediaeval and Renaissance Texts and Studies, ed. W. L. Brackman, 6. Brussels: OMIREL, UFSAL, 1982.
82. "University of Kansas." In *Survey of Instructional Resources in History of Pharmacy Undergraduate Courses*, ed. Robert J. Buerki, 13-14. Columbus: Ohio State University Press, 1982.
83. "Botanical Data and Late Medieval Rezeptliteratur." In *Fachprosa-Studien: Beiträge zur mittelalterlichen Wissenschafts- und Geistesgeschichte*, ed. Gundolf Keil et al., 371-395. Berlin: Erich Schmidt, 1982.
84. "Folkloristic Elements in Medieval Herbalism." In *Actes [du] XXVIe Congrès international d'histoire de la médecine, Plovdiv (Bulgaria), 20-25 août 1978*, 2:203-205. Sofia: n. p., 1981.
85. "The Multiple Uses of Dill (*Anethum graveolens* L.) in Medieval Medicine." In *"Gelêrter der arzenîe, ouch apotêker": Beiträge zur Wissenschaftsgeschichte; Festschrift zum 70. Geburtstag von Willem F. Daems*, ed. Gundolf Keil, 411-424. Würzburger Medizenhistorische Forschungen, 24. Pattensen, Hanover: Horst Wellm, 1982.
86. "Medicinal Plants and Folk Remedies in Pliny, *Historia naturalis*." *History and Philosophy of the Life Sciences* 4 (1982): 3-23
87. *Selected Medieval Garden Plants and their Uses*. Lawrence: University of Kansas, 1983.
88. "*...findet man in den apotecken*: Notices Concerning the Availability of Medicamenta in Medieval Fachliteratur." In *Perspektiven der Pharmaziegeschichte: Festschrift für Rudolf Schmitz zum 65. Geburtstag*, ed. Peter Dilg, 356-376. Graz: Akademische Druck- und Verlagsanstalt, 1983.
89. Marilyn Stokstad and Jerry Stannard. *Gardens of the Middle Ages*. Lawrence: Spencer Museum of Art, University of Kansas, 1983.
90. "Medieval Gardens and Their Plants." In *Gardens of the Middle Ages*, 37-69. Lawrence: Spencer Museum of Art, University of Kansas, 1983.
91. "Medieval Gardens in Modern Kansas." In *History at Kansas* 2, no. 2: 1. Lawrence: Department of History, University of Kansas, 1983.
92. "Botany." In *Dictionary of the Middle Ages*, ed. Joseph R. Strayer, 2:344-349. New York: Scribner's, 1982.
93. "Selections from Medieval Herbals and Related Medical Texts." *Würzburger medizinhistorische Mitteilungen* 1 (1983): 167-171.
94. "Medieval *Arzneitaxe* and Some Indigenous Plant Species." In *Orbis Pictus: Kultur- und pharmaziehistorische Studien; Festschrift für Wolfgang-Hagen Hein zum 65.*

Geburtstag, ed. Werner Dressendörfer and Wolf-Dieter Müller-Jahncke, 267-272. Frankfurt a. M.: Govi-Verlag, 1985.

95. "The Theoretical Bases of Medieval Herbalism." *Medical Heritage*, 1 (1985): 186-198.
96. "Linnaeus, Nomenclator Historicusque Neoclassicus." In *Contemporary Perspectives on Linnaeus*, ed. John Weinstock, 17-35. Lanham, Md.: University Press of America, 1985.
97. "Aspects of Byzantine Materia Medica." In *Symposium on Byzantine Medicine*, ed. John Scarborough, 205-211. Dumbarton Oaks Papers, no. 38. Washington, D.C.: Dumbarton Oaks Research Library and Collection, 1985.
98. "Herbal Medicine and Herbal Magic in Pliny's Time." In *Pline l'Ancien, temoin de son temps: Conventus Pliniani Internationalis, Namneti 22-26 Oct. 1985 Habiti Acta*, ed. J. Pigeaud and J. Oroz, 95-106. Bibliotheca Salmanticensis, 87. Salamanca and Nantes: n.p., 1987.
99. "Alimentary and Medicinal Uses of Plants ." In *Medieval Gardens*, ed. Elisabeth B. MacDougall, 69-91. Washington, D.C.: Dumbarton Oaks Research Library and Collection, 1986.
100. "Medieval Herbalism and Post-Medieval Folk Medicine." In *Folklore and Folk Medicines*, ed. John Scarborough, 10-20. AIHP Publication, no. 10. Madison, Wis.: American Institute of the History of Pharmacy, 1987.
101. "Medieval Hepatic Therapy and Some Folk Medical Survivals." *Würzburger medizinhistorische Mitteilungen* 6 (1988): 207-223.
102. Sherry Kay, Jerry Stannard, and Richard Kay. "Medieval Cookery for Credit." *Teaching the Middle Ages* 3 (1989): 160-171.
103. Robert J. Smith and Jerry Stannard, eds. *The Folk, Identity, Landscapes and Lores*. University of Kansas Publications in Anthropology, 17. Lawrence, Kans.: Department of Anthropology, 1989.
104. "Belief and Practice in Medieval Folk Medicine." In *The Folk, Identity, Landscapes, and Lores*, ed. Robert J. Smith and Jerry Stannard, 121-131. University of Kansas Publications in Anthropology, 17. Lawrence, Kans.: Department of Anthropology, 1989.
105. "The Scientific Revolution." In *Patterns in Western Civilization*, ed. James Woelfel and Sarah Chappel Trulove, 224-242. Needham Heights, Mass.: Ginn Press, 1991.
106. "Diseases of Western Antiquity." In *The Cambridge World History of Human Disease*, ed. Kenneth F. Kiple, 262-270. Cambridge, Eng.: Cambridge University Press, 1993.

I

HIPPOCRATIC PHARMACOLOGY

Traces of ancient Greek pharmacology still linger on in the present day: the bland diet during a fever, a purgative such as castor oil, a soothing ointment flavored perhaps with aqua rosae—all testify to practices elaborated over two millennia ago. It is no argument to say that current medical practice seldom resorts to these measures, for they are encountered as often in the idle conversations of the sophisticated as they are still current in the more remote of our rural areas. Thus, although official medical practice and the larger social structure have drastically changed in the interval, the presence of these older practices is evidence of a tradition which still exerts considerable influence. Such a tradition does not, usually, remain viable unless it possesses some value. It is the early stage of this tradition of Western European pharmacology—its development, its values, and its contributions, that are to be discussed in the following pages.

The beginnings of Greek pharmacology were probably so unpretentious that we can be certain of very little save the names of a few simples, generally of botanical origin, and how they were employed. One example must suffice. In the eleventh book of the Iliad, Homer recounts how the Trojans wounded Odysseus. Perceiving this, Machaon, the reputed son of Asclepius, is quickly taken to the tents of the Greeks in order to be on hand when Odysseus returns. Homer relates the value of Machaon in these words:

> For a leech is of the worth of many other men for the cutting out of arrows and the spreading of soothing simples.[1]

Later in the same book, a wounded soldier addresses Patroclus:

> But me do thou succour, and lead me to my black ship, and cut the arrow from my thigh, and wash the black blood from it with warm water and sprinkle thereon kindly simples of healing power. . . . Then Patroclus made him lie at length and with a knife cut from his thigh the sharp-piercing arrow, and from the wound washed the black blood with warm water, and upon it cast a bitter root, when he had rubbed it between his hands, a root that slayeth pain, which stayed all his pangs; and the wound waxed dry, and the blood ceased.[2]

---

[1] Homer, *The Iliad* XI 514-515. Trans. A. T. Murray (Loeb Library), London, 1954, vol. I, p. 518.

[2] Homer, *op. cit.* XI 828-831, 844-848. Trans. Murray, vol. I, pp. 540-542.

The question of whether there was a Near-Eastern basis of Greek pharmacology is a controversial one.[3] It is not easy to find an alternative solution to the question of the origin of Greek pharmacology. A probable answer, however, is the belief that there was a crudely empirical basis for the aforementioned passage, typical of many of the early references.[4] This empirical basis was, in turn, probably the distillate of archaic or folk medicine practiced for many centuries.[5] Folk medicine such as this probably depended more on local drug lore than on any extensive or systematic borrowing from the then culturally more advanced Eastern nations. It is, of course, impossible to prove this hypothesis because folk medicine by its very nature is never committed to writing at the time it is *en vogue*. Traces of folk medicine occur in Greek writings of all periods: e. g., Thucydides (II 47), Plato (Symp. 201 D), and others testify to the belief than an epidemic could be averted by prayers and the consultation of oracles.[6]

Despite these and other references to folk medicine, a system cannot be reconstructed from them. There was no explicitly formulated system of medicine, including pharmacology, until the Hippocratic writings were composed, though again indications of pre-Hippocratic medical theories are well-known.[7] It is thus with Hippocrates that a study of Greek pharmacology must begin. Incidentally, the term "Hippocrates" here means the anonymous author of any one of the writings comprising the Hippocratic Corpus. We know very little of the historical Hippocrates; even the ancient tradition that he was born ca. 460 B. C. on the island of Cos is not above suspicion.[8]

Today, pharmacology is divided into three disciplines: pharmacognosy, pharmacodynamics, and pharmacotherapy. The ancient writers on drug lore clearly did not possess the refined techniques practiced today in the clinic and in the laboratory. Consequently, it is difficult to avoid projecting our practices and vocabulary on the ancient writings in the belief that we

[3] The attribution of Greek pharmacological practices to Egyptian sources, despite its many defenders, requires much further study. For an attempt to account for the origin of Greek pharmacological practices without recourse to Near Eastern sources, cf. Georg Sticker, "Entwickelungsgeschichte der spezifischen Therapie." *Janus*, 1929, *33*: 131-190.

[4] P. Saintyves, "L'Art de Guérir est-il d'origine empirique?" *Janus*, 1917, *22*: 372-390.

[5] Owsei Temkin, "Beiträge zur archaischen Medizin." *Kyklos*, 1930, *3*: 90-135.

[6] Hugo Magnus, "Die Volksmedizin, ihre geschichtliche Entwickelung und ihre Beziehungen zur Kultur." *Abhandlungen z. Gesch. d. Med.*, 1905, *15*: 1-112.

[7] Charles Daremberg, *État de la médecine entre Homère et Hippocrate*. Paris: Didier, 1869.

[8] Ludwig Edelstein, "The genuine works of Hippocrates." *Bull. Hist. Med.*, 1939, *7*: 236-248.

I

are thereby explaining them. Nevertheless, the triple division of pharmacology offers a convenient method of pulling together a wide array of material bearing on Hippocratic pharmacology.

I. Pharmacognosy

A study of Hippocratic pharmacognosy is largely confined to an examination and identification of those simples which were employed for therapeutic purposes. The first class to be examined is that derived from the vegetable kingdom, that is, the botanicals of older medical terminology. Botanicals supplied by far the largest number of drugs for the ancient materia medica and they retained this position up to the middle of the last century. The preeminent position accorded plants is due to several factors among which may be mentioned their common occurrence, widespread distribution, ease of identification, and the relative simplicity with which they may be prepared.

In order to reduce to manageable size a discussion of botanicals, the following procedure has been adopted. After some remarks on the identification of the plants occurring in the Hippocratic writings, several of them will be examined under the following headings: 1. Mode of administration. 2. The specific purpose for which the drug was prescribed. 3. The Hippocratic explanation of, or rationale for, its specific action. 4. A comparison of the ancient explanation with that found in the modern literature on pharmacognosy.

If there is any thesis to be maintained in this paper, it is simply that a comparative study of the type indicated in the last point is very important for the evaluation of ancient pharmacology. But clearly, if the findings of modern pharmacognosy are to be utilized, it is necessary that the identification of the plant be established with as much certainty as possible. It is because it aids in the identification of ancient plant names, that it is necessary to include under pharmacognosy a few remarks on the Hippocratic explanation of a drug's specific action. Hence, we must first ask the question: What are the means by which plants may be identified? Of the more than 200 plant names which occur in Hippocrates, the great majority are capable of at least a generic identification and many can also be given a specific identification with a fair degree of certainty.[9]

The first step in the identification of an ancient plant is to search for some clue provided in its description. Frequently, a few well-chosen

[9] J. H. Dierbach, *Die Arzneimittel des Hippokrates*. Heidelberg: Gross, 1824. Though old, this work has not yet been superseded.

descriptive epithets are more valuable for purposes of identification than are all the imprecise and conflicting references taken together. Our most important source for the description of plants is the " History of Plants," written by Theophrastus, the disciple and slightly younger contemporary of Aristotle. The nine books which constitute this work have justly earned for him the title, " the first systematic botanist." For our purposes, the importance of Theophrastus consists in the precise descriptions of plants which he studied at first hand. These descriptions, which have enabled historians of botany to identify most of the plants mentioned by him, are doubly important for the student of Hippocratic botany. For Theophrastus, writing within a century of the death of Hippocrates, often uses the same plant names as those in Hippocrates. Moreover, the flora described by Theophrastus—mainland Greece, the islands of the Archipelago, and the coast of Asia Minor, is, ecologically speaking, identical to the flora known to Hippocrates.[10] If a Hippocratic plant name also occurs in Theophrastus, there is a good likelihood that it is the same plant that is in question. A good example is the plant called *teutlon* by Theophrastus. This plant is usually accepted to be the wild beet (*Beta maritima* L.). The same term occurs in Hippocrates.[11] From the analysis given of this plant in Hippocrates, plus its therapeutic uses, it seems safe to conclude that it is the same plant as Theophrastus' *teutlon*.

The second criterion for the identification of ancient drug plants is that supplied by modern phytogeography. For instance, having provisionally identified a plant in terms of descriptive evidence, we then ask: Has this plant ever been recorded as occurring in Greece, the islands, or the coast of Asia Minor? [12] This procedure must be used with caution, however, for although the absence of any record is presumptive evidence that the plant did not occur there in antiquity, the converse is not true. The fact that a plant now occurs in Greece, etc., does not necessarily mean it is indigenous. For plants are frequently introduced for commercial reasons, seeds are carried by birds, and finally some plants are able to flourish, after escaping from cultivation, in a state indistinguishable from their originally indigenous habitat.

[10] Martin Rikli, *Das Pflanzenkleid der Mittelmeerländer*. 3 vols. Bern: Huber, 1943-1948.

[11] Hippocrates, *On Regimen,* 54 (VI 560 Littré). Unless otherwise specified, all references to Hippocrates are to the edition of Emile Littré, *Oeuvres complètes d'Hippocrate*. 10 vols. Paris: Baillière, 1839-1861.

[12] The definitive work on Greek phytogeography is E. von Halácsy, *Conspectus florae graecae*. 4 vols. Leipzig: Engelmann, 1901-1912.

A third criterion is supplied by iconography or the pictorial representation of plants, trees, and their fruit. Pliny, in the first century A. D., refers to a herbal with colored illustrations which is, unfortunately, no longer extant.[13] Our earliest illustrated herbal dates from the year 512, when a manuscript of Dioscorides was presented as a wedding present to Juliana Anicia, daughter of the Emperor of the West. But much older than the representations supplied by herbals are those found on Greek coins and vases. Among the plants found on ancient coins are myrtle, barley, quince, grapes, storax, and several different species of conifers.[14] Honeysuckle and lotus are among the most common floral decorations on Greek vases,[15] while on the walls of the palace of Cnossus on Crete, ivy, crocus, and lilies can still be discerned.[16]

A fourth criterion, that supplied by philology, is of especial value when used in conjunction with phytogeography. Since it is common, even in our times, for the same plant to have different local names, the correlation of local name and scientific identification is important.[17] In some cases, local names are recognizable variants of ancient plant names which have disappeared and form an essential link with the past. The investigation of the etymology of Greek plant names is an entire field in itself, and its results are important corroborative evidence for the correct identification of a plant.[18] Thus *triphyllon,* literally three-leafed, is generally accepted to be *Psoralea bituminosa* L. and closely related to our alfalfa.[19]

Finally, mention should be made of later glossaries or dictionaries. This is a common literary genre, especially between the sixth and fourteenth centuries. Frequently an entry in a glossary will shed light on the identity of an ancient plant name by defining it in terms of later or more common plant names.[20] Elsewhere, in an article to be published shortly,

[13] Pliny, *Naturalis historia* XXV 4, 8.

[14] F. Imhoof-Blumer und Otto Keller, *Tier- und Pflanzenbilder auf Münzen und Gemmen des klassischen Altertums.* Leipzig: Teubner, 1889, pp. 54-63 and Tafeln IX-X.

[15] Cf. R. M. Cook, *Greek Painted Pottery.* London: Methuen, 1960, p. 118.

[16] Charles Seltman, *Approach to Greek Art.* New York: Dutton, 1960, p. 17 and Plate 11a.

[17] For the correlation of classical Greek plant names and their Byzantine synonyms, cf. Bernhard Langkavel, *Botanik der spätern Griechen vom dritten bis dreizehnten Jahrhunderte.* Berlin: Berggold, 1866.

[18] A. Carnoy, *Dictionnaire étymologique des noms grecs de plantes.* Louvain: Publications universitaires, 1959.

[19] *Triphyllon* is mentioned in *Diseases* III 16 (VII 150 L.) as one of the ingredients in an expectorant eclegma. On its identity, cf. Carnoy, *op. cit.,* p. 270 and C. Fraas, *Synopsis plantarum florae classicae.* München: Fleischmann, 1845, p. 62.

[20] For Byzantine botanical glossaries, cf. A. Delatte, *Anecdota atheniensia et alia.* Liége: Vaillant-Carmanne, vol. II, 1939, pp. 273-454.

I have confirmed, with the aid of Byzantine and Arabic botanical glossaries, the identity of a Hippocratic plant name as our common garlic.[20a]

The last criterion is what may be called comparative pharmacognosy. By comparing the physiological action of chemically-recognized ingredients obtained from an accurately identified plant with the ancient descriptions of a plant and its therapeutic uses, a provisional identification may be corroborated. Thus, Moïssidés, in studying the plants employed as abortifacients, noted many cases where a rationale can be given for their use in antiquity.[21] Other classes of drugs, including diuretics, emetics, and expectorants have been studied by von Grot and others.[22] Mention must be made, finally, of two plants used as " psychopharmaca." [23] The root of mandragora (*Atropa Belladonna* L.) was recommended as a drink for " sick persons who, in their sadness, wish to hang themselves," [24] while poppy (*Papaver somniferum* L.) was used several times for relief of pain.[25] Comparative pharmacognosy also plays a valuable role by demonstrating that many botanical drugs were probably ineffective. An example is provided by the juice of common lettuce or its wild form. This milky-sap, well known to all gardeners, was believed to have been an anodyne by Hippocrates.[26] Despite the extraction of a bitter crystallizable substance, lactucin, tests have failed to indicate any sedative properties of *Lactuca scariola* L. or *L. sativa* L.[27]

With these as the means for identifying Hippocratic medical plants, we can resume our discussion by considering some examples of botanical drugs and the uses for which they were intended. As the following examples will indicate, the particular action of plant drugs was recognized

[20a] Cf. my paper, forthcoming in *Osiris,* entitled " The plant called Moly."

[21] M. Moïssidés, " Contribution à l'étude de l'avortement dans l'antiquité grecque." *Janus,* 1922, *26*: 59-85, 129-145.

[22] Rudolf von Grot, Über die in der hippokratischen Schriftensammlung enthaltenen pharmakologischen Kenntnisse," in Rudolf Kobert, ed., *Historische Stud. a. d. pharmakol. Inst. d. k. Univ. Dorpat,* 1889, *1*: 58-133; T. Beck, " Haben die Hippokratiker das Secale cornutum therapeutisch verwendet? " *Arch. f. Gesch. d. Med.,* 1908, *2*: 279-284; H. Fischer, " Helleborus im Altertum und bei Paracelsus." *Schweiz. med. Wchnschr.,* 1936, *66*: 484-489.

[23] Anne E. Caldwell, *Psychopharmaca.* Washington: U. S. Dept. of Health etc., 1958. Psychopharmaca are defined on p. 1 as " drugs that primarily affect the mental state."

[24] *Places in Man* 39 (VI 238 L.).

[25] *Fistula* 7 (VI 456 L.) ; *Diseases of Women* 206 (VIII 400 L.).

[26] *Diseases of Women* 78 (VIII 196 L.) ; *ibid.* 101 (VIII 224 L.).

[27] F. A. Flückiger and D. Hanbury, *Pharmacographia.* London: Macmillan, 1874, p. 357.

in antiquity, though, of course, the ancients had inadequate explanations for their use.

In Hippocrates and all Greek medical writers after him, the juice of a plant called *skammōnia* is prescribed as a drastic purge.[28] This plant has been identified as *Convolvulus Scammonia* L., a member of the morning glory family and closely related to the common bindweed as well as to jalap. Older chemical analysis indicated that there is no essential difference between the active ingredients of scammony and Mexican jalap (*Ipomoea orizabensis* Led.).[29] The active principle, jalapin, is a glucosidal resin. It is owing to this substance that scammony, which last appeared in U. S. P. IX, is classified as a hydragogue cathartic.

Common flax (*Linum usitatissimum* L.) occurs frequently in Hippocratic writings under the name *linon* from which our word "linen" is derived. It is used internally as a drink for "uterine suffocation"[30] and as a detersive or emollient clyster for uterine disorders,[31] and externally in the form of cataplasms.[32] Modern uses center around its laxative, demulcent, and emollient action. It is perhaps odd that there is no evidence of its being used as a laxative in Hippocrates, but later, in the time of Dioscorides, its laxative properties are explicitly mentioned.[33] The Hippocratic use, especially as cataplasms, is borne out by analysis of the emollient effect of linseed oil. The oil is also used in an ointment, and Hippocrates recommends it to soften the uterine orifice in the belief that it will correct sterility.[34] Although this use was probably ineffective, it clearly points to a recognition of linseed oil as a lubricant.

Another drug of common occurrence in Hippocratic pharmacology was *sikuos agrios* or the squirting cucumber (*Ecballium Elaterium* Rich.) which yielded the juice called *elaterion*. This juice is known to contain elaterin, a crystallizable substance which possesses a powerful action as a hydragogue cathartic.[35] The purgative property of *elaterion* was known to Hippocrates, who employed it also as an oxytocic and abortifacient.[36]

[28] *Diseases of Women* 16 (VIII 54 L.).

[29] Flückiger and Hanbury, *op. cit.*, p. 396.

[30] *Diseases of Women* 129 (VIII 276 L.).

[31] *Ibid.* 64 (VIII 132 L.); *ibid.* 78 (VIII 190 L.).

[32] *Ibid.* 198 (VIII 382 L.).

[33] Dioscorides II 103, ed. Max Wellmann, *Pedanii Dioscuridis Anazarbei De materia medica libri quinque*. Berlin: Weidmann, vol. I, 1907, p. 177.

[34] *Superfetation* 32 (VIII 500 L.).

[35] H. W. Youngken, *Textbook of Pharmacognosy*. 6th ed., Philadelphia: Blakiston, 1948, p. 861.

[36] Oxytocic: *Superfetation* 4 (VIII 478 L.); abortifacient: *Diseases of Women* 91 (VIII 220 L.).

I

Although there appears to be no chemical action which would justify this usage, it is well known that a drastic purge may be sufficient to perform the specific action desiderated by Hippocrates.

Finally, mention should be made of another instance where modern pharmacognosy supplies a rationale for the Hippocratic usage. Today, common dill (*Anethum graveolens* L.) is recognized in the form of aqua anethi as a carminative. Among the many uses of dill or *anethon,* in Hippocrates, including a mouth wash in which dill was probably used as a flavoring agent,[37] it is explicitly mentioned as effective for " relaxing the stomach," which is, after all, the *raison d'être* of a carminative.[38]

Many other instances could be given of the Hippocratic use of botanicals which we know today to be justified, e. g., the use of the rind of the pomegranate (*Punica Granatum* L.) as an astringent.[39] The high tannin content of the rind is evidence that, as an astringent, it was probably quite effective.[40]

The next class of simples to be examined is that derived from the animal kingdom. While there is some reason to suppose that the ancient therapeutic use of many botanicals can be justified in modern terms and that they offered some relief, the case is different as regards drugs of animal origin.[41]

At least ten animal phyla can be identified as the source of these drugs: mammals, birds, fish, arthropods, including insects and crustacea, and molluscs are the most common, though amphibia, reptiles, coelenterates, echinoderms and porifera are also represented. For example, in Chapter 48 of " *On Regimen*," 17 fish, 10 mollusca, 5 crustacea, 1 echinoderm (a sea urchin), and 1 coelenterate (a sea anemone) are listed as having specific medicinal properties. Since marine animals are rather less adaptable than plants to a changing environment, the marine fauna of modern Greece probably differs little from that of Hippocrates' time.[42] But while zoologists are in general agreement as to the identification of the vast majority of these organisms, specific therapeutic data are lacking, because

[37] *Diseases of Women* 185 (VIII 366 L.),

[38] *Affections* 43 (VI 252 L.).

[39] *Ibid.* 54 (VI 264 L.); *Diseases of Women* 110 (VIII 238 L.).

[40] Youngken, *op. cit.*, p. 591 estimates the tannin content to be 22-28 percent.

[41] For the ineffectiveness of drugs of animal origin, cf. W. B. McDaniel, " The medical and magical significance in ancient medicine of things connected with reproduction and its organs." *J. Hist. Med. & Allied Sc.*, 1948, *3*: 525-546.

[42] E. g., cf. D'Arcy W. Thompson, *A Glossary of Greek Fishes*. London: Oxford Univ. Press, 1947, passim.

most of these animals have long since been dropped from the pharmacopoeias.

Various portions of these animals were employed for a wide range of complaints but with little evidence that they were superior to botanicals or, indeed, that they were more effective than no treatment at all. Such substances included flesh, blood, skin, eggs, horn, etc.[43] One may speculate as to the effectiveness of such remedies as cuttle fish eggs, administered as an emmenagogue suppository or the use of deer horn, powdered or calcined, for sterility.[44] An explanation, however, can probably be given for the use of the lung of a sheep when inserted in the nostrils for repair of a fracture.[45] Likewise, the effectiveness of goose-grease and seal-oil as lubricants or as the vehicle for ointments and suppositories can be recognized.[46] Goat's liver, eaten raw for metrorrhagia is an instance where, in the absence of any but the simplest data of enzymology and physiological bio-chemistry, silence seems judicious.[47]

Closely connected with the foregoing is a class of substances of animal origin used particularly in mixtures of various simples. This class includes butter, wax, marrow, whey, rennet, etc. Butter, for instance, is used frequently in the preparation of suppositories but is also recommended to be drunk as a galactagogue.[48] Marrow, likewise, was used in suppositories and also as an ointment for local skin eruptions.[49]

No discussion of animal substances would be complete without mention of two drugs which retained their place in pharmacopoeias until recent times. I refer to the cantharis, variously known as blister beetle or Spanish Fly, and castoreum.

Cantharis vesicatoria L., a small beetle, is the source of cantharidin, a crystalline anhydrid of cantharidic acid. Official in U. S. P. XI, it was used principally as a vesicant. In Hippocratic writings it was considered

[43] A full account of drugs of animal origin and their uses will be found in Max Höfler, *Die volksmedizinische Organotherapie und ihr Verhältnis zum Kultopfer.* Stuttgart: Union deutsche Verlagsgesellschaft, [1908].

[44] Cuttlefish eggs; *Diseases of Women* 84 (VIII 208 L.); deer horn: *Sterility* 13 (VIII 434 L.).

[45] *Joints* 38 (IV 168 L.). The passage continues by noting that it is superior to sponge in that it does not absorb fluids which would cause it to expand.

[46] *Diseases of Women* 203 (VIII 390 L.); 90 (VIII 218 L.).

[47] *Ibid.* 110 (VIII 236 L.).

[48] On suppositories: *Diseases of Women* 78 (VIII 180 L.); galactagogue: *ibid.* 44 (VIII 102 L.).

[49] Emollient suppositories: *Diseases of Women* 205 (VIII 394 L.); *Superfetation* 33 (VIII 500 L.); an ointment for genital *aphthai: Diseases of Women* 34 (VIII 82 L.).

especially valuable for dropsy and as an emmenagogue suppository.[50] Castoreum has now disappeared completely from medical practice. But this substance, obtained from the preputial follicles of the European beaver, was once thought to be a powerful anti-spasmodic. This belief, now rejected, apparently owes its origin to the Hippocratic writings in which castoreum was supposed to have a special affinity for the uterus—and hence, to be useful in uterine complications generally.[51]

Before we turn to the third class of simples—drugs of mineral origin—it is well to note a further, rather heterogeneous class of substances whose full importance will emerge later in this paper. In the language of the pharmacist, these substances would be termed either correctives or excipients rather than the basis or adjuvant. This class, comprising both botanical and animal-derived drugs, can perhaps be sub-divided into three sub-classes. There are, first of all, substances which appear as items of food in the normal diet, e. g. alphita (barley groats), olive oil, honey, wine, etc. Medicinally, the substances of this group are used mainly to ensure the proper consistency of a drug or to overcome some undesirable property of the basis. Thus, we find that honey is used in the preparation of emollient ointments, e. g., following cauterization of haemorrhoids,[52] and and as a corrective it is used to sweeten the bitter flavor of water of lentils drunk as an emetic.[53] A second group includes substances whose major uses are cosmetic as much as pharmacological, e. g., *muron*—a sweet smelling oil used in unguents or in perfumes.[54] A third group is exemplified by gum and resin. Resins of various types are used, e. g., in compresses for compound fractures.[55] Whether there was a disinfectant or bacteriostatic property imparted to the dressing may be doubted because of the other ingredients including wine and beet leaves.[56] Gums, like myrrh

[50] For dropsy: *Diseases of Women* 59 (VIII 118 L.); emmenagogue suppository: *ibid.* 84 (VIII 208 L.).

[51] Ointment to soften uterine orifice: *Superfetation* 32 (VIII 500 L.); in suppository for sterility: *Sterility* 9 (VIII 428 L.); drunk with wine after succussion: *Excision of the Foetus* 4 (VIII 516 L.).

[52] *Haemorrhoids* 2 (VI 438 L.).

[53] *Diseases of Women* 146 (VIII 322 L.).

[54] *Ibid.* 201 (VIII 384 L.).

[55] *Fractures* 29 (III 514 L.).

[56] The use of beet leaves, with and without wine, is mentioned several times, *Joints* 63 (IV 272 L.), *Affections* 38 (VI 248 L.), *Use of Liquids* 5 (VI 128 L.). F. B. Lund, "Hippocratic Surgery," *Ann. Surg.*, 1953, *102*: 537, suggested that the wine served as an antiseptic. On the other hand, as Garrison has pointed out, the aseptic technique of dry dressings is something quite different from antiseptic technique, F. H. Garrison, *An Introduction to the History of Medicine.* 3rd. ed., Philadelphia, London: Saunders, 1921,

I

(from *Balsamodendron myrrha* Nees), were frequently used in fumigations. References to both nasal and vaginal fumigations for uterine disorders leave some question, however, as to the effectiveness of the therapy as well as to Hippocratic anatomy.[57]

A last class of simples to be mentioned, and again as much botanical as animal-derived, includes the dung of various animals, both domesticated and wild. Thus, cow dung is used, chiefly as a fumigant, for "uterine suffocation," as an anodyne for uterine pain, and as an emmenagogue,[58] while more exotic excrement, like hawk dung, is drunk with wine as a cure for sterility.[59]

We come now to the third class of simples, those of mineral origin. The number of minerals is restricted owing, no doubt, to the technological inability to obtain water soluble salts and purified samples of those minerals which are usually found in combination with other insoluble or inert minerals. Nevertheless, some 11 chemical elements can be counted among the mineral drugs. Two elements, sodium and copper, account for a large share of the mineral drugs which also include aluminum, antimony, arsenic, iron, lead, mercury, sulphur, and probably calcium and potassium salts in smaller amounts. Sodium occurs in the form of common salt, NaCl or soda, Na_2CO_3. In the form of common salt, its uses are numerous, e. g., in a drink following a purge, if the patient is feverish.[60] It is tempting to explain this use in modern terms as an effort to re-establish the fluid level and to maintain a correct electrolyte balance. While I think such an interpretation is correct, it is quite certain that this was not the explanation given in antiquity. I mention this because it is a particularly good example of one of the chief difficulties in the history of pharmacology—the ease with which the fallacy of historical projection can be committed.[61] The vast difference between the ancient and modern explanation of the physiological action of mineral drugs will be seen by considering another use of salt, viz., that it dries the body. Whatever

pp. 90-91. On the possibility of volatile oils of herbs as antiseptics, cf. L. Laloy, "Die Antisepsis bei den Alten." *Janus,* 1900, *5*: 535.

[57] Nasal fumigation: *Diseases of Women* 137 (VIII 310 L.); vaginal fumigation: *Superfetation* 32 (VIII 500 L.).

[58] For "uterine suffocation": *Diseases of Women* 203 (VIII 388-390 L.); as an anodyne: *ibid.* 206 (VIII 400 L.); as an emmenagogue: *Places in Man* 47 (VI 346 L.).

[59] *Diseases of Women* 89 (VIII 214 L.).

[60] *Affections* 41 (VI 250 L.).

[61] In *Diseases* II 59 (VII 92 L.) where pleurisy *may* be in question, Hippocrates advises that air from an inflated bladder be injected into the chest cavity. Lund, commenting on this passage, states, "Here is the ancient germ of nitrogen injections in phthisis . . .," F. B. Lund, *op. cit.*, p. 536.

be the modern explanation of this, there is no doubt about the ancient explanation. As we shall see later, the ancient explanation is formulated in terms of the theory of the elements. The other common sodium salt was **νίτρον**—the *nitrum* of Latin writers, the *neter* of the Hebrews, and the nitre of the English translations of the Old Testament. The dispute, going back to the Renaissance, whether this meant soda or saltpetre (KNO_3), has just recently been put to rest by the detailed study of Partington, who has shown conclusively that **νίτρον** etc., always meant soda.[62] It was used internally, but more often externally in the form of cataplasms and wet dressings. The internal uses are, however, of more interest historically. Formerly, it was believed that the internal use of minerals was the direct result of the Paracelsian revolution of the sixteenth century. But, as a few examples may indicate, the internal use of minerals was already well established in the Hippocratic Corpus, and an equal frequency of occurrence can be found in succeeding generations of Greek medical writers. Thus, soda is drunk with honey for flatulence or used in clysters for ulcers or lesions.[63] Or consider the following recipe:

> Soda, 2 drachmas
> Castoreum or sagapenum (*Ferula persica* Willd.) 1 obol
> Asphalt, i. e., liquid natural petroleum, 1 drachma
> Drink in an oil-wine mixture to expel the afterbirth.[64]

Other minerals included alum, probably impure aluminum sulphate, and sandarach, disulphide of arsenic, both of which were used in ointments and in suppositories.[65] *Misy,* an impure antimony sulphide, was likewise used, and there is one recipe in which it is to be drunk with water as a contraceptive.[66] Copper salts are probably the most difficult to identify since copper enters into the composition of many common minerals in many different chemical forms. Copper salts in the form of an acetate were used in suppositories, while the oxide was used in plasters, and the carbonate was especially recommended as an ointment for fistula in ano.[67]

[62] J. R. Partington, *A History of Greek Fire and Gunpowder.* Cambridge: Heffer, 1960, pp. 298-339.

[63] For flatulence: *Diseases of Women* 177 (VIII 360 L.); uterine clyster: *ibid.* 78 (VIII 192 L.).

[64] *Ibid.* 78 (VIII 184 L.).

[65] Alum: *Fistula* 3 (VI 450 L.); *ibid.* 10 (VI 460 L.); Sandarach: *Superfetation* 32 (VIII 500 L.); *ibid.* 33 (VIII 504 L.).

[66] *Disease of Women 76* (VIII 170 L.).

[67] Suppositories: *Haemorrhoids* 8 (VI 442 L.); *Sterility* 11 (VIII 432 L.); plasters: *Diseases of Women* 104 (VIII 226); fistula in ano: *Fistula* 3 (VI 448). On Hippocratic mineral drugs cf. J. Berendes, *Die Pharmacie bei den alten Culturvölkern.* Halle: Tausch & Grosse, 1891, pp. 196-204.

I

Lastly, we must mention water. Three types were usually distinguished, excluding aqueous solutions—ordinary water, sea water, and brine.[68] Naturally, they were used most importantly in drinks, especially in the form of decoctions of leaves.

The last remarks concern dosage. The entire problem of metrology and posology is so complex that little more can be done here than call attention to a few miscellaneous matters. It is of interest that the quantity of each of the several ingredients is often specified in the preparation of mixtures. One might conclude that, the required ingredients being available, it would be possible to duplicate the recipe. But that would be an unsafe inference. Since we have no knowledge of what the Hippocratic ingredients would assay, we could not duplicate a recipe volumetrically according to modern standards of exactness, despite our ability to duplicate it qualitatively. While there is probably a variation no wider than that permitted by modern standards between the botanicals of ancient time and the same plants found today, this is not true as regards mineral drugs or such excipients as vinegar or wine. Indeed, unless the wine came from the same jar, the possibility of even a rough assay is probably precluded.

Although our inability to duplicate a recipe volumetrically somewhat lessens the importance of exact decimal equivalents of the dosage of ancient drugs, we can in some cases transcribe them in modern notation: e. g., 1 obol = ⅙ drachma = 1.05 grams = ca. 15 grains. In other cases, however, we find recipes typified by the following, in which neither precise equivalents nor volumetric analysis seems possible:

> For a woman suffering during a difficult delivery
> She will drink:
> 4 dark seeds of peony
> 1 cheramys of a decoction of elder fruit (= .009 **L.**)
> castoreum as large as a bean.[69]

Beyond the difficulties of interpreting Hippocratic weights and measures, other references to dosage seem as straightforward as they are sensible: e. g., dosage *ad libitum* is frequently mentioned, while in one case the dosage is specified to be " in accordance with the severity of the complaint." [70]

[68] *Use of Liquids* 1 (VI 118 L.). Cf. also, *In the Surgery* 13 (III 316 L.) and Dierbach, *op. cit.*. pp. 1-5.

[69] *Diseases of Women* 136 (VIII 308 L.).

[70] *Ibid.* 175 (VIII 358 L.); *Sterility* 18 (VIII 440 L.); *Diseases of Women* 78 (VIII 184).

II. Pharmacodynamics

There are scattered throughout the Hippocratic writings theories and speculations that shed valuable light on the earliest period of Western pharmacodynamics. But because these passages are not systematically connected, a more probable beginning can be found by examining the relation between drugs and diet. It is well known that dietary and other forms of regimen bulked large in Hippocratic writings.[71] It is probably safe to assume that some of the distinctive features of Hippocratic pharmacology are derived, more or less directly, by observing the gains, unspectacular but certain, of a controlled regimen.[72]

Let us begin by briefly comparing two of the more famous Hippocratic writings from the standpoint of drug administration and diet. *Epidemics* I is so well known that there is no need to rehearse the evidence that it is one of the earliest examples of the now-familiar case-method. And here we meet a strange thing. Despite the writer's insistence that the patient's regimen must be regulated by considering his history,[73] in only one or two instances is there any mention of drugs, and even there the composition of the prescription is not specified. This case describes a woman suffering perhaps from puerperal fever. " She began," so runs the text, " by feeling pain in the stomach and the right hypochondrium. Pains in the genital organs. The discharge stopped. A suppository relieved these conditions." [74] With the exception of a second reference to this suppository some ten lines later, no other drugs are mentioned in the fourteen cases with which *Epidemics* I closes. One might conclude from this that the writer of the *Epidemics* was guilty of a callous, inhuman devotion to scientific accuracy; and in order to attain this goal, the patient's welfare was neglected. This is, I think, unlikely for the following reasons.

There is no simple, botanical or otherwise, which is effective against a highly infectious disease. Quite probably the Hippocratic writer, through

[71] It has recently been argued that for Hippocrates as well as for succeeding generations of Greek physicians, " scientific medicine was identical with dietetic medicine," Owsei Temkin, " Nutrition from classical antiquity to the baroque," in Iago Galdston, ed. *Human Nutrition, Historic and Scientific.* New York: International Univ. Press, 1960, p. 83. Cf. also Ludwig Edelstein," ΠΕΡΙ ΑΕΡΩΝ und die Sammlung der hippokratischen Schriften." *Problemata,* 1931, *4*: 100 sq.

[72] Cf. *Regimen in Acute Diseases* 9 (II 280 L.) and such empirical findings as, " In every disease it is a good sign when the patient's mind is sound and his appetite good; the contrary is a bad sign," *Aphorisms* II 33 (IV 480 L.). Cf. *ibid.* VII 66 (IV 598 L.).

[73] *Epidemics* I 3, 11 (II 676 L.). Cf. also *Nature of Man* 9 (VI 54 L.) and *Aphorisms* I 2 (IV 458 L.).

[74] *Epidemics* I 13, case 4 (II 690 L.).

numerous failures, knew this also. For this reason, the most effective measures he could take were palliative. He would try to keep the patient as comfortable as possible and await either the crisis that spelled recovery or the famous prognostic sign, the *facies hippocratica,* the face of death. There are suggestions in *Epidemics* I that the patient's comfort was a matter of concern to the physician; he was given water when thirsty, cooled when feverish, etc.[75] The old concept of "let nature take her course" was not a counsel of despair; it was rather the admission that there was no effective rational treatment available.[76]

Now if we turn to another writing and consider the treatise, *On Regimen,* we shall find something quite different. We may, in fact, be in a better position to understand the virtual absence of drugs from *Epidemics* I. *On Regimen* is concerned largely with two matters: a list of precepts designed to promote or maintain health and, secondly, a similar list of food stuffs classified, in terms of their useful properties, for a variety of different complaints.[77] The interesting feature for us is the analysis of foodstuffs. This analysis will provide a good example of early pharmacodynamics and also help to explain *Epidemics* I. Foodstuffs are classified in three ways: (i) in terms of their physiological action, e. g., diuretics or laxatives, (ii) in terms of their therapeutic properties, supposed or real, e. g., cholagogues, and (iii) in terms of their relative nutritive value. The question then arises how to distinguish foodstuffs from drugs for, as we have seen, the same items may appear as either. This is, indeed, a difficult problem, as we know from Galen's

[75] Case 7 of *Epidemics* I (II 700-702 L.) describes a feverish patient. On the second day, "after frequent drinking, he had a copious stool," and on the fifth day "his head was bathed." Case 2 (II 688 L.) states, "on the tenth day [the patient] received no drink." Elsewhere attendants are mentioned (II 692 L.) who presumably took care of such matters and perhaps supervised the diet. Ackerknecht summed up the point succinctly: "For better or worse Hippocrates observed sick people, not diseases." Erwin H. Ackerknecht, *A Short History of Medicine.* New York: Ronald Press, 1955, p. 112. For a detailed examination of the physician-patient relation, cf. Ludwig Edelstein, "The professional ethics of the Greek physician." *Bull. Hist. Med.,* 1956, *30*: 391-419 and H. N. Couch, "The Hippocratean patient and his physician." *Tr. Am. Philological A.,* 1934, *65*: 153 sq.

[76] The belief that the physician "assists nature by drugs" is contained in such aphorisms as, "Nature is the physician of diseases," *Epidemics* VI 5 (V 314 L.); "Nature (lit. natures) is untaught in all things," *On Nourishment* 39 (IX 112 L.). For an attempt to explain "natures" in terms of modern medical science, cf. August Bier, "Hippokratische Studien." *Quellen u. Stud. z. Gesch. d. Naturw. u. d. Med.,* 1932, *3*: 51-78.

[77] Cf. Adolf Palm, *Studien zur hippokratischen Schrift* Περὶ Διαίτης. Diss., Tübingen: Tübinger Chronik, 1933.

remarks on the subject.[78] The distinction between drugs and foodstuffs is, I believe, a purely pragmatic one: If an item is administered to a sick person with a curative aim, regardless of its mode of administration and the nature of the complaint, it is *ipso facto,* a drug. Thus, *cardamon,* probably our cress (*Lepidium sativum* L.), is said to be heating and to melt the flesh; however, since it congeals white phlegm, so as to produce strangury, its indiscriminate usage must be controlled and it is positively prohibited for phlegmatic persons.[79] Coriander (*Coriandrum sativum* L.), another common confectionery and flavoring agent, also possesses medicinal value. It is analyzed in the following manner: it is hot and astringent; it stops heartburn and when eaten last causes sleep.[80] Calaminth, perhaps our catnip (*Nepeta cataria* L.), warms and purges,[81] while blite (*Amaranthus blitum* L.) warms without passing well by stool.[82] Thus, by a combination of common sense and observation, based perhaps on his own dietary habits, the writer presents us with a theory of how botanical drugs work. But in virtue of the fact that these plants also form part of the normal diet, the viewpoint of the dietician is never wholly absent in Hippocratic discussions of simples. This comes out particularly well in another treatise, *On Ancient Medicine,* in which the inelegant but accurate word " slops " designates barley gruel, a basic item in the normal diet. Depending on the patient's condition as well as on certain theories to be mentioned presently, the gruel is either thickened, diluted with water, honey, or wine, or flavored by various herbs, e. g., dill or anise.[83] It is not surprising, then, to find few explicit references to drugs in *Epidemics* I. From the standpoint of that writer, botanicals are part of the ordinary dietary regimen which is, presumably, carefully regulated during any illness. The language and logic of the *Epidemics* suggest that attendants,[84] or the physician himself, supervised the diet. Thus the statement " They were not hungry, in fact, food was able to be given " [85] and such observations as " a complete lack of appetite " [86] and " They gladly took all sorts of food, after which their fever abated," [87] indicate, at the very least, that food was given during the course of the illness and perhaps it formed part of the treatment. On the other hand,

[78] Galen, *De alimentorum facultatibus* in *Galeni opera omnia,* ed. C. G. Kühn, Leipzig: Cnobloch, 1821-1830, VI, 468. For a discussion of this point, cf. Temkin, " Nutrition . . ." (cf. n. 71), p. 87.

[79] *On Regimen* 54 (VI 558 L.).

[80] *Ibid.* 54 (VI 558 L.).

[81] *Ibid.* 54 (VI 558 L.).

[82] *Ibid.* 54 (VI 560 L.).

[83] Anise flavored wine: *Diseases of Women* 34 (VIII 82 L.).

[84] Cf. note 75.

[85] II 610 L.

[86] II 628 L.

[87] II 632 L.

from the strongly taxonomic and analytic standpoint of the writer of *On Regimen,* these same items are considered as drugs and classified accordingly.

Before concluding this section on pharmacodynamics, three further points must be mentioned. The first concerns the role of experiment. In our sense of the term, experiments, with all the conceptual apparatus of prediction, hypothesis, controls, statistical evidence, and precise instrumentation, are of course wanting. But experiment, etymologically derived from the Greek *peiraomai,* "to attempt," need not be taken so strictly. There is little doubt that experiment in the sense of utilizing first-hand experience, as opposed to theorizing about how things ought to be, underlay many of the references to simples, e. g., the taste or odor of a certain plant.[88] Moreover, the Hippocratic writer must have been impressed by observing favorable or unfavorable results of his therapeutic practice. On the other hand, there are instances of experiments in the stricter sense of testing. In the treatise, *On the Heart,* it is asserted that some liquid, instead of passing into the stomach, passes into the trachea. As proof of this, some colored water is given to a thirsty pig. If its throat be cut immediately, the trachea will be found colored.[89] What is important here is the attempt at substantiating a claim by empirical *and experimentally-produced evidence.*

This brings up a second point whose mere mention is of considerable historical importance. There is, in the Hippocratic Corpus, no evidence of the practice of magic medicine. The absence of incantations and amulets is a sufficiently strong reminder of the darker days that were to follow.[90] Dreams, it is true, are mentioned as an aid to prognosis, but this is not, so far as I know, ever actually utilized in medical practice.[91] Bags of aromatic herbs are worn about the neck,[92] but again, this is

[88] The surprisingly large amount of botanical information scattered throughout the Hippocratic Corpus suggests that the botanical resources were known to the Hippocratic physician through practical experience. A study of this material is in preparation and will appear elsewhere.

[89] *On the Heart* 2 (IX 80 L.). There is an English translation of this short treatise. F. R. Hurlbutt, Jr., "Peri Kardies. A treatise on the heart from the Hippocratic corpus." *Bull. Hist. Med.,* 1939, *7*: 1104-1113. For a discussion of this and other related passages, cf. G. Senn, "Über Herkunft und Stil der Beschreibung von Experimenten im Corpus Hippocraticum." *Arch. f. Gesch. d. Med.,* 1929, *22*: 217-289, and W. A. Heidel, *Hippocratic Medicine.* New York: Columbia Univ. Press, 1941, p. 105.

[90] A. R. Mills, "Peasant remedies from the Greek islands." *Bull. Hist. Med.,* 1948, *22*: 441-450.

[91] *On Regimen,* Book IV (VI 640 sq.) is subtitled, "On Dreams." For the use of dreams in diagnosis, cf. *Humors* 4 (V 480 L.).

[92] *Fistula* 10 (VI 460 L.).

only ineffective not magical, especially so, when it is remembered that the beneficial action of the inhalation of the volatile oils of many herbs is a belief that persisted up through the seventeenth century.

The last item to be mentioned is the role of toxicology. There is little evidence that poisons, either natural or introduced by artifice, formed a very important part of Hippocratic pharmacodynamics. *Peganon,* or rue (*Ruta graveolens* L.) is stated to be a prophylactic against poisons, while in the Hippocratic *Oath,* the physician is enjoined from engaging in the administration of poisons.[93] It was only later, when poisoning became a widespread social threat that major importance was given to antidotes and the *Giftbuch.*

III. Pharmacotherapy

This final section raises some difficult questions concerning the interpretation of Hippocratic pharmacology. Two such problems are the Hippocratic concept of disease and the important role played by non-medical theories in the formulation of the disease concept.

There is a tradition going back to antiquity that the Coan, or Hippocratic school, so named for the island of Cos, disagreed wth the rival Cnidian school over a fundamental question. At issue allegedly was the relative weight to be attached to prognosis and to diagnosis. The Hippocratic school, we are told, came out strongly in favor of prognosis.[94] However this may be, much of Hippocratic pharmacotherapy is concerned with the treatment of symptoms. Because of the elaborate symptomatology, it is sometimes possible to identify a disease and thus to evaluate the particular therapy. A noticeable characteristic of Hippocratic writings is that a drug is usually prescribed for a definite complaint. Although purges, for example, are often recommended for specified complaints, these complaints are, more often than not, symptoms rather than identifiable clinical entities. Other complaints for which drug therapy is recommended include various types of discharges and gastrointestinal disturbances, often marked by biliousness, headache, fever, and chills. For the treatment of such symptoms there was, as indicated above, a full stock of drugs available. The extent of available drugs may

[93] Rue as a prophylaxis: *On Regimen* 54 (VI 558 L.). For the passage in the Hippocratic Oath on poisons, cf. Ludwig Edelstein, *The Hippocratic Oath. Supplements to Bull. Hist. Med.,* no. 1, Baltimore: The Johns Hopkins Press, 1943, pp. 6 sq.

[94] Theodor Meyer-Steineg, "Die Bedeutung der Prognose in den hippokratischen Schriften." *Arch. f. Gesch. d. Naturw. u. d. Technik*, 1913, *6*: 258-262. On the motives for, and social significance of, prognosis in the Hippocratic Corpus, cf. Edelstein, ΠΕΡΙ ΑΕΡΩΝ (cf. n. 71), pp. 60-88.

I

by measured by the fact that no less than 53 different simples are recommended as emmenagogues.

In discussion of the clinical background of Hippocratic pharmacotherapy, one must bear in mind that theory often determined the choice of therapy. We have observed above that an analysis of a drug included the specification of its properties, e. g., its tendency to warm and dry, or to evacuate bile, or its reputed effectiveness as a galactagogue or soporific. In order to more fully understand such claims, it is necessary to say a few words about the influence on Hippocratic medicine of two important theories. The theory of the humors and the associated theory of the four elements probably had their origin in pre-Socratic natural science of the fifth century.[95] We find, for example, traces of these theories in Empedocles, Alcmaeon, and other early writers. The best known version of the doctrine of the humors, the one which became classical, states that man is composed essentially of four vital humors or juices—blood, phlegm, yellow bile, and black bile. When the humors are in equilibrium, bodily health is maintained, while a disturbance, depending on its severity, causes a proportionately severe disturbance or disease. Moreover, and as a consequence of this general statement, each disease is characterized by the relative excess or depletion of one or more of the humors. Here, then, in crudely quantitative terms, is the foundation of Hippocratic physiology, pathology, and aetiology. In order to illustrate this, I shall cite the account of epilepsy found in the treatise *The Sacred Disease.* " The disease," the author states,

> attacks the phlegmatic but not the bilious or choleric. Its origin begins in the embryo while it is still in the womb, for like the other parts, the brain too is purged and has its impurities expelled before birth. In this purging, if the action be thorough and regulated, and if there flow away neither more nor less than is proper, the infant has a perfectly healthy head. But if the flux from all the brain be too abundant, and a great melting take place, he will have as he grows a diseased head, and one full of noise. . . .[96]

Here we have the aetiology of a disease stated in terms of a theory which, if accepted, necessarily determines the therapy. The object of therapy will be to re-establish the equilibrium of the humors. It will now be possible to see more clearly why the writer of *On Regimen* was so careful to specify the properties of foods and drugs.

[95] Charles Kahn, *Anaximander and the Origins of Greek Cosmology.* New York: Columbia Univ. Press, 1960, pp. 119-165.

[96] *The Sacred Disease* 8, in *Hippocrates.* Trans. W. H. S. Jones (Loeb Library), London, 1952, vol. II, pp. 155-157.

The famous medical precept of *contraria contrariis curantur,* " opposites are cured by opposites," has its origin here.[97] The rational grounds of drug therapy are found within the framework supplied by the twin theories of the humors and the four elements. Since phlegm, for instance, is thought to partake equally of the cold and moist, a phlegmatic person is characterized by the predominance of cold and moist humors in his body. Any drug that promised to restore the warm-cold balance of the body was thereby indicated: e. g., *thymos* (*Thymbra capitata* L.) evacuates phlegm because it heats and dries.[98]

Another crucial matter pertains to the old belief, going back to Plato, that Hippocratic medicine is concerned with the " whole man." [99] An extension of the relation between regimen and drug therapy can now be mentioned. Regimen covered much more than diet in the modern sense. It included precepts designed to regulate man's relation to his environment, not only with respect to diet, but also with respect to exercise and rest, clothing, shelter, a favorable location for one's dwelling, sexual activity, and the like.[100] If a sick person is to be treated effectively, he must be viewed as a part of a larger context. The writer of the short work, *Regimen in Health,* summed up the thought nicely, " It is necessary to establish regimen with regard to age, season, habit, land, and physique, and to counteract the prevailing heat or cold." [101] It is as important in the prognosis and treatment to regulate exercise and rest as it is to institute a supervised diet. The case-method is thereby justified in the attempt to tabulate all of the data relevant to the prognosis. This information concerning the individual determines the choice of therapy fully as much as the theories mentioned above and probably more than the ready availability of specific drugs. Thus, a person suffering from an overabundance of bile is, according to the theories current in Hippocratic writings, hot and dry. Therapy may be the administration of a cho-

[97] This doctrine is stated in various ways in the Hippocratic writings, e. g. *Nature of Man* 9 (VI 52 L.); *On Regimen* 40 (VI 538 L.).

[98] *On Regimen* 54 (VI 560 L.). For the identification of *thymos,* cf. Alfred C. Andrews, " Thyme as a condiment in the Graeco-Roman era." *Osiris,* 1958, *13*: 150-156.

[99] The relevant passage in Plato (*Phaedrus* 270 C) reads, " Do you think it possible to obtain knowledge of the nature of the soul without knowing the nature of the whole? " Whether this means " whole of man " or " the whole (sc. universe)" cf. Ludwig Edelstein, " The role of Eryximachus in Plato's Symposium." *Tr. Am. Philological A.,* 1945, *76*: 97 sq., F. Steckerl, " Plato, Hippocrates and the Menon Papyrus." *Classical Philology,* 1945, *40*: 166-180, and my paper, " Socratic eros and Platonic dialectic." *Phronesis,* 1959, *4*: 120-134.

[100] Cf. e. g., *On Regimen,* chapters 57-66 (VI 570-588 L.).

[101] *Regimen in Health* 2 (VI 76 L.).

lagogue, e. g., beet juice or marjoram.[102] On the other hand, any drug is indicated which increases the cold and moist elements in the body and, so to speak, neutralizes the hot and dry. Barley juice is a good example, for it is explicitly stated to be cold and moist.[103] But, equally important, the patient must forego heating and drying foods; furthermore, he must avoid any environmental factors or any activity which would promote the specific therapy is justified at three different levels: (i) establishment of the proper balance of humors; (ii) establishment of the proper relation of the four elements—the hot, dry, cold, and moist; (iii) the availability of specific drugs.

Several miscellaneous topics must round out this discussion of Hippocratic pharmacotherapy. It has been mentioned above that it is difficult to identify the diseases whose symptoms are described and treated. There are relatively few instances where a disease can be identified with some certainty, e. g., mumps in *Epidemics* I 1, epilepsy in the *Sacred Disease,* and a few other conditions such as apoplexy and Pott's Disease have been recognized. It is not intended here to catalogue the diseases nor even to offer tentative identifications.[105] It is more germane to our interests to note what sort of diseases and disorders were treated by the Hippocratic physician. For this purpose it would be invaluable to possess statistics on the incidence of diseases complete with all the tabulated data of our Public Health surveys—age, sex, racial and geographical indices, etc. Barring this, we must be content with generalities. From the pharmacotherapeutic standpoint, the most important group of diseases were probably gastro-intestinal, followed closely by upper-respiratory complaints. It must not be forgotten that malaria was endemic in ancient Greece.[106] It is, therefore, probable that a latent or chronic malarial infection colored the clinical manifestations of many disorders. The frequent references to enlarged or painful spleen and the periodical chills and fevers are ample evidence.[107] Other common complaints for which drugs were prescribed included localized lesions and infections, a wide variety of dermatological conditions, and a host of gynaecological dis-

[102] Beet juice: *Diseases of Women* 109 (VIII 232 L.); marjoram: *On Regimen* 54 (VI 560 L.).

[103] *On Regimen* 40 (VI 536 L.).

further increase of his already hot and dry constitution.[104] Thus, the

[104] For a general statement, cf. *Regimen in Health* 1 (VI 72-74 L.).

[105] E. W. Goodall, "On infectious diseases and epidemiology in the Hippocratic collection." *Proc. Roy. Soc. Med.,* Hist. Sect., 1934, *27*: 524-534.

[106] W. H. S. Jones, *Malaria and Greek History.* Manchester: Univ. Press, 1909.

[107] References to swollen spleen are frequent: *Airs, Waters, Places* 7 (II 26-28 L.); *Affections* 20 (VI 228 L.); *Humors* 8 (V 488 L.); *Aphorisms* III 22 (IV 496 L.).

orders whose differential diagnosis is a matter of dispute.[108] The Greek term *phthisis,* current even in the nineteenth century as a synonym for consumption, undoubtedly included pulmonary tuberculosis.[109] However, since the presence of the tubercle bacillus clearly cannot be demonstrated, a positive diagnosis in the technical sense is hardly possible.[110] Infectious diseases, probably including diphtheria and mumps, yielded to no therapy save palliative.[111] At the other extreme, there are few references to battle wounds, bites of reptiles or insects, and remarkably few eye complaints. Although fractures are treated in a series of very capable treatises, they are of more interest to the history of orthopedic surgery than to pharmacology.

Anyone familiar with more than half a dozen of the Hippocratic treatises must recognize profound differences in content and in style, sometimes reaching the point where contradictory assertions can be adduced with equal authority. Multiple authorship is proven beyond question. Moreover, it is likely that different audiences were intended by the various Hippocratic writings.[112] Some of the books appear to be written for the professional physician, others for the layman, and perhaps a third group ought to be added comprising midwives, leeches, medical assistants, and apprentices. In those books with complicated recipes and equally complicated preparation and administration, a professional physician is likely to be the intended reader. He, or his assistant, is probably expected to collect the simples and prepare them as stipulated. On the other hand, in those works in which the therapy is largely dietary or " regiminal " a lay person may be the intended reader. This cannot, of course, be proven, but it does offer a partial explanation of the otherwise surprisingly wide range of pharmacotherapeutic methods and substances.

With mention of the wide range of durgs available, we have come full circle in our discussion of Hippocratic pharmacology and the place of drugs in Hippocratic medicine. Western medicine got off to an auspicious start with such a rational application of drugs and their administration.

[108] Cf. H. Fasbender, *Entwickelungslehre, Geburtshülfe und Gynäkologie in den hippokratischen Schriften.* Stuttgart: Enke, 1897.

[109] Cf. Hans Moritz, " Die Behandlung der Lungentuberkulose im Altertum." *Ztschr. f. Tuberk,* 1934, *69*: 28-38, 100-113, and A. S. Pease, " Some remarks on the diagnosis and treatment of tuberculosis in antiquity." *Isis,* 1940, *31*: 380-393.

[110] Zambaco, as late as 1912, concluded that " nos constatations démontrent mathématiquement " the presence of tuberculosis, etc. in antiquity. D. Zambaco, " L'antiquité de la léprose, de la syphilose et de la tuberculose." *Arch. f. Dermat. u. Syph.,* 1912, *111*: 193-211.

[111] Diphtheria is described in *Dentition* 20 (VIII 546 L.). Cf. H. Haeser, *Lehrbuch der Geschichte der Medicin.* 3. Aufl., Jena: Dufft, vol. III, 1882, pp. 428-429.

[112] On this matter, cf. Owsei Temkin, " Greek medicine as science and craft." *Isis,* 1953, *44*: 213-225.

II

Pliny and Roman Botany

IF HISTORY CAN be personified, then it may be said that she has been unkind to Pliny. There was a time, indeed, when his influence equaled that of Aristotle, Galen, and Dioscorides; and throughout the Middle Ages, it surpassed that of Theophrastus. With the reaction to classical science, beginning in the late fifteenth centry, his fortunes suffered an eclipse. The long shadows cast by the Renaissance continue to this day to obscure his place in the sun.[1]

It is generally agreed that the thirty-seven books comprising his encyclopedic *Historia naturalis* share the defects of their author. He was gullible and uncritical, he lacked great originality, and he was not possessed of sufficient training always to understand the details of the scientific and technical problems he discussed. These charges cannot be put aside altogether, nor is that my intention. Rather, I shall suggest that a study of the botanical portions of his writings, long overdue, will reveal not only that Pliny is an invaluable source for tracing the development of early botany, but also that he made important contributions of his own.

From Pliny's nephew we learn that a servant read aloud to him at the dinner table as well as in a sedan chair being carried through Rome. Excerpting, and quickly translating if the text were in Greek, Pliny dictated to another servant.[2] Because of these less than ideal working conditions, necessary for the busy administrator that he was, it is not surprising that inaccuracies and errors were introduced. It is possible, for example, that the plant name *helxine* is an error for *ixine*.[3] There is this much truth in the assumption: the name *helxine* occurs in an almost literal translation of a passage from Theophrastus where *ixine* is discussed.[4] The cause for this may have been nothing more serious than an auditory lapse on Pliny's part, analogous to

* University of Colorado. This paper was delivered at the annual meeting of the History of Science Society, Philadelphia, 28 December 1963. Support from U.S.P.H.S. G.M.–9361–02 is acknowledged.

1 It is worth noting that the initial criticism of Pliny came at the hands of lexicographers who, like Hermolao Barbaro and Nicolò Leoniceno, were more adept at philological exegesis than at botany. For a general statement of the problem, cf. Bernhard Milt, "Schweizerische Theophrastforschung und schweizerische Theophrasteditionen im 16. Jahrhundert und ihre Bedeutung," *Gesnerus*, 1946, *3*:76, and Jerry Stannard, "Dioscorides and Renaissance Medical Botany" (in press).

2 Plinius Caecilius Secundus, *Epistularum libri*, III 5. Ed. R. C. Kukula (Leipzig: Teubner, 1912), pp. 65–66.

3 Pliny, *Natural History*, XXI 56, 96. Ed. H. R. Rackham, *et al.* (Loeb Classical Library) (10 vols. London: Heinemann, 1938–1962). All references to Pliny are to this edition and the translations are those of the Loeb editors.

4 Theophrastus, *Enquiry into Plants*, VI 4, 9. Ed. Sir Arthur Hort (Loeb Classical Library) (2 vols. London: Heinemann, 1948–1949).

Reprinted from *Isis: Journal of the History of Science Society*, Winter 1965, Volume 56, No 186, by kind permission of the University of Chicago Press.

II

a typographical error by a not too attentive secretary.[5] I mention this to serve notice to historians who, in lumping together inaccuracies and botanical errors, have only confounded the issue.

Of course, Pliny made errors; that cannot be denied. Yet even here, one must read the text before pontificating, as a well-known historian of science has recently done. To say that Pliny "added nothing [to ancient botany] beyond the cult of marvels to the existing texts which he pillaged" [6] betrays as profound an ignorance of Pliny's text as it does of ancient biology.

Anyone who has seen a venerable cork oak with its scarred and mutilated trunk must be struck by the fact that it endures an injury fatal to other trees.[7] "All trees perish," wrote Theophrastus, "if the bark is stripped off all round. . . . However, one should perhaps except the cork-oak; for this, they say, is all the stronger if the bark is stripped off." [8] Its vitality, in point of fact, is *not* improved by removing the outer bark.[9] Yet, to Theophrastus, this appeared to be a natural deduction based upon the fact that this process left the cork oak unimpaired – so much so that it could be repeated time and again. Pliny duly reports this belief.[10] Yet it has sometimes been forgotten that, in another passage, he described the cork oak in a straightforward manner, and he enumerated some dozen uses for the cork without once mentioning Theophrastus' claim.[11] The moral is easy to draw: if error is chargeable, let the charge be exact.

The last example points up another area of controversy. Pliny, we know, relied heavily upon the writings of others, but for all that, he was no slavish copyist. When Pliny and his source disagree, it certainly does not follow that the former's report must be inferior. Thus, in several instances, what appears to be a mere extract from Dioscorides turns out to be on closer inspection a more detailed account.[12] Both writers, in one instance, agree that portions of the plant called *androsaimon* stain the fingers red and, for that reason, obtained its name, literally "man's blood." Despite the overall similarity between the two passages, Pliny made one significant alteration. While Dioscorides is content with stating that the fruit "when crushed, make the fingers, as it were, bloody," Pliny more precisely declared that the stain proceeds from the crushed *comae*.[13] Commenting on this passage, Andrews has written: "It looks as if Pliny were being more specific than Dioscorides, localizing the staining substance in the filament (the *comae* of

5 Cf. Jacques André, "Pline l'Ancien botaniste," *Revue des Études Latines*, 1955, *33*:308.

6 A. R. Hall, *The Scientific Revolution* (Boston: Beacon, 1954), p. 276.

7 Knowledge of the debarking process of the cork oak must have been fairly common even before Theophrastus' time, judging from references to corks as floats for fishing nets in such writers as Aeschylus, *Choephoroi* 506, and Pindar, *Pythian Odes*, II 80.

8 Theophrastus, *ed. cit.*, IV 15, 1.

9 The tree as a whole appears not to benefit from the debarking process though the quality of the cork produced is greatly enhanced. Cf. F. L. Sargent, *Plants and Their Uses* (New York: Holt, 1913), p. 279.

10 Pliny, *op. cit.*, XVII 37, 234.

11 *Ibid.*, XVI 13, 34.

12 Cf. Max Wellmann, "Sextius Niger, eine Quellenuntersuchung zu Dioscorides," *Hermes*, 1889, *24*:530–569. Wellmann argues that parallels between Pliny and Dioscorides are to be explained by assuming that they borrowed from a common source – Sextius Niger.

13 Dioscorides, *De materia medica*, III 155. Ed. Max Wellmann (3 vols. Berlin: Weidmann, 1907–1914); Pliny, *op. cit.*, XXVII 20, 37.

our text), the hair-like part of the stamen that supports the anther." [14] It may be concluded that Pliny has incorporated material independent of that in Dioscorides. The correctness of this addition, it may be noted, was confirmed in 1911, when Černý succeeded in isolating a red pigment called hypericin from the floral parts of *Hypericum perforatum* L., the *androsaimon* of Pliny and Dioscorides.[15]

Nowhere has Pliny been more maligned than in connection with the transmission of unverified, and apparently unverifiable, reports — what Blainville once termed his " empiricisme pur." [16] Yet, with few exceptions, scholars have ignored the fact that some of the most valuable portions of Pliny's botany are derived not from the texts of others but from personal experience. As a young man in Rome, he tells us that Caecina Largus showed him the large specimens of the nettle tree (*Celtis australis* L.) growing on his estate.[17] This is not, perhaps, a particularly illuminating observation, but it testifies to the fact that he had eyes for something besides the written word. In another passage, he mentions that mint, green in summer, nevertheless turns yellow in the autumn.[18] A homely remark such as this is deserving of attention if for no other reason than that, while known to every gardener, it is not the sort of information that often finds its way into books.

A more critical observation concerns Solon of Smyrna, an erstwhile botanist of repute known to us principally through Pliny. Judging from the context, Solon had claimed that orache (*Atriplex* spp.) would not grow in Italy. At least, this is what is implied by Pliny's retort: " I am at a loss to understand why Solon has stated that orache is difficult to grow in Italy." [19] The chapter in which this passage occurs opens with the words " atriplex et silvestris est," meaning, I take it, that both the cultivated and wild forms were found in Italy.[20]

An unparalleled ethnobotanical observation relates that the Chauci, a tribe inhabiting the littoral around the mouth of the Elbe, live in houses raised on turf hillocks in order to escape the tides. Pliny, who may have actually had dealings with this vassal tribe while he was cavalry commander in the Germanies between 47 and 57 A.D., also enumerated some of the items they manufactured out of reeds and sedges.[21] Referring to an even more northerly locale, the Borkum Islands in the North Sea, he tells us how they obtained their then-current name — *Fabaria* — the Bean Islands.[22] Despite

14 Alfred Andrews (in) Pliny, *op. cit.*, Vol. VII, p. 411.

15 C. Černý, " Über das Hypericin (Hypericumrot)," *Zeitschrift für physiologische Chemie*, 1911, *73*:371–382.

16 On a so-called unverified account, which has in fact been authenticated by a combination of field work and literary research, see Jerry Stannard, " The Plant Called Moly," *Osiris*, 1962, *14*:272–273.

17 Pliny, *op. cit.*, XVII 1, 5.

18 *Ibid.*, XIX 47, 159.

19 *Ibid.*, XX 83, 220.

20 The cultivated species, *Atriplex hortense* L., is widely distributed over most of Italy; the wild form is probably *A. nitens* Reb. but perhaps *Chenopodium album* L. Both are reported by Giovanni Arcangeli, *Compendio della flora italiana. . . .* Ediz. seconda (Torino: Loescher, 1894. Pp. 210, 213), though the former has only a local distribution.

21 Pliny, *op. cit.*, XVI 1, 4.

22 *Ibid.*, XVIII 30, 121.

the unlikely name, this is one of several observations which botanists have taken seriously.[23] Pliny's explanation is not only correct, it is the first reference in the literature to *Pisum maritimum* L., a species of pea not found, in the wild state, in the Mediterranean Basin.[24]

There are good reasons for asserting that if Theophrastus was the Father of Botany, to Pliny belongs the honorific Father of the History of Botany. Early developments in agricultural and related technological practices and the dates of introduction into Italy of foreign plants are but two of the subjects he discussed from an historical point of view. As he states in the prefatory letter to Titus: "There is not one person to be found among us who has made the same venture, nor yet one among the Greeks who has tackled single-handed all departments of the subject." [25] Much has been written on the subject of illustrated herbals, and from these, as he tells us, he learned to associate many Greek names with native plants.[26] Yet herbals were not his only source of information. "I at least," he states, "have enjoyed the good fortune to examine all but a few plants through the devotion to science of Antonius Castor, the highest botanical authority of our time; I used to visit his special garden, in which he would rear a great number of specimens." [27] Were it not for this passage, we should never know about the prototype of the modern physic garden. Nor would we know the names or writings of men like Amphilochus, said to have written a book on *medica*,[28] or Moschion, credited with one on radishes.[29] Even Gaius Valgius, a favorite of Augustus', is nowhere else reported as having left unfinished at the time of his death a book on the therapeutic properties of plants.[30]

It is surely worth something to find a phytogeographer like Alphonse De Candolle citing the *Natural History* as evidence that the pistachio (*Pistacia vera* L.) entered the West in historical times from Syria.[31] Likewise, it attests

[23] The outstanding example is the determination of a new species, *Faba Pliniana* Tr., on the strength of Pliny, *op. cit.*, XVIII 30, 121: "nascitur [faba] . . . in Mauretania silvestris passim, sed praedura et quae percoqui non possit." The author of this species reports that, notwithstanding Pliny's account, "il est néanmoins récolté par les femmes indigènes." Louis Trabut, *Répertoire des noms indigènes des plantes spontanées, cultivées et utilisées dans le nord de l'Afrique* (Algiers: Imprimeries "La Typo-Litho" et Jules Carbonel réunies, 1935), p. 110.

[24] Cf. Johannes Hoops, *Waldbäume und Kulturpflanzen im germanischen Altertum*, (Strassburg: Trübner, 1905), p. 465. In his *Species plantarum*, Linnaeus did not specify the region for his type specimen beyond stating "Habitat in Europae borealis littoribus maris arenosis"; however, he cited six lines above his own earlier *Flora suecica* (1746) whose ecological zones would include the Borkum Islands. Carl Linnaeus, *Species plantarum*, A facsimile of the first edition, 1763. Ed. W. T. Stearn and J. L. Heller (2 vols. London: Ray Society, 1957–1959), Vol. 2, p. 727.

[25] *Praefatio*, 14.

[26] For Pliny's use of an illustrated herbal, with accompanying descriptions, see especially XXV 4, 8. This, and the few other testimonia to Crateuas' herbal, have been examined by Max Wellmann, "Krateuas," *Abhandlungen der königlichen Gesellschaft der Wissenschaften zu Göttingen, Philol-Hist. Klasse*, 1897–1899, N.F. 2, no. 1:1–32 (Berlin: Weidmann, 1899).

[27] Pliny, *op. cit.*, XXV 5, 9.

[28] *Ibid.*, XVIII 43, 144. *Medica* has been identified as *Medicago sativa* L. or lucerne.

[29] *Ibid.*, XIX 26, 87.

[30] *Ibid.*, XXV 2, 4.

[31] Alphonse De Candolle, *Origin of Cultivated Plants* (New York: Appleton, 1885), p. 316. Cf. Pliny, *op. cit.*, XIII 10, 51, and XV 24, 91.

to Pliny's descriptive abilities to observe that Naudin, who wrote a monograph on the Cucurbitaceae or melon family, stating,

> il [Pliny] précise bien . . . les caractères du Melon, sa forme obovoïde, sa couleur jaune, son odeur et sa séparation spontanée d'avec le pédoncle, bien qu'il s'arrondisse à terre, et ne soit pas suspendu. Ces deux derniers caractères suffiraient pour caractériser le Melon à l'exclusion de toute autre espèce.[32]

In addition to the historicity and accuracy of his reports, the *Natural History* is a storehouse of information concerning contemporary practices, without whose proper understanding the history of early botany is lacunose. On the subject of statistics, Book XVIII provides the best evidence available for the extraction rates of flour. The latest writer on the subject has stated that Pliny "provides statements enabling us to relate given quantities of ground products to the quantity of grain used in their production, and this in turn makes it possible to know something about the quality of the various products."[33] A paraphrase of the principal passage runs like this: From one *modius* (1 peck or 8 quarts) of wheat from Campania, one may expect 4–5 *sextarii* (4–5 pints) of ordinary flour, ½ a *modius* of fine flour, 4 *sextarii* of *secundarium* (seconds), and an equal amount of bran.[34] Moritz, using a variety of wheat thought to most closely resemble that of antiquity, actually ground by hand the seeds in a first-century Roman quern found in Sussex.[35] The results strikingly confirmed Pliny's figures — for they were within the predictable range of the extraction rates just mentioned. It is pleasant to reflect how a Roman mill, dating perhaps from Pliny's time, has helped to rehabilitate his reputation.

There remains much more that could be written about Pliny's services to science and his immense influence. When it is recalled that the encyclopedic tradition, represented by Isidore of Seville, Vincent of Beauvais, Bartholomew the Englishman, and others, found its mainstay in his writings, it is not to be wondered at that something of Pliny's choice of language should find its way into the technical vocabulary of later botany. Terms such as *radix, folium,* and *fructus* were not, of course, the invention of any one writer. Yet, by studying Pliny's vocabulary, a key is provided for tracing the evolution of technical terms before they became standardized. Some of these terms occur for the first time in the literature in Pliny and, though used in a nonbotanical sense, their appropriateness as descriptive terms is at once apparent. For example, *pistillum* is still a pestle, while *stigma* refers to the brand by which slaves were identified.[36] On the other hand, some terms occurred with the same meaning then as now. *Internodium,* for

32 Charles Naudin, "Revue des cucurbitacées cultivées au muséum, en 1859," *Annales des sciences naturelles,* 1859, 4 sér. *12*:112. The passage to which Naudin refers is XIX 23, 67. Later commentators have followed Naudin in accepting Pliny's description as referring to *Cucumis Melo* L.

33 L. A. Moritz, *Grain-Mills and Flour in Classical Antiquity* (Oxford: Clarendon Press, 1958), p. 168.

34 Pliny, *op. cit.,* XVIII 20, 86–87. Cf. the detailed analysis in Moritz, *op. cit.,* pp. 170–171.

35 Moritz, *op. cit.,* p. 152.

36 Pliny, *op. cit.,* XXXIV 50, 169 (*pistillo lapideo*); XXX 10, 30 (*stigmata*).

example, is used to describe the space between the joints of a giant reed – probably bamboo – precisely what we call the internodal space or internodium.[37]

I have left for last a passage which epitomizes the direct connection between ancient and modern botanical terminology. *Lilium candidum* L., the beautiful and well-known madonna lily, is described thus:

> The flower is of an exceeding whiteness, fluted (*striati*) on the outside, narrow at the bottom, and gradually expanding in width after the fashion of a basket. The lips (*labrum*) curve outwards and upwards all round; the slender pistil (*pilus*) and stamens (*stamina*), the colour of saffron, standing upright in the center. So the perfume of the lily, as well as its colour, is twofold, there being one for the corolla (*calix*) and another for the stamens (*stamina*), the difference being slight.[38]

This passage is one of the most noteworthy descriptions in all of ancient botany. Both from the point of view of fidelity to nature and succinctness, it is unrivaled, and the identification of the plant so described is made absolutely certain. Moreover, five technical terms occur in this passage which are in use today, and one of them, *stamen*, appears for the first time in the literature in its correct botanical denotation. By recognizing that the pistil is a separate organ, distinct from the corolla, another giant step was taken in plant morphology. That Pliny should have described the sexual organs of a plant more carefully than Theophrastus may help historians to revise their opinions concerning his powers of observation and description.

Buffon, not known for his undue reverence for the past, once said of the *Natural History* that " cette compilation contient des choses rassemblées d'une manière si neuve, qu'elle est préférable à la plupart des livres originaux qui traitent de cette matière." [39] A case in point would be, I submit, Pliny's description of the lily.

[37] *Ibid.*, VII 2, 21.

[38] *Ibid.*, XXI 11, 23.

[39] Quoted by E. Littré, ed., *Histoire naturelle de Pline* (2 vols. Paris: J. J. Dubochet, 1850–1851), Vol. 1, p. vi.

III

MEDICINAL PLANTS AND FOLK REMEDIES IN PLINY, *HISTORIA NATURALIS*

It is well known that folk societies have always practiced some kind of medicine. In the process, those societies have accumulated a body of customs and lore traditionally associated with beliefs about and uses of plants for therapeutic purposes. As a result of many studies and investigations it has become increasingly apparent that many of those beliefs and uses rested on evidence that was not obtained, and perhaps cannot now be verified, by methods normally followed today such as bioassays, clinical trials, pharmacological screening and the like. It follows that the beliefs regarding the therapeutic uses of many of the plants known to have been employed in folk medicine depended upon evidence of a sort that would not be considered acceptable today.

The purpose of the present study, therefore, is to examine some of the grounds for the belief in and use of plants for therapeutic purposes independent of the kinds of scientific data mentioned above. In order to keep this study to a manageable size, I shall concentrate on the reports found in Books XII-XXVII of the *Historia naturalis*, written by Gaius Plinius Secundus (23-79 A.D.), better known as Pliny the Elder.[1]

Although Pliny has often been criticized, and in some instances rightly so, for his lack of critical judgment and his misunderstanding

---

[1] All references to Pliny are to his *Historia naturalis* eds. H. Rackham *et alios*. London, 1938-1962. 10 vols. (Loeb Classical Library).

of technical details, there are several reasons for choosing him for this particular study. First of all, because of his great command of the written sources, some of which subsequently have been lost,[2] the *Historia naturalis* provides a wide range of beliefs about medicine in general and medicinal plants in particular.[3] Equally important is the fact that Pliny included not only the claims of the better known Greek physicians and scientists, but folk beliefs and oral reports as well. These he gathered from rustics, tradesmen, and a wide assortment of those whom he termed *inperiti*.[4] A careful reading suggests, however, that Pliny was not always advocating or even accepting some of the grosser superstitions and magical practices about which he wrote. It is rather the case, that like a modern anthropologist, he was reporting what he had learned from his informants.[5] This is clearly indicated by his use of such terms as *aiunt*, *dicitur*, and *traditur* which frequently introduce his remarks concerning a fabulous report, superstition, or magical practice. Consequently, such passages do not necessarily represent Pliny's personal opinion.

While many historians of folklore and religion have recognized the importance of Pliny as a historical source, historians of science and medicine have not always appreciated the fund of information concerning folk science and folk medicine contained in his *historiae*.[6] It may be true that Pliny's findings are not of the calibre of Hippocrates,

---

[2] J. Stannard, *Lost Botanical Writings of Antiquity* in *Actes du XIe Congrès Intern. d'Hist. des Sciences*, Varsovie, 1968 t. V, pp. 319 sqq.

[3] On Pliny's botany in general, including medical botany, cf. J. André, *Pline l'Ancien botaniste*. « Revue des Etudes Latines », *33*, 1955, 297-318; M. Brosig, *Die Botanik des älteren Plinius*. Progr. Graudenz, 1883, and J. Stannard, *Pliny and Roman Botany*, « Isis », 1965, 420-425.

[4] XXV, 69; cf. XVII, 176 and XVIII, 206.

[5] With reference to *arsenogonon* and *thelygonon*, Pliny reports beliefs held by unspecified persons (*narrant*) and adds, *huius potu mares generari, si credimus*. XXVI, 162. Cf. also XVII, 58; XXI, 153; XXVII, 141 etc.

[6] Fewer studies exist on the folkloristic elements of Pliny's *phytotherapia* than on his *zootherapia*. Noteworthy are L. Moulé, *Le folklore médical de Pline*, « Bull. Soc. franc. d'hist. méd. », *17*, 1923, 71-85 and W. B. McDaniel, *Some Roman Remedies in Italian Folk medicine*, « Trans. College of Physicians of Philadelphia », *12*, 1944, 72-87. Cf. also W. H. S. Jones, *Ancient Roman Folk Medicine*, « Journal of the History of Medicine », *12*, 1957, 459-471 and E. Stemplinger, *Der Heilmagnetismus bei Plinius*, « Ztschr. f.d. deutschösterr. Gymnasien », *69*, 1919, 1-20.

Aristotle, or Theophrastus, all of whom he had read thoroughly, if sometimes too quickly. But it must be borne in mind that his purpose was different from theirs. Like the early naturalists, many of whom, incidentally, borrowed extensively from the *Historia naturalis*, Pliny's task was to collect, describe and record. With respect to the therapeutic uses of plants and plant products, he recorded what he found, not because he believed it, but because it added another dimension to the natural world whose objects he was describing.

Finally, mention should be made of Pliny's enormous influence, second only to that of Aristotle and Galen, in providing information and misinformation alike to succeeding generations of scholars, encyclopedists, physicians and others. This was especially true of the Middle Ages where Pliny's influence can be detected in many texts dealing with the world of nature and its history.[7]

* * *

The majority of the more than one thousand herbs or trees mentioned and/or described in the *Historia naturalis* served one or more medicinal uses. Those uses ranged from the topical application of macerated leaves, eating a portion of the root, or drinking an aqueous decoction to *conpositiones*, designed for internal ailments, containing several different ingredients, botanical as well as non-botanical. In addition to these simple, natural remedies, for which parallels can be found in the folk medical practices of many different cultures,[8] Pliny described a variety of rituals, charms, and magical practices, many of which involved the use of plant material for therapeutic purposes. Belief in these rituals and magical practices was widespread among all classes of Roman society, as we know from contemporary texts. Moreover, they were accepted as part of traditional medicine

[7] K. Rück, *Die Naturalis Historia des Plinius im Mittelalter*, « Stzb. Bayr. Akad. Wiss., philos-hist. Kl. », 1898, pp. 203-318.

[8] For example, cf. W. Bonser, *The Medical Background of Anglo-Saxon England*, London, 1963, *passim* and M. Höfler, *Volksmedizinische Botanik der Germanen*. Wien, 1908, *passim*. There are numerous references to Pliny in G. Ferraro, *Botanica Popolare di Carpento d'Acqui*, « Archivio per lo Studio delle Tradizioni popolari », 4, 1885, cf. pp. 129, 130, 131, 187 etc.

just as fully as the natural remedies noted above. And like the poultices and decoctions, parallels can also be found in other cultures for superstitious beliefs and magical practices. It may be noted in passing, that such beliefs continued until the recent past in some regions and may, in fact, still continue here and there today.

In order to understand even a fraction of the complex relations that existed between these different kinds of remedies, it will be necessary to examine some of the beliefs about medicinal plants in general. For, as Pliny noted, one can fully understand the nature of plants only in terms of their medicinal effects.[9]

Perhaps the single, most prevalent belief about plants was that many of them shared a double or even a treble use, each of which was associated in some way with the healing arts.

Many plants served as a foodstuff in the normal diet, for example, *abellana* (*Corylus avellana* L., hazel nut); *brassica* (*Brassica oleracea* L., cabbage); *cepa* (*Allium cepa* L., onion); *hordeum* (*Hordeum vulgare* L., barley); *cotoneum* (*Cydonia oblonga* Mill., quince); *siser* (*Pastinaca sativa* L., parsnip), etc. Depending, however, upon the method of preparation, the inclusion of other ingredients, the form of administration, dosage, and frequency, each of the above species also served one or more therapeutic purposes.

Abellanae, for example, were eaten dried as a remedy for catarah. When pulverized, they were added to hydromel for cough or taken in raisin wine.[10] The medicinal uses of *brassica* were, and still are, innumerable. It was used externally, mixed with various herbs or barley meal, as a topical application for gout while the water in which it had been boiled was used as a fomentation for muscular pains.[11] Macerated cabbage leaves were applied to wounds and *carcinomata* and the juice of freshly pounded leaves was dropped into the ears as a cure for deafness.[12] Internally, it was alleged to be good for the spleen, stomach and the chest and for colic.[13] The virtues of

---

[9] *... vera ... cuiusque natura [sc. plantae] non nisi medico effectu pernosci potest,* XIX, 189.

[10] XXIII, 150.

[11] XX, 81.

[12] XX, 83.

[13] XX, 80; XX, 82.

cepae rivalled those of cabbage for they were used internally and externally, raw or cooked, with and without other ingredients. Among the properties attributed to *cepa* was that it clarified the eyes, promoted sleep, healed abrasions and ulcers, and was beneficial for dog bite, spider bite, and the sting of scorpions.[14] As might be expected of such a panacea, Pliny noted that physicians tended to disagree about the medicinal properties of *cepae*.[15] As one of the basic staples, *hordeum* was widely used as a tisane in restorative diets while its meal formed the basis of poultices.[16] In addition, the ashes were applied locally for burns and employed as a dentrifice.[17] Pliny concluded his account of the medicinal uses of *hordeum* with two reports: that eating barley bread prevented gout and that, as a part of a ritual, nine barley grains thrown into a fire with the left hand, will immediately cure carbuncles.[18] The fully ripened *cotoneum*, eaten raw, was claimed to be effective for spitting of blood, dysentery, and coeliac complaints.[19] Externally, they were applied, raw or cooked, for stomach ache and, to the chest, for fever.[20] The blossoms also were used for inflamed eyes and as an emmenagogue.[21] Finally, the root of the tree was worn as an amulet as a cure for scrofula.[22] Lastly, *siser* was listed as a diuretic and aphrodisiac, an antidote for poisoning and for diarrhoea.[23]

A third use of plants is associated with their function as *condimenta*. Plants used for that purpose were usually not eaten by themselves, at least in an uncooked state. But cooked or uncooked, they served in a variety of ways as seasonings, relishes, and garnishes. Examples of such plants are *anetum* (*Anethum graveolens* L., dill); *cuminum*

---

[14] XX, 39.
[15] XX, 42.
[16] XXII, 136; cf. XVIII, 75.
[17] XXII, 134.
[18] XXII, 135.
[19] XXIII, 100.
[20] XXIII, 101.
[21] XXIII, 103.
[22] XXIII, 103.
[23] XX, 34.

(*Cuminum cyminum* L., cumin); *feniculum* (*Foeniculum vulgare* Mill., fennel); *ruta* (*Ruta graveolens* L., rue); *thymum* (*Thymus vulgaris* L., thyme) etc. All of these herbs were widely used as *medicamenta* when taken separately, usually raw. But they were also extensively used as ingredients in composita for a wide range of complaints. *Anetum*, for example, was used in a variety of ways, most of which probably can be attributed to the pungently-pleasant odor of its foliage and seeds. Thus, it was combined with other herbs as a drink for women in labor or in a vinous decoction with cumin and rue.[24] *Cuminum* was added to a decoction of *heliotropium* in order to dissolve urinary calculi, presumably on the grounds that it, like other members of the *Umbelliferae*, increased diuresis.[25] The seed was also mixed with other ingredients and applied externally for colic which seems strange since elsewhere Pliny remarked that it served as a carminative when taken internally for the relief of flatulence.[26] *Feniculum* is another old favorite that is still used in contemporary folk medicine. Along with horehound and honey, it served as a cough medicine and its ashes were used by some as a dentrifice.[27] Finally, on the basis of a fable that snakes when shedding their skin eat fennel to improve their eyesight, it was added, along with frankincense (a gum-resin derived from *Boswellia sacra* Flueck. and probably other species of the same genus) to an ointment made of viper flesh designed to improve the vision.[28] Pliny began his account of the medicinal properties of *ruta* by stating that it is one of the principal medicinal plants.[29] This is borne out by the large number of recipes in which it is included. The sharp, penetrating odor of the crushed foliage doubtless determined most of its uses for he stated that as a remedy for the sting of a scorpion, an aqueous decoction of *ruta* and *phlomis* was as beneficial as it was bitter.[30] Elsewhere it is recorded as an

---

[24] XX, 101; XXIX, 46.

[25] XXII, 59; XXVI, 87.

[26] XXVIII, 205; XX, 159.

[27] XX, 242; XXX, 27.

[28] XX, 254; XXIX, 119.

[29] *In praecipius autem medicaminibus ruta est*, XX, 131.

[30] *Contra scorpiones bibitur radix (sc. phlomis) cum ruta ex aqua, magna amaritudine sed effectu pari*, XXV, 121.

ingredient in external applications for spasms and burns, for carbuncles, and for spreading sores.[31] Its use in remedies to be taken internally include a decoction for intestinal worms, as a cough syrup, and in a medicament to eliminate an excess of yellow bile.[32] *Thymum*, like other genera of the *Labiatae*, was widely used in both cooking and in medicine, a fact which Pliny explicitly noted, while elsewhere he reported that it was macerated and served with salt as an appetizer.[33] The pleasant odor of the foliage, due principally to the presence of an essential oil, governed most of its medicinal uses: as a tonic to promote digestion, as a flavoring for wine, and for old cheese, while thyme honey was expecially valued for ulcers and complaints of the eyes.[34]

Some plants and plant products even played a further role in folk medical practice. In this small but heterogeneous class are several plants which indirectly contributed to the preparation or administration of medicaments. For example, the dried, unperforated rind of *malum granatum* (*Punica granatum* L., pomegranate) served as a vessel to mix and heat liquid substances.[35] A branch of the fig tree (*Ficus carica* L.) was used to stir a mixture of various substances, possibly, though Pliny does not state this, because the latex of the branch might help to thicken the mixture or add a whitish color.[36] The culm of *harundo* (*Arundo donax* L., bulrush) was employed as a pipe through which fumes were inhaled for medicinal purposes while a sharpened splinter was used to cut or puncture the ripe fruit of *cucumis silvestris* (*Ecballium elaterium* (L.) A. Rich., squirting

[31] XXIII, 4; XXII, 103; XXIV 25.

[32] XXII, 155; XXVI 31 et 29; XX, 217.

[33] Writing with reference to two kinds (*duo genera*) of thymum, Pliny adds, « *utraque* [...] *existimantur et in cibo et in medicamentis* », XXI, 154. As an appetizer, cf. XXI, 157.

[34] XIX, 92; XIV, 111; XI, 242; XI, 38.

[35] *cum rosaceo in calyce punici calfactus*, XXIV, 57; cf. XXII, 90 et XXIV, 77.

[36] The recipe for preparing *schiston*, a milkish medicament: « id fit hoc modo; fictili novo fervet caprinum maxime, ramisque ficulneis recentibus miscetur additis totidem cyathis mulsi quot sint heminae lactis », XXVIII, 126; cf. XXVIII, 186. The use of the latex for purposes of curdling milk, which Pliny notes, XXIII, 117, is already known to Homer: ὡς δ'ὅτ' ὀπὸς γάλα λευκὸν ἐπειγόμενος συνεπηξὲν ὑγρὸν ἐόν ... *Iliad*, V, 902-903.

cucumber) whose pulp was then used as a drastic purge.[37] The ever-present cabbage leaf served as a bandage to retain a poultice or as a simple dressing for sores.[38] And finally, the seeds of various species served as a metrological unit.[39] These and other domestic uses point to the extent to which the common folk relied on plants in order to satisfy their daily needs.[40]

Generalizing from the above, it was the exceptional plant that did not serve in some medicinal capacity. In fact, sometimes Pliny stated explicitly that a particular plant had no medicinal use.[41] Elsewhere he remarked that it is only a specific portion of a plant (root, leaves, seeds, etc.) that was used in medicine.[42] But the opposite statement is frequently made also, viz. that a specific portion or organ had no medicinal use.[43] Statements of the latter sort seem to imply that the remaining portions or organs had some medicinal use. In one notable passage, he wrote with reference to *consiligo* (*Pulmonaria officinalis* L., lungwort) that no medicinal use had yet been found for its above-ground portions.[44] This passage may be taken to suggest that the folk were continually searching for new or better sources of medicine.[45] It is unlikely that the passage pertaining to *consiligo*

---

[37] XXVIII, 230; XX, 3.

[38] XX, 127; XXX, 30.

[39] Seeds and fruits used as units of weight and measure include: *capparis*, *Capparis spinosa* L., XV, 15; *cicer*, *Cicer arietinum* L., XII, 131; *corylus*, *Corylus avellana* L., XII, 32; *faba*, *Vicia faba* L., XVI, 27; XIII, 105; *juglans Junglans regia* L., XX, 7; *lens*, *Lens culinaris* Medik., XX, 233; *lupinus*, *Lupinus* spp., XX, 190; *malus*, *Malus pumila* Mill., XVI, 27; *papaver*. *Papaver somniferum* L., XX, 20; *pirus*. *Pyrus communis* L., XIII, 104.

[40] *Mille praeterea sunt usus earum* (*sc. arborum*) *sine quis vita degi non possit*, XII, 5. Elsewhere, in reference to the uses of the pitch pine (*Pinus silvestris* L.), he states, « *materies vero praecipua est tradibus et plurimis vitae operibus* », XVI, 42.

[41] *Blitum* (*Amaranthus lividus* L.) *iners videtur ac sine sapore aut acrimonia ulla*, XX, 252. Cf. XXIV, 83; XXVII, 142 etc.

[42] Roots: XXII, 24; XXV, 96; leaves: XXV, 92; seeds: XXV, 116; sap: XXV, 69.

[43] XXV, 92; XXV, 120; XXVII, 122.

[44] *Superficies eius herbae an sit in aliquo usu adhuc incertum est*, XXVI, 38.

[45] Cf. Pliny's remarks about the recent discovery of *herba cantabrica* (*Convolvulus cantabricus* L.). The passage concludes: « Nec alias defuere Hispaniae herbis exquirendis, ut in quibus etiamnunc hodie in more sit laetiore convictu potionem e centum herbis mulso additis credere saluberrimam suavissimamque. Nec quisquam genera earum iam novit aut multitudinem, numerus tamen constat in nomine », XXV, 85.

referred to a consciously planned or deliberately designed experiment of the kind that we associate with scientific medicine today.[46] Yet the fact remains, that folk remedies did change.[47] Moreover, in the evolution of folk medicine, several effective remedies were discovered, later to be transmitted to other cultures where traces of earlier uses can still be found. The comparatively slow rate of change in folk medicine doubtless must be attributed to many factors. But, in order to meet new diseases successfully and to take advantage of newly introduced plants, folk medicine responded with new remedies.[48] At the same time, it must be noted, new rituals and magical practices were being introduced from the East, especially from Egypt.[49] These so quickly became a part of Roman medicine that they threatened to supplant some of the older, native practices.[50]

Pliny was echoing an old, popular belief when he stated that plants were created by nature for their usefulness.[51] There can be little doubt that herbs and trees were valued by the folk as much for utilitarian as for aesthetic purposes. Colorful and sweetly-perfumed flowers, no less than bitter tasting and malodorous plants, were as highly regarded for their medicinal properties as for decorative purposes.[52] Wreaths, garlands and chaplets provide a good example. While *coronamenta* made of many different species played an important role in civic ceremonies, religious rites and military triumphs, they

---

[46] Cf. Pliny's refence to the accidental discovery (*fortuito experimento*) of a remedy for elephantiasis, XX, 144. Another fortuitous discovery is reported in XXII, 120.

[47] Perhaps new remedies were sought for as antidotes or, as Pliny suggests, perhaps even poisonous substances may have some medicinal worth, « adeo nullo omnia experiendi fine ut cogerent etiam venena prodesse », XXV, 37.

[48]. For new diseases, cf. XXVI, 1; for newly-introduced species for medicinal purposes, cf. XXI, 48; for a new remedy, cf. XXX, 44.

[49] Cf. G. Heuten, *La diffusion des cultes égyptiens en Occident*, « Revue de l'Hist. des Religions », *104*, 1931, 409–416.

[50] Pliny concludes his description of a complex medicament, containing many exotic ingredients, by stating, *haec ostentatione; simplex quidem et antiqua illa salubritas pavere et melle constat*, XX, 208; cf. also the reference to a remedy *ex barbaris tradidit*, XXVIII, 200. Native plants were regarded as useful as exotica, XXII, 118. For references to specifically Roman medicaments, cf. XXII, 128; XX, 78.

[51] XXI, 2; XXII, 1.

[52] ...*florum odorumque generi est magnus usus in medicina*, XXI, 116.

eventually acquired various medicinal and apotropaic properties.[53] Headache, for example, was said to be relieved by wearing a garland of *viola* (*Viola odorata* L., violet), *spina alba*, *puleium* (*Mentha pulegium* L., pennyroyal) and others.[54] A chaplet of *crocus* (*Crocus sativus* L., saffron) was even recommended to alleviate intoxication.[55]

* * *

On the basis of the foregoing remarks, it is necessary to ask two questions: (1) Why was it believed that certain species of plants were therapeutically useful? and (2) Why were certain plants believed to be especially (*privatim*) effective for specified complaints or illnesses? [56] It will make little difference at this point whether or not later investigators have confirmed the medicinal properties attributed to particular species. The fact remains that folk remedies constitutes a large part of folk medicine and that the principal ingredients in most folk remedies were of plant origin.

Grounds for the general belief that certain plant species were medicinal (and indirectly, why others were not so regarded) are of several different types. Naturally, they cannot be sharply distinguished because folk beliefs, unlike scientific data, do not lend themselves easily to categorization. But supplemented by analogies to the folk medical practices of other cultures and the findings of modern pharmacological studies, the following types of reasons are exemplified in the *Historia naturalis*.

There are, first of all, those plants whose ingestion led to predictable, overt physiological responses, for example, *scammonia* (*Convolvulus scammonia* L.; scammony) as a drastic purge or *veratrum* (*Veratrum album* L., white hellebore) as an emetic.[57] How these plants were

---

[53] For the different kinds of *coronamenta*, cf. XXI, 1 sqq.; on their preparation, cf. XXI, 51 sqq. For the use of chaplets and garlands by the *herbarii*, cf. XXVII, 117.

[54] XXI, 130; XXIV, 108; XX,152; etc.

[55] XXI, 138.

[56] The adverb *privatim*, originally a legal term, frequently is used by Pliny in the technical sense of « specific (for) » e.g., *verbascum privatim tonsillis in aqua potum*, XXVI, 24 or *hippolapathi radix privatim ungues scabros detrahit*, XX, 232.

[57] For *scammonia*, cfr. XXVI, 59; for *veratrum*, cfr. XXV, 51.

first discovered is not our concern here. It is sufficient to note that the early indiscriminate use of plants and their products for alimentary purposes, for stock animals and humans alike, undoubtedly contributed to the recognition that some plants produced an immediate physiological response.[58] As evidence of this supposition, it may be remarked that Pliny, like most of those who have written about the folk uses of plants, has emphasized the toxic, sometimes lethal properties of fungi.[59] On the other hand, Pliny praised the useful properties of other fungi, for example, edible mushrooms which were, and still are, prized as gastronomic delicacies.[60] Other species, for example Pliny's *agaricum*, were widely used in folk medicine, both externally and internally.[61] Since the chemical compositions of fungi were unknown, the distinction between lethal, toxic, inert and edible varieties can only have been based on trial and error and custom. The results of many generations of folk use were then transmitted by those forms of communication peculiar to each society such as proverbs, popular songs, myths and the like.[62]

---

[58] Pliny sometimes accepted the widespread folk belief that because animals seek out a certain plant, therefore it is useful for humans, cf. XX, 37; XX, 254. On the other hand, he reports that because animals will not eat *ocimum* (*Ocimum basilicum* L., basil), humans should also avoid it, XX, 119.

[59] XXII, 92 sqq. Several folk beliefs are reported in XXII, 94–95 regarding signs for determining whether a fungus is edible or not. For a similar belief. cf. HORACE, *Sat.* II iv, 20.

[60] XVI, 31; XIX, 33. For *boletus* (*Boletus* spp.) as a delicacy, cf. MARTIALIS, *Epigr.* I, XX; for tubera, cfr. *Juvenal*, *Sat.* V, 116.

[61] XXV 103, 119; XXVI, 42 etc. Cf. also MARCELLUS EMPIRICUS, *De medicamentis liber* ed M. NIEDERMANN. Leipzig, 1916, I, 166; XX, 13, 19, etc. and DIOSCORIDES, *De materia medica* ed. M. WELLMANN, Berlin, 1906, III, 1.

[62] As a good example, cf. the lines from what may have been a popular song. Pliny is writing about reseda (*Reseda alba* L.) and the uses of this well known plant:

Reseda, morbis reseda,
scisne, scisne quis hic pullus egerit radices?
nec caput nec pedes habeat. (XXVII, 131).

For other examples, cf. XIX, 144; XX, 252. An interesting parallel from the nineteenth century is the following proverb: An funzos, An carstos

(Anno fungato, Anno tribolato)

recorded by B. MAVENA, *Alcune nozioni del calendario agricolo*, « Rivista delle Tradizioni Popolari Italiane », *1*, 1893/1894, 453–455.

Once it was discovered and made known that specific plants, the recognition of which was not always public information, produced an overt response or a particular reaction, a basis was provided for the belief that other plants of similar appearance might possess similar properties. That belief, in turn, was reinforced, both positively and negatively, by other beliefs some of which will be mentioned below.

A second basis for the general belief in the healing properties of plants was indirectly suggested above. This is the so-called « Doctrine of Signatures » which, in both its primitive and sophisticated forms, played an enormously important role in the history of therapeutics. Logically, the belief in the signature(s) of a plant substance (or animal substance) depends upon the acceptance of two premises, though seldom are both made explicit.[63] First, is the admission of a recognition of a similarity, fanciful or real, between the morphological appearance of a plant (color, habit, armature etc.) or some portion thereof (seed capsule, root, leaf etc.) and a symptom or abnormal condition of the human body. Then, once that similarity is asserted, it is further assumed that *because* of that similarity, the plant (or specified portion thereof) is specifically and therapeutically effective for that symptom or abnormal condition. The second of these two premises of *Signaturenlehre* is thus one of several variants of the well-known axiom *similia similibus curantur* that is found to underlie many « sympathetic » cures and remedies.[64]

Numerous passages in the *Historia naturalis* indicate that such a belief was a significant factor, not only in the selection and use of particular plants for specified complaints, but for an explanation of their activity as well. The following is a selection of typical examples.

A simple instance of how a plant's signature determined its use concerns various orchidaceous plants. Several genera, for example, *Orchis*, *Ophrys*, *Serapias*, and probably *Aceras*, are referred to by

---

[63] Such a belief underlies Pliny's comments on *natura* and her gifts to man: « Dederat quas diximus molles cibisque gratas, pinxerat remedia in floribus visuque ipso animos invitaverat etiam deliciis auxilia permiscens », XXII, 16.

[64] Cf. TH. WEIDLICH, *Die Sympathie in der antiken Litteratur*. Progr. Karls-Gymnasium, Stuttgart, 1894 and J RÖHR, *Der okkulte Kraftbegriff*, « Philologus », Supplm. Bd. XVII/1 1925, pp. 34-76. Pliny explicitly mentions sympathia et antipathia as a Greek doctrine, XX, 1.

Pliny. The underground portions of each of these genera is characterized by two swollen, ellipsoid tubers, resembling testicles or *orchides.* As a result of their shape, some orchidaceous species were used for complaints connected with the genitalia.[65] For similar reasons, they were also regarded as *aphrodisiaca.*[66] Finally, the larger of the two tubers was believed to promote the production of male children, the smaller, female children.[67] Another example of the pervasive influence of the belief in signatures concerns those plants which possess a milky sap or latex. Not unnaturally, those plants were especially valued as galactagoga, a function which they retained in both school medicine and folk medicine for many centuries.[68] A last example pertains to remedies for the sting of a scorpion.[69] The basis for selecting specific plants for that purpose rested upon a similarity in shape of the segmented body of the scorpion to the seed capsules of same genera of the *Leguminosae.* The maturing capsules of some species of *Coronilla, Scorpiurus,* and *Ornithopus,* each of which has several species indigenous to Italy, roughly resemble, in fact, a scorpion because of the constricted, slightly curved, and sometimes sharply terminating point.[70] For this reason, such plantes were thought to be useful for the scorpion's painful sting. It is not so easy, however, to determine whether this use preceded or followed the folkname *scorpio* < σκορπίον or *scorpiuron* < σκορπίουρον.[71] At any rate, once name and use were correlated, the plant was assured of a potential therapeutic role in any region in which scorpions could be expected.

---

[65] XXVI, 96.

[66] XXVI, 95.

[67] XXVI, 97. For a modern folk variation of this belief, cf. A. Treichel, *Volksthümliches aus der Pflanzenwelt, besonders für Westpreussen,* « Altpreussische Monatsschrift », *31,* 1894, p. 467 *sub voce* Orchis oder Knabenkraut.

[68] Among the lactiferous plants used as galactagoga, Pliny mentions *soncus* (*Sonchus oleraceus* L.; XXII, 88); *iasine* (*Convolvulus arvensis* L.; XXII, 82) and *polygala* (*Polygala venulosa* Sbth.; XXVII, 121).

[69] For the many references to plants allegedly useful for the sting of a scorpion, cf. XX, 133, 145, 155, 162, 164, 182, 185, 223, 236 etc.

[70] Cf. Eugenio Baroni, *Guida Botanica d'Italia.* 4th ed. Milano, 1963 pp. 135-136 et fig. 105.

[71] Ex argumento nomen accepit scorpio herba. Semen enim habet ad similitudinem caudae scorpionis, XXII, 39.

Before we turn to some of the superstitious and magical grounds for believing in the therapeutic properties of plants, it will be useful to examine several miscellaneous claims. Like the belief in signatures, these claims were widespread and appealed to the popular mind because of their simplicity and easy application. Although they may appear to us to be irrational and unsupported by empirical data, these simple and easily understood claims about plants were formulated in naturalistic terms. As such, they exerted a great influence and were easily communicated. But because they did not involve any occult powers or violate laws of nature, they may be distinguished from superstitions and magical practices which will be examined below.

A good example of these claims is furnished by the simple assertion that the bitter taste of a plant, or alternatively its penetrating aroma, is evidence of its medicinal properties.[72] From this it follows that the more bitter the juice of a plant is, the better medicine, it is.[73] Another example concerns the distinction between the medicinal properties of domesticated and wild forms of a given species. Leaving aside whether the distinction is taxonomically valid, the claim was apparently a common one, judging from the numerous instances of it in Pliny and in the latin nomenclature conserved by him.[74] Thus, the wild form of *staphylinus* is said to be more effective than the cultivated form, though both forms can probably be identified as the common carrot (*Daucus carota* L.).[75] On the other hand the distinction

---

[72] Elatine ... nascitur in segete, acerba gustu et ideo oculorum fluctionibus efficax foliis cum plenta tritis et inpositis, XXVII, 74; cf. also XIX, 80; XXV, 68. As regards aroma, cf. Pliny's description of *sinapi* (or mustard (*Brassica* spp.): magis in nares et cerebrum penetret ... comitiali morbo aut vulvarum conversione suffocatas excitat odore aceto mixto, XX, 236, 238; cf. also XXIV, 21; XXI, 15 etc.

[73] Lactucae (*Lactuca* spp.; lettuce) ... in cholera quoque coctas patinis dederunt, ad quod utilissimae quam maximi caulis et amarae, XX, 67; cf. XXV, 121; XXI, 58 and XXVIII, 134 where the same reasoning underlies the use of rancid butter.

[74] Differentiating between ruta domestica (*Ruta graveolens* L.; garden rue) and r. silvestris (*R. montana* L.), Pliny states, silvestris horrida ad effectum est et ad omnia acrior, XX, 131; cf. also XX, 76; XX, 124 etc.

[75] Alterum genus est staphylinus, quod pastinacam erraticam vocant ... Ad reliqua et sativa pollet. Efficacior tamen silvestris magisque in petrosis nata, XX, 30 et 32; for the identification, cfr. J. André, *Lexique des Termes de Botanique en Latin*, Paris 1956 p. 240 *sub voce* Pastinaca.

between *asparagus* (*Asparagus officinalis* L., asparagus) and *corruda* (*Asparagus acutifolius* L., wild asparagus), while taxonomically valid, seems not to be reflected in any different pharmacological activity, despite Pliny's claims for the superiority of *corruda*.[76] Another instance of these naturalistic claims is very revealing of the structure of folk medicine because the opposite claim is made elsewhere. With reference to several species of plants, Pliny stated that the darker colored form is more effective than the lighter colored form.[77] Yet with respect to at least two other species, the opposite is maintained, one of which *scilla* (*Urginea maritima* (L.) Baker, squill) has been the subject of numerous studies in which the red and white forms show no difference in activity.[78]

We come now to an assortment of superstitions and magical practices insofar as they relate to the therapeutic uses of plants. No less than the materials assembled above, superstition and magic were closely connected with the practice of medicine in all strata of Roman society. Insofar as they helped to shape beliefs about the use of plants for medicinal purposes, they were as much a part of Roman life as was illness itself.

Since it is not my intention here to try to explain the origin and nature of magic, on which there exists a voluminous literature, it may simply be noted that many of the beliefs which we could term magical or superstitious were debased, secularized forms of ceremonies institutionalized, at an earlier date, by various state, municipal and religious bodies. For example, the great popularity of a plant variously termed *verbena*, *verbenaca*, or *hiera botane* (ἱερὰ βοτάνη) (*Verbena officinalis* L., verbena) in folk remedies can be traced back to its use in certain solemn ceremonies conducted, in part, by the *verbenarius*.[79]

---

[76] Silvestrem asparagum aliqui corrudam, aliqui libycum vocant, Attici orminum. Huius ad supra dicta omnis efficacior vis, et candidiori maior, XX, 110. For the identification, cf. ANDRÉ, *op. cit.*, p. 102 *sub voce* corruda.

[77] Git (*Nigella sativa* L.) ... optimum quam excitatissimi odoris et quam nigererrimum, XX, 182; cf. XX, 69; XX, 76, etc.

[78] Cf. J. STANNARD, *Squill in Ancient and Medieval Materia Medica, with Special Reference to its Employment for Dropsy*, « Bull. New York Acad. Med. », *50*, 1974, 684–713.

[79] Cf. J. STANNARD, *Magiferous Plants and Magic in Medieval Medical Botany*, « Maryland Historian », *8*, 1977, 33–46.

A similar example concerns *sagmen*, a sacred piece of turf which, as Pliny stated, was used, along with *verbena*, to cure the ills of state.[80]

Because of the sacred nature of official rituals and ceremonies, many of which incorporated or were associated with some plant or plant product, it was not difficult to extend the range of useful properties of those ceremonial plants. Moreover, one could attribute to other plants similar properties with a potential for magical as well as medicinal use. The attribution of those properties, it should be obvious, went far beyond what could be deduced from an examination of the plants themselves; in this respect the belief in what Pliny called *vanitates magicae* differs radically from a belief in a plant's signature.[81]

Sometimes this took the form of accepting an older tradition whose origin, possibly etruscan or italic, is matter of conjecture.[82] At any rate, Pliny has recorded some details, mostly marvels and *secreta*, connected with these *plantae magicae*.[83] Because their identification is far from certain, the rationale underlying their uses is not easy to determine nor is the mechanism (physical or chemical) by means of which they functioned as they did. In the few instances where the plant in question seems not to have been purely imaginary, a few tentative remarks might be made. For example, the plant called *minyas* or *corinthia*, which may possibly be identified as *Smyrnium olusatrum* L., alexanders or horseparsley, is said to be a cure for snake bite, in the form of a fomentation, but also that it causes death if one steps on it or if it be sprinkled over one's body.[84] Despite this poisonous property, Pliny concludes by reporting that it is an antidote

---

[80] XXII, 5. A similar belief is mentioned by HORACE, *Odes* II XV, 17; III viii, 3-4.

[81] For Pliny's criticism of the *magicae vanitates*, cf. XXVI, 18.

[82] As an example, cf. the charm « Luxum siquod est, hac cantione sanum fiet », described by Cato. *De agri cultura*. 160. The apparently meaningless words of the charm are neither Latin nor Greek. For a later parallel, cf. T. ZACHARIAE, *Abergläubische Meinungen und Gebräuche des Mittelalters in den Predigten Bernardinos von Siena*, « Ztschr. f. Volkskunde », *22*, 1912, 126.

[83] Cf. the extended discussion in XXIV, 156-168.

[84] XXIV, 157; cf. XIX, 162 where the same plant may be in question. For the identification of Pliny's minyas-corinthia, cf. J. ANDRÉ, *Notes de Lexicographie Botanique Grecque*, Paris 1958 p. 36 *sub voce*.

to other poisons. Another of the eighteen magical plants, listed *seriatim* by Pliny,[85] was *ophiusa*, literally a snake plant, ὀφιοῦσσα < ὄφις As its name suggests, it was associated with snakes, but not as a remedy following snake bite. Rather, Pliny reported, that if taken as a drink it caused mental derangement by producing visions of snakes.[86] Like *ophiusa*, *thalassaegle* and *gelotophyllis*, cannot be identified, but to each of the latter two are attributed psychotropic powers.[87] It is no more than a conjecture, but these, and possibly others of the magical plants, may have been hallucinogenic plants used in various eastern cults.[88] It is probably no coincidence that many of Pliny's *plantae magicae* also occur in Graeco-Egyptian magical papyri.[89]

There is, however, another class of plants whose effects were no less marvellous. But unlike the magical plants discussed above, these were real plants, and, as such, their identification is reasonably certain. Such plants may be termed magiferous plants. When certain plants were subjected to rituals or associated with *incantamenta*, they were believed to acquire extraordinary properties. That included, *inter alia*, greatly enhanced therapeutic properties. However, despite the incredible properties attributed to those plants, they were prepared and administered in the same fashion as the ordinary medicinal plants of the garden or fields. Examples of magiferous plants are common throughout the history of medical botany and a few examples from Pliny will demonstrate their general character.[90] A striking example of a magiferous plant is *vettonica* (*Stachys officinalis* (L.) Trev., betony).[91] Typical of such plants is the wide range of uses for common

---

[85] XXIV, 157-167.

[86] Ophiusam ... qua pota terrorem minasque serpentium observari ita ut mortem sibi eo metu consciscant, XXIV, 163.

[87] XXIV, 164.

[88] I owe this idea to H. Brugsch, *Die Alraune als altägyptische Zauberpflanze*, « Ztschr. f. ägypt. Sprache », *29*, 1891, 32.

[89] Cf. C. Wessely, *Synopsis Florae Magicae*, « Bull. Inst. francais d'Archéol. Orient », *30*, 1930, 17-26.

[90] Cf. J. Stannard, *Maryland Historian* (cited above n. 79).

[91] Vettonica is one of the most frequently mentioned medicinal plants in the *Historia naturalis*; cf. especially XXV, 84; XXVI, 40-42, 112-114, 116-119 *et passim*.

complaints: colic, cough, gout, indigestion, inflamed eyes, looseness of the bowels, sciatica etc. For such complaints as these, *vettonica* was used either by itself or mixed with one or two liquid substances, for example, raisin wine, hydromel, or squill vinegar. In this respect, then, *vettonica* served the same needs in the same fashion as many other common plants. But, there is another side to its many uses and it is to this that *vettonica* achieved its importance in the history of materia medica. Among these latter uses of a plant which Pliny described as *ante cunctas laudatissima*, is its remarkable herpicidal effect.[92] Snakes, so Pliny reported, whip themselves to death when enclosed by a circle made of fresh plants.[93] Another property no less remarkable was that it protected a home against all misfortunes when planted nearby.[94] In powdered form, taken every day, it assured protection against any poison.[95] Other, less dramatic uses, included a remedy for the sting of a scorpion, for improving a lead-gray complexion, for the after-effects of wine, and as a remedy for epilepsy and *insania*.[96]

It is not an easy matter to explain why so many unusual properties were attributed to *vettonica* for outside of its attractive, crenate leaves and its reddish-purple, spicaform inflorescence, there is little to set it apart from many other equally attractive plants that possess, in addition, some conspicuous characteristic. Since it was unknown to Theophrastus, one cannot trace its history before Pliny. But there must have been a Greek tradition because Pliny cites as synonyms, *psychrotrophon* < ψυχρότροφον and *cestros* < κέστρον.[97] That all of these names denote the same plant is clear from the opening sentence of Dioscorides' account, an account which in some places is nearly identical to various phrases used by Pliny.[98] But since neither Pliny

---

[92] XXV, 84.

[93] XXV, 101.

[94] XXV, 84.

[95] XXV, 128.

[96] XXV, 122; XXVI, 102, 112, 113, 118.

[97] Vettones in Hispania eam quae vettonica dicitur in Gallia, in Italia autem serratula, a Graecis cestros aut psychrotrophon, XXV, 84.

[98] Κέστρον τὸ καλούμενον ψυχρότροφον, ἐπειδὴ ἐν τοῖς ψυχροτάτοις εὑρίσκεται τόπου, ὃ 'Ρωμαῖοι βεττονίκην ἢ ῥουσμαρίναν καλοῦσι κτλ, Dioscorides IV, I Wellm.

nor Dioscorides cite the other, a common source has been postulated.[99] Whatever may be the final answer as regards the extraordinary properties ascribed to *vettonica*, later writers have left no doubt that it was one of the principal magiferous plants.

The belief in the magico-medical efficacy of certain plants was reinforced by popular tales, fables, and legends with their mythological association. By conjoining these traditional accounts and plants which, like the *laurus* (*Laurus nobilis* L., laurel) also played some role in ritual, the popularity of those plants was increased.[100] And with that went an increased potential for therapeutic use. Thus, to introduce the plant called *hieracion*, literally, hawk-plant (< ἱέραξ, hawk), Pliny stated that accipiters (which would normally include hawks along with falcons) rip the plant apart and moisten their eyes with the juice (a milky latex), and thereby restore their vision. Then, he concluded, that the juice, mixed with mother's milk, heals various eye diseases, and especially, dimness of vision.[101] The identification of *hieracion* as *Urospermum picroides* Schm. (no common English name known) is tentative but possibly correct: it is a lactiferous plant, it fits Pliny's description, and it is indigenous to Italy.[102] An example of how a popular legend helped to promote the medicinal uses of a plant is provide by *nymphaea* (*Nymphaea alba* L., white water lily). According to tradition, wrote Pliny, Nymphaea was the daughter of a nymph who died out of jealousy for Hercules, whence its synonym, *heracleon*. Then, the passage concluded: therefore, those who have drunk *nymphaea* for twelve consecutive days are incapable of sexual intercourse.[103] Whether, in fact, *nymphaea* was used as a remedy to insure purity, we have no way of knowing. But judging from the

---

[99] M. WELLMANN, *Krateuas*, « Abh. Gesellsch. f. Wissenschaft zu Göttingen », philol-hist. Kl. N.F. II/1, 1897/1899, 1-32.

[100] For the medicinal uses of *laurus*, cf. XXIII, 152-158. In 158 Pliny refers to his account, XV, 127-138; these latter include symbolic and ritual uses, historical anecdotes, some lexicographical data, and a variety of folk beliefs pertaining to state cults and private superstitions.

[101] XX, 60-61.

[102] For the identification, cf. ANDRÉ, *Lexique* p. 162 *sub voce* hieracion (unfortunately misspelled!). As an Italian indigene, cf. BARONI, *op. cit.* p. 333 and fig. 182.

[103] XXV, 75.

numerous plants which were believed to repress sexual desires or to produce sterility, the claims made for *nymphaea* are not unusual at all.[104]

Finally, mention must be made of another common use of plants for therapeutic purposes. This involves the belief that certain plants, or often merely portions thereof, possessed one or more apotropaic properties in themselves, and independent of any ritual or ceremony to which they might be subjected. In the case of some plants that were said to possess such a property, probably there was originally some magical or ritualistic reason for believing in their capacity to avert sickness, misfortune or potential danger. But whether that reason is obvious or not, a large number of common plants were thought to provide security by being carried or worn on one's person as a pendant, knot, or amulet, or indeed, even planted by one's house. These uses include placing a sprig of *puleium* behind the ear as protection against the heat of the sun or tying springs of *artemisia* (*Artemisia vulgaris* L., mugwort) to oneself in order to avert fatigue.[105] *Andrachle* (*Portulaca oleracea* L., purslane) by merely being carried prevented the sting of a scorpion.[106] Twigs of various plant species, if held in the hand, prevented chafing of the thighs while *condurdum*, suspended from the neck, was said to repress scrofulous sores.[107] Lastly, and in order to observe the full range of apotropaic properties, Pliny reported that it was an old custom to plant *asphodelus* (*Asphodelus* sp., asphodel) in front of the gate in order to avert the noxious effects of sorcery while wearing a simple wreath of *laurus* provided protection against thunder.[108]

---

[104] Albertus Magnus records among the uses of *nenufar albus*: affert somnum et aufert dolores capitis frigidos; sed cum hoc debilitat caput, minuit pollutiones et desiderium coitus, *De vegetabilibus*, edd. E. Meyer et C. Jessen. Berlin, 1867. Lib. VI, sectio 395 (pp. 542-543).

[105] XX, 152; XXVI, 150.

[106] XXV, 164; cf. XX, 232.

[107] Virgam populi (*Populus alba* L.) in manu tenentibus intertrigo non metuatur, XXIV, 47; cf. also XXIV, 63; XXVI, 91; XX, 151. On condurdum (? *Saponaria* sp.) cfr. XXVI, 26.

[108] Tradunt et ante portas villarum satum (sc. asphodelum) remedio esse contra veneficiorum noxiam, XXI, 108. On *laurus*, cf. XV, 135.

* * *

In summary, it may be said that there were many different bases for believing in the medicinal properties in general as well as for the assignment of particular properties to specific plants. The preceding pages are a summary of those beliefs as they are recorded in Pliny's *Historia naturalis*. It is not intended to be exhaustive, rather the aim has been to examine passages representative of folk medicine insofar as plants are concerned. That many of the beliefs noted by Pliny became a part of Western European medicine is evidence not only of the role of plants in history but of the value of Pliny as a historical source for the early stages of a long tradition.

IV

LUCIANIC NATURAL HISTORY

Interpretations of Lucian are many and diverse, ranging from satirist to moralist, and imaginative romancer to critical historian of his own times. Doubtless he is all of these, and more. While I have no intention of preparing a *catalogue raisonnée* of the many facets of his genius, it is, I believe, time to call attention to one aspect of his work that has been somewhat neglected. Lucian was not a naturalist, still less was he a scientist, but that does not mean that his writings are devoid of interest to the student of ancient science.[1] In calling attention to some passages of scientific interest, I shall try to show how Lucian made use of his knowledge of natural history. It will be suggested, moreover, that despite his readiness to admit to the charge of ψευδολογία some of his observations must be taken seriously.

Lucian's references to natural history are intentionally subordinated to the demands imposed by genre, style, meter, and vocabulary. But when one looks beneath the literary artifices which have constituted a not unimportant part of his enduring appeal, there can be found a wealth of materials relating to and drawn from the world of nature. The entire corpus of Lucian's writings, when looked at in this fashion, indicates that he was a master in the art of utilizing such material in the most effective fashion. They indicate, also, how sensitive he was to details, selecting precisely those which emphasized the matter at hand or pointed a moral in a telling manner. The conclusion that is forced upon a reader attentive to details is that Lucian took pains to be exact.

[1] A parallel is provided by a study of the medical references in Lucian by J. D. Rolleston, "Lucian and Medicine," *Janus*, xx (1915), 83-108. Cf. Adolf Braun, "Die Tragopodagra des Lukian," *Münch. med. Wchn.*, Nr. 36 (4 Sept. 1936), pp. 1477-78.

Exactness and fidelity to nature, stemming perhaps from boyhood when he modeled waxen figures of animals (*Somn.* 2), should not, of course, be confused with truth. Lucian's humor greatly depends upon his ability to fabricate and to populate the skies, the waters, and the nether world with bizarre creatures: ὧν μήτε εἶδον μήτε ἔπαθον μήτε παρ' ἄλλων ἐπυθόμην, ἔτι δὲ μήτε ὅλως ὄντων μήτε τὴν ἀρχὴν γενέσθαι δυναμένων (*Ver. Hist.* I. 4). Yet one must take seriously his earlier claim κἂν ἓν γὰρ δὴ τοῦτο ἀληθεύσω λέγων ὅτι ψεύδομαι (*loc. cit.*) No reader is apt to confuse his ἀεροκώνωπες (*Ver. Hist.* I. 16) with real insects or to suppose that the κύων addressed as ὦ βέλτιστε ἰχθύων (*Pisc.* 48) is really a dog fish. But it might be forgotten, momentarily at least, that Lucian's motives entail that *he* should not confuse the real with the imaginary; after all, says Momus, we all know that θεοὺς μὲν τοὺς θεούς, κυνοκεφάλους δὲ τοὺς κυνοκεφάλους (*Deor. Conc.* 11). The success of these fabled creatures lies in their author's tight control upon his fantasies. He exaggerated, but did not lose sight of those objects whose dimensions he playfully stretched, he compounded but without ignoring the structural details of the component pieces and, if he had traffic with legend, it was to smile at those who took it seriously. This was made possible by the fact that he, like his latter-day admirer Rabelais, first looked at Nature and acquainted himself with her uneven hues before he painted her in unseemly colors. To borrow a passage from him to whom this volume is dedicated, one might say of Lucian, as it was said of Chariton, " By being closer to legend [he] is closer to nature and reality." [2]

Much of the natural history contained in Lucian's writings consists in passing allusions to the world about him, πάνθ' ἁπλῶς ὁπόσα τρέφει ζείδωρος ἄρουρα (*Icar.* 12). Mention of the better-known members of the fauna and flora is often accomplished through the agency of analogy and other forms of comparison. Their meaning depends partly upon the reader's ready grasp of the point being expressed and this in turn depends upon Lucian's use of plants and animals which were known to all, e.g. καὶ ἀνεμώνη ῥόδῳ καὶ ἀνθρώπῳ πίθηκος (*Pro Merc. Cond.* 11). But they also depend upon the context and their appropriateness to or within that context.

The false followers of Aristotle, in an extended example of Lucian's art, are first likened to certain unidentified fish, described

[2] B. E. Perry, "Chariton and His Romance from a Literary-Historical Point of View, *AJP*, LI (1930), 101.

as ἀκανθώδεις καὶ τὴν ἐπιφάνειαν ἐκτετραχυσμένους. This analogy is realistically reinforced by adding ἐχίνων δυσληπτοτέρους—"harder to grasp than sea urchins" (*Pisc.* 51). While this is only one of many specimens of Lucian's technique, it demonstrates his ability to put data from the real world to the service of the story teller's art. Since sea urchins were regarded as delicacies and hence known to all, Lucian made his point without unnecessary elaboration.[8] By prolonging the analogy, a less skillful artist might run the risk of boring his audience—certainly an anathema to be avoided by a satirist. Examples could be multipled of Lucian's ability to incorporate those details, and only those details, which further the story without interfering with the plot or the reader's almost intuitive grasp of the situation. Capitalizing on his ability to portray graphically a situation in succinct terms, Lucian did not hesitate to include homely references to matters of common knowledge. The passing remark that a goose runs along the ground in order to gain sufficient momentum for flight (*Icar.* 10) is an example. There was, in this case, no essential reason to note such a mundane matter. But in doing so, a mood and associated images were invoked, based on Lucian's own observations as well as those of his audience.

It was not Lucian's purpose to describe animal life except as incidental to the descriptions of his fellow men. Not all men, in fact very few, lived the blameless life of Demonax. Some less-than-fatal flaw gave Lucian the opportunity of finding further defects. He was aided in his search by the proclivity of the human species to act like animals. Anacharsis, more bemused than shocked, wonders at the sanity of those athletes who ἐν τῷ πηλῷ συναναφύρονται κυλινδούμενοι ὥσπερ σύες (*Anach.* 1). Analogy is piled upon analogy in depicting the animal-like behavior of wrestlers: τὰ μέτωπα συναράττουσιν ὥσπερ οἱ κριοί . . . ὥσπερ αἱ ἐγχέλυες ἐκ τῶν χειρῶν διολισθαίνοντες (*loc. cit.*). Animal traits supply Lucian with a repertory limited only by his knowledge of natural history and the antics of those who may fittingly be so described. Gaping expectation is graphically illustrated by the action of birds ὥσπερ τὴν χελιδόνα προσπετομένην τετριγότες οἱ νεοττοί (*Timon* 21) while a milling crowd is depicted ὥσπερ ἐπὶ τὴν γλαῦκα τὰ ὄρνεα (*Harm.* 1).

Enough has been said to demonstrate the appropriateness of Lucian's analogies and comparisons. That they rest upon a fund

[8] For edible sea urchins, cf. Aristotle, *Historia Animalium* IV. 4, 530b2; V. 12, 544a16-18.

of observational evidence cannot be questioned. It remains to note, however, that information of this sort need not derive solely from personal observation. Lucian himself admits his indebtedness to Homer, Herodotus, and others, including those whose names have been lost through the accidents of history, and those anonymous writings which he may have consulted as sources (e.g., Αἰγύπτιαι βίβλοι, *Philops.* 31). To the Poet, not unnaturally, he most often refers and while he can "laugh at Homer and his nonsense" (ψυχρολογία, *Dial. Mort.* XVI. 5) he is not slow to exploit it. At the outset of *True Stories*, he states that he will avail himself of ψεύσματα as the most suitable means of replying to the monstrous fables and myth-mongering τῶν παλαιῶν ποιητῶν τε καὶ συγγραφέων καὶ φιλοσόφων (*Ver. Hist.* I. 2). This does not mean that the "ancient classics" were valueless for Lucian's purposes. A Celt, surprisingly well-read in Greek literature, describes the persuasive powers of the Trojan leader in flowery (εὐανθῆ) terms. Then, as an afterthought λείρια γὰρ καλεῖται, εἴ γε μέμνημαι, τὰ ἄνθη (*Hercules* 4), referring evidently to the *Iliad* III. 152. But Homer was no more exempt from criticism than was any other writer. Nigrinus' vaunted wisdom eclipsed Homer's sirens, εἴ τινες ἄρα ἐγένοντο, καὶ τὰς ἀηδόνας καὶ τὸν Ὁμήρου λωτόν (*Nigrinus* 3).[4] Liberties were taken with Homer's verse when, with a deft alteration, a comical effect could be produced. Lucian's favorite targets, the sham philosophers, are made more ridiculous by having their grave comments interspersed with mock-heroic and mock-Homeric verses:

Πρόσθε κύων, ὄπιθεν δὲ λέων, μέσση δὲ χίμαιρα.
(*Fug.* 30; cf. *Iliad* VI. 181)

Lucian's parody of the χίμαιρα or she-goat resembles his parodies of other writers whom he read. Herodotus' εἵματα ἀπὸ ξύλων (VII. 65), worn by the Indians and usually identified as cotton garments, are turned into ὑαλίνη μαλθακή—malleable glass (*Ver. Hist.* I. 25).[5] More of Herodotus' fabled creatures will turn up later. Exactly how much Lucian owed to the writings of his predecessors is difficult to determine. A case in point occurs in his *Dipsades*. In half-serious tones he concludes that his motive for writing was one of friendship for an unnamed person, but ταυτὶ οὐ μὰ Δία πρὸς Νίκανδρον τὸν ποιητὴν

[4] For the problems involved in identifying the plants of antiquity, cf. Jerry Stannard, "Hippocratic Pharmacology," *Bull. Hist. Med.*, XXXV (1961), 497-518.

[5] Identified as cotton (*Gossypium herbaceum* L.) by Friederich Kanngiesser, "Die Flora des Herodot," *Arch. f. Gesch. d. Naturw. u. d. Techn.*, III (1910), 88. Cf. Herodotus III. 47 and III. 106.

φιλοτιμούμενος διεξῆλθον (*Dips.* 9). Whether this means that Lucian depended exclusively on Nicander must be left to the industry of the *Quellenforscher*. But there is no doubt that he availed himself of the former's *Theriaca*, as the list of venomous beasts indicates (cf. *Dips.* 3).[6] Occasionally, citations from other writers contain stray bits of natural history and appear to have been chosen advisedly. A verse from Archilochus on the cicada is chosen for a twofold purpose—the proverb itself points to the difficulty Lucian's critic may expect, but secondly, Archilochus' own penchant for invective will supply a precedent in keeping with a version of *la Querelle des Anciens* (*Pseudol.* 1).[7] More strained, perhaps, but none the less ingenious is Lucian's citation of a verse from Pindar concerning ποντίου θηρὸς πετραίου, the cuttlefish, and how it supplies a model for the mimic (*Salt.* 67).[8] That these and other passages testify more to Lucian's reading habits than to field work in natural history can be granted. What must not be overlooked, however, is that Lucian knew enough natural history to turn these passages to his own, original purposes.

Midway between Lucian's knowledge of natural history, based upon observation and everyday experience, and the literary exercises just discussed, stands another literary genre, the proverb. A few words are required on this topic for, as several scholars have shown, rudiments of ancient natural history survived well into the Middle Ages.[9] Frequently they were transmitted by the exempla which lightened many a medieval sermon even if they failed to enlighten the devout.

Proverbs, however mechanically they may be used, usually rest upon a factual basis. Even those whose meaning rests upon an inversion of nature, exaggerate but do not obliterate this basis; otherwise such a *cliché* as ὁ νεβρὸς τὸν λέοντα (*Dial. Mort.* VIII. 1) would be pointless. Instances of both types, those with straightforward and those with oblique meanings, occur in Lucian. They tend, moreover, to substantiate the claim regarding Lucian's knowledge of natural history. A miscast proverb is grotesque; yet, in order to illustrate gross stupidity, what could be more economical than

[6] All of the creatures mentioned in *Dips.* 3, save the φύσαλος (cf. *Philops.* 12) occur in the *Theriaca*. For their identification, see Nicander, ed. Gow and Scholfield (Cambridge, 1953), pp. 19-23.

[7] Archilochus fr. 143 Bergk[3].

[8] Pindar fr. 208 Turyn; cf. Turyn's notes *ad loc.* for further references.

[9] Cf. G. R. Owst, *Literature and Pulpit in Medieval England*, 2nd ed. (Oxford, 1961), pp. 190-209.

"mute as fish," (*Adv. Indoct.* 16) or "senseless as grasshoppers" (*Jup. Trag.* 31)?[10]

Whether it is analogy, parody, or proverb, Lucian's ability to use them nicely permits at least one generalization: he was alert to any information which had potential value, whatsoever its source or degree of credibility. A telltale example occurs in some questions Zeus casually asked regarding things on earth. He inquired first, said Menippus, "about the price of wheat nowadays in Greece and whether last winter's rainfall was violent and whether the vegetables require more rain" (*Icar.* 24). Mundane questions though these may be, they commanded a widespread interest because they were the common concern of all. We may imagine Lucian, in his rambles about town, alive to the trivia of shopkeepers, sailors, and peasants. κἂν ἰδιώτης ᾖς, he once remarks, ἤτοι σκυτοδέψης ἢ ταριχοπώλης ἢ τέκτων ἢ τραπεζίτης (*Vit. Auct.* 11), that does not preclude the possibility of a vital piece of information. Conversations with farmers, perhaps, lie in back of a series of observations about agricultural practices (*Hesiodus* 7). Lucian's willingness to listen, as well as to speak, gave him access to information which he could later weave into the texture of his compositions. There is, for instance, no awkwardness in putting into a boatman's mouth the rectification of one of the oldest fables of literary ornithology, viz. the harsh fact that swans have equally harsh voices (*Electr.* 5). Discussions with professionals from all walks of life undoubtedly opened Lucian's eyes to superstitions passing all belief (cf. *Philops.* 7).[11] But there were other sources too.

Religious rites, for instance, provided an opportunity for a rudimentary course in comparative anatomy. *On Sacrifices*, diatribe though it may be, contains sufficient material to reconstruct, in part, some of the sacrificial techniques and to identify some of the victims. In addition to the ox, goat, and lamb (*Sacr.* 12), Lucian records the unusual practice of sacrificing a boar (*Syr. Dea* 54) and even a horse (*Luctu* 14).[12] The actual procedure followed by the priest

10 For the ichthyological background of the former proverb, see D'A. W. Thompson, *A Glossary of Greek Fishes* (Oxford, 1947), p. 62 s.v. ἜΛΛΟΨ.

11 Superstitious beliefs about animals were often transferred to the drugs derived therefrom. For almost exactly contemporary uses, cf. Stannard, "Materia Medica and Philosophic Theory in Aretaeus," *Sudhoffs Archiv f. Gesch. d. Med. u. d. Naturw.*, XLVIII (1964), 27-53.

12 Whether the sacrifice of a boar is based on idolatrous or Semitic practices is a moot question. Cf. F. S. Bodenheimer, *Animal and Man in Bible Lands* (Leiden: Brill, 1960), p. 212, and *Hori Apollinis Hieroglyphica*, ed. Francesco Sbordone (Napoli: Loffredo, 1948), p. 120.

was to cut the throat, eviscerate the animal, remove the heart and then either burn the carcass (*Sacr.* 13, cf. *Necyom.* 9) or simply bury the animal without removing the entrails (*Sacr.* 15). How much Lucian learned from observing or participating in sacrifices is problematic. But that he had some experience with sacrificial equipment is proved by one of those realistic details necessary to support the superficial persiflage. Mnesitheus is censured by Zeus for failing to live up to his promise of a hecatomb. Instead, he sacrificed a single old cock and "four pieces of frankincense so rotted with mildew (εὐρωτιῶντας) that they failed to burn upon the coals and produced so little smoke that they could not be smelled at all" (*Jup. Trag.* 15). Lucian's word-magic is nowhere better illustrated than in the deft touch by which such a detail, insignificant in itself, spells the difference between a *tableau vivant* and an overworked commonplace. The point, however, is not Lucian's dramatic ability but the fact that he must have known that a mildewed piece of incense was worthless.

One other source deserves mention because its obvious modern parallel has provided similar information for the amateur naturalist. Athletic spectacles and bouts, no doubt, provided more amusement than edification. But when one recollects his own boyhood days and the circus, it is easy to recall that some natural history insinuated itself into the mind, and perhaps the nostrils as well. The craze for horse racing is mentioned several times by Lucian (*Nigrinus* 29, *Timon* 50) and may have provided the basis for his remarks about a horse's gait (*De domo* 10) and whether driving a team is a τέχνη or not (*Paras.* 8). A limited knowledge of horses obtained in this way, and perhaps from a horse-breeder as well (cf. *Timon* 23), will help us to understand why Lucian praised Zeuxis' hippocentaur as realistic. "The Centaur herself," he begins,

> is depicted lying on fresh young grass with all the horse part of her on the ground. Her feet are stretched behind her. The human part is slightly raised up on her elbows. Her fore-feet are not now stretched out, as you might expect with one lying on her side; one foot is bent with the hoof drawn under like one who kneels, while the other on the other hand is beginning to straighten and is taking a grip on the ground, as is the case with horses striving to spring up. She holds one of her offspring aloft in her arms, giving it the breast in human fashion; the other she suckles from her mare's teat like an animal. (*Zeuxis* 4, trans. by K. Kilburn, *Lucian*, Loeb Edition, Vol. VI)

Gladiatorial shows must have supplied a further glimpse into the ways of strange or angered animals, but in the one passage where

Lucian describes such a spectacle, he merely states that he saw θηρία κατακοντιζόμενα καὶ ὑπὸ κυνῶν διωκόμενα (*Tox.* 59).

A complete study of those sources from which even a passive bystander might profit would, however, fail to do justice to Lucian's command of natural history. There are, that is, far too many passages that can be explained only by supposing that he actually consulted Nature as the final arbiter.

That *tour-de-force*, the epideictic *Muscae Encomium* is, as the title suggests, devoted to the most improbable of all subjects. The mockery of current rhetorical practices should not disguise the fact that Lucian's description of the fly ranks as one of the finest pieces of descriptive entomology in antiquity. After describing its flight and comparing the peculiar buzzing with the sound produced by other insects, he continues:

> Regarding the remainder of the body, her head is lightly connected to the neck and easily turned about, not united as it is in the case of grasshoppers. The eyes are prominent and resemble horn. The thorax (στέρνον) is firm and compact and the legs grow out from the waist (τῇ ἐντομῇ) which is not constricted as it is in wasps. Its abdomen is armored, even as wasps, and resembles a corselet (θώραξ) by its flattened segments and scales. She defends herself, however, not by means of her tail-end, as wasps and bees do, but by her mouth and her proboscis (τῇ προβοσκίδι). As in the case of elephants, she has the means of foraging, seizing and holding fast in virtue of an organ attached at the end, resembling a sucker (κοτυληδόνι). Projecting from it is a tooth by which she bites and drinks blood (for she drinks blood as well as milk) though her bite is not very painful. She has six feet but uses only four in walking, while the two in front are used as hands. You may see her standing on four feet holding in her two hands a piece of food, very much like a person. (*Musc. Enc.* 3) [13]

Quite apart from the purpose behind the encomium, the details could only have been ascertained by a direct, personal examination.

It cannot be doubted that Lucian closely examined various objects. A simple yet effective experiment is recorded when it is noted that a decapitated fly continues to exhibit signs of life, including breathing (*Musc. Enc.* 6). It would be unfair to tax Lucian for failing to recognize this behavior as a reflex phenomenon, yet it is to his credit that he does not go beyond his evidence as Albertus Magnus did, a millennium later, when repeating a similar experiment on a cicada.[14] Another sample of personal in-

[13] I have translated Harmon's text (Loeb Library ed., Vol. I, p. 84), as Jacobitz' Teubner text (Vol. III, p. 135) appears to be defective.

[14] Albertus Magnus, *De Animalibus*, XXVI. 14, ed. H. Stadler (Münster, 1920), II, 1585.

spection concerns the habits of the social insects. Well before Lucian's time, the communal activities of ants had served as a model for Stoic moralizing.[15] And while Lucian, too, is unable to resist the temptation, the following description reads as if it derives from an actual observation. The organized way of life of ants is introduced by these words: "One ant carries out dung while another, having seized somewhere a bean skin or a piece of cracked wheat, is running, bearing it off" (*Icar.* 19). Then, armed with a ready excuse to animadvert upon the similarities between ants and humans, he adds, "Doubtless, among them there are, according to their way of life, builders and demagogues, councilmen, musicians, and philosophers" (*Icar.* 19). This extrapolation in no wise lessens our opinion of Lucian's curiosity and visual acuity. While poking fun at Pythagoras and his notorious taboo against beans, the Pythagorean spokesman attempts to justify the master's *ipse dixit* regarding the divinity of beans.[16] They are, he announces, the generative principle (τὸ πᾶν γονή), "and if you should open a bean while it is yet green, you will observe its resemblance to the *membrum virorum*" (*Vit. Auct.* 6). The structure in question, ἡ φυή, is probably the hypocotyl of a typical dicotyledonous plant. Its recognition is not surprising in virtue of the important place of legumes in the ancient diet.[17] Acquaintance with such staples explains other botanical observations such as the tough integument of the lupine seed (*Ver. Hist.* I. 14). Finally, but still within the vegetable kingdom, comes a remarkably prescient hint of things to come. Demonax, always ready with a quick retort, was once asked how much smoke would result from burning a thousand pounds of wood. The answer, said he, little realizing the prophetic step he was taking in plant chemistry, was simple: "Weigh the ashes, the remainder will be smoke" (*Demonax* 39).[18]

A noteworthy feature of Lucian's natural history consists in the numerous passages dealing with insects or, more precisely, Arthropoda, for Lucian, like the ancients in general, did not distinguish

15 Cleanthes fr. 515 ed. H. von Arnim, *Stoicorum Veterum Fragmenta,* I (Leipzig, 1905), 116.

16 Cf. Alfred C. Andrews, "The Bean and Indo-European Totemism," *American Anthropologist,* LI (1949), 274-92.

17 Cf. Jacques André, *L'alimentation et la cuisine à Rome* (Paris: Klincksieck, 1961), p. 39.

18 The later history of the subject is treated by L. F. Kebler, "Dr. Francisco Redi, A Pioneer in Plant Ash Studies," *J. Am. Pharm. Assoc.,* XXVI (1937), 240-44.

the classes Insecta and Arachnida. We have already commented on his description of the fly and the social behavior of ants.

It remains now to call attention to a few other passages pertaining to insects. No other insect, save possibly the bee, is mentioned so often by the Greeks as the τέττιξ or cicada. The frequency of citation is an index of its common occurrence and widespread distribution. For this reason, a large literature could be assembled on its life-habits as seen by the ancients. Lucian adds his share to *tettigologia,* not all of which is accurate, but then Aristotle shared the mistaken belief that the cicada feeds on dew.[19] The majority of Lucian's references pertain to its song which was so appreciated that it gave rise to proverbs and similes and reached a peak in the fad of keeping cicadas imprisoned in little cages (cf. *Bacch.* 7, *Imag.* 13, *Peregr.* 41 etc.).[20] More significant, however, is Lucian's observation on the membranous wing (ὑμενόπτερος), shared alike by grasshoppers (τὰς ἀκρίδας), cicadas, bees, and flies (*Musc. Encom.* 1; *Dips.* 3). Among the laity there has always existed some confusion between the cicada and the grasshopper, yet Lucian is clear enough on the distinction. The flight pattern (πτῆσις) of the fly is " not, as in the case of bats, a smooth, continuous motion of the wings, nor, as in the case of grasshoppers, a bounding spring, nor, as in the case of wasps, a whirring motion, but rather so flexible as to permit her to fly wherever she may wish (*Musc. Encom.* 2). The dozen or so other insect types mentioned by Lucian are of interest chiefly in indicating that no detail was too small or too trivial to warrant a place in his writings (e.g., *Sacr.* 9).

Proverbs referring to insects likewise are cited where the same point could be made, though perhaps not so economically, by describing some aspect of insect life. A minimal knowledge of insects on the part of his audience left Lucian freedom to refer to insects for illustrative purposes. The famous strong man Milo will, like all mortals, reach that time when he will not be able to lift a mosquito (ἐμπίς, *Contempl.* 8). Lynceus' vision is so penetrating that he can descry a mosquito's nest (*Icar.* 12). In order to understand passages such as these, no especial information is required nor is there any reason for supposing that Lucian was a field entomologist. It is evident, nonetheless, that he had an eye for detail and was capable of using empirical evidence, witness his remark about the dung beetle and its curious habit of " pill-rolling "

[19] Aristotle *Hist. An.* IV. 7, 532b14. . . . τῇ δρόσῳ τρέφεται μόνον.
[20] For insects kept in cages, cf. *Anthologia Palatina,* VII. 189-98.

(*Pseudol.* 3). That moths eat woolens (*Saturn.* 21), coleopterous larvae destroy papyrus rolls (*Adv. Indoct.* 1, 17), and weevils infest granaries (*Saturn.* 26) are but some of the pieces of common information that found their way into Lucian's writings. Other passages betray a literary source, though the Indian ants of Herodotus (III. 102) do not deceive Lucian (*Saturn.* 24; cf. *Gallus* 16). On the other hand, he lends support to a belief that persisted, in one form or another, to the time of the famous debate between Pasteur and Pouchet on spontaneous generation. A grim future is predicted for Simon the miser and his hoarded wealth, for μετ' ὀλίγον πάντα ταῦτα καταλιπόντα σίλφην ἢ ἐμπίδα ἢ κυνόμυιαν γενέσθαι (*Gallus* 31). If the belief in the spontaneous generation of insects is only implicit here, it is spelled out elsewhere. Regarding the fly, he writes, "it is not generated in the form just described, but rather as a larva (σκώληξ), either from dead men or animals. Then, in stages, the legs emerge, the wings grow and from a creeping animal comes a flying animal; it becomes pregnant and gives birth to a little larva that is the future fly" (*Musc. Encom.* 4).

Turning from insects to other flying creatures, it may be said that Lucian's ornithology follows much the same pattern as his entomology. His bird-lore, that is, is a blend of popular knowledge and fable, strengthened by observational bits and scraps, to which a dash of the fabulous adds flavor. There is one major difference, however. Lucian's birds are either demonstrably real, and hence capable of identification, or they are fictitious, invented for the purpose at hand. In the latter category a creature such as the γρὺψ ὑπόπτερον θηρίον (*Navig.* 44) may be relegated to myth in the same fashion as Herodotus' φοῖνιξ is dismissed (*Navig.* 44, cf. *Peregr.* 27).

When we turn from the world of fantasy to the real avifauna, we find that Lucian refers to and illustrates his writings with some twenty different species. Not all can be identified to the species rank because such a term as γύψ denotes not only the vulture but several other species as well. The common birds, such as the crow (κορώνη), jackdaw (κολοιός) and raven (κόραξ) account for nearly half of Lucian's references (cf. *Eun.* 8, *Electr.* 5, *Jup. Trag.* 31, *Gallus* 21, etc.).

Other species are cited only once or twice but always in such a manner as to suggest that Lucian knew enough about their appearance or habits to refer to them when they answered to his specific needs. Thrasycles' table manners, προαρπάζων ὥσπερ ἴκτινος τὰ ὄψα

(*Timon* 54) allude to the ferocious grasping habits of the common kite. Lucian's sojourn in Egypt was not without its rewards, for it added to his fund of information. Describing the desolate wastes of Libya, he calls attention to the few animal species capable of withstanding a xerophytic area. The monotony and fear of venomous serpents, scorpions, and the dreaded *βούπρηστις* is relieved by the ostrich (*Dips.* 2-3). Not only is it hunted for food, but its eggs provide nourishment and utensils alike for the nomadic Garamantes (*Dips.* 7). Egypt was also the home of the sacred ibis, pictured so often in tomb reliefs.[21] The half-human, half-animal Egyptian gods are held up for ridicule in several passages (*Jup. Trag.* 42, *Deor. Conc.* 10). Lucian takes delight in contrasting the anthropomorphic Greek gods with the zoomorphic Egyptian gods and the endless possibilities provided by a belief in human transformations (*Sacr.* 14). In one passage he ceases his raillery long enough to list some gifts sought from Egypt which are sufficiently credible to suppose that Lucian was acquainted with the desiderata: τὰ Νειλῷα ταῦτα ταρίχη τὰ λεπτὰ . . . ἀπ' Αἰγύπτου ἢ μύρον ἀπὸ τοῦ Κανώπου ἢ ἶβιν ἐκ Μέμφιδος (*Navig.* 15).

One could, of course, go on to enumerate all the mammals, fish, and plants, perhaps emboldened by relying on Lucian himself—κόρος οὐδεὶς τῶν καλῶν (*Dips.* 9). But enough has been said, I believe, to support the claim that Lucian knew what he was talking about.

[21] Cf. Otto Keller, *Die antike Tierwelt* (Leipzig: Engelmann, 1913), II, 198-202.

V

Materia Medica and Philosophic Theory in Aretaeus

Except for HIPPOCRATES, with whom ARETAEUS has been favorably compared, there is no major medical writer of antiquity whose biographical details are more uncertain. Not only is his birthplace unknown, as is the land where he practiced, even the century of his birth has been the subject of widely differing opinions[1]. He may have studied in Alexandria and there is some probability that he resided in Italy[2]. Concerning the date of his birth, the little evidence available derives from his own writings. The use of the Ionic dialect and his reliance on Stoic-inspired pneumatic theories are compatible with either a first or second century date. He refers several times to passages from the Hippocratic Corpus, but to no writer by name unless the Philo mentioned in connection with a salve is the PHILO OF TARSUS known in antiquity as the inventor of a famous eye salve[3]. ARETAEUS, in turn, is mentioned only by such late authors as ALEXANDER OF APHRODISIAS, AETIUS OF AMIDA and PAUL OF AEGINA[4]. It is generally accepted that he flourished in the second

[1] R. KOSSMANN, Wann lebte Aretaeus von Cappadocien? Münch. med. Wschr. 49 (1902), pp. 1265—67 argued that he was a pupil of NICANDER OF COLOPHON who died ca. 130 B.C. On the other hand, ROBERT FUCHS thought that he might have lived as late as the 4th century A. D., Puschmanns Handbuch der Geschichte der Medizin ed. MAX NEUBURGER und JULIUS PAGEL. Vol. I (Jena 1902) p. 366.

[2] The best evidence for an Italian residence is the use of *φάρ*, from Latin *far*. VI 2 p. 125.16 ed. C. HUDE, Aretaeus. Editio altera. Berlin 1958 (*Corp. Med. Graec.* t. II). All references to ARETAEUS are to this edition; for the form of the reference cf. n. 6. Italian wines are mentioned in VI 3 p. 128.13 but since they were exported abroad (cf. LUCAN, Pharsalia X 163) this does not necessarily mean an Italian residence; nor does the description of diphtheria as *helkea Syriaca*, I 8 p. 11.27. The reference to Egyptian plants, barley wine and the water of the Nile (I 8 p. 11.22—26) may indicate that ARETAEUS travelled to Egypt.

[3] The compound drug of PHILO's is mentioned only in connection with gonorrhoea (VIII 5 p. 164.25) and it may not be the salve mentioned by CELSUS, De Medicina VI 6,3. For other references to a *drug of Philo's* cf. Rufus of Ephesus, ed. CH. DAREMBERG et C. E. RUELLE. Paris 1879 p. 397. For the references to HOMER and HIPPOCRATES, cf. J. H. and H. N. COUCH, The Literary Illustrations of Aretaeus of Cappadocia. *Canad. Med. Assoc. Jrnl.* 33 (1935) 556—59.

[4] Cf. MAX WELLMANN, Die Pneumatische Schule bis auf Archigenes. Philologische Untersuchungen 14 (1895) pp. 23—24.

V

half of the first century, A.D., and was thus slightly younger than Galen (131—200 A.D.)[1]. The failure of Galen to mention Aretaeus by name is, however, difficult to reconcile with an early pre-Galenic date. To avoid this dilemma, Adams suggested that Aretaeus and Galen were contemporaries and out of professional jealousy refrained from mentioning the other by name[2]. The complexity of a satisfactory dating has been increased by Wellmann's demonstration of a series of parallels between Aretaeus and Archigenes of Apameia, another pneumatist[3]. From the resemblances between these passages, he concluded that Archigenes was the primary source from whom Aretaeus borrowed. However, since Archigenes' date is also uncertain, one could with equal justice reverse the argument[4].

Whatever our uncertainty on the date of Aretaeus, his position in the front ranks of ancient medical writers appears secure. Although not a prolific writer, as judged by the titles attributed to him, a good estimate can be formed by what has survived[5]. Despite the little attention given him in antiquity, his reputation since the editio princeps of 1554 has steadily grown[6]. This can be measured,

[1] Arturo Castiglioni, A History of Medicine. Translated and edited by E. B. Krumbhaar. New York 1941 p. 215. Heinrich Haeser, Lehrbuch der Geschichte der Medicin. 3. Aufl. Vol. I (Jena 1875) p. 341.

[2] *The Extant Works of Aretaeus, the Cappadocian.* Edited and translated by Francis Adams. London, The Sydenham Society, 1856. Preface p. V. Traditionally, Aretaeus' main work has been divided as follows: On the Causes and Symptoms of Acute Diseases, 2 books; On the Causes and Symptoms of Chronic Diseases, 2 books; On the Therapy of Acute Diseases, 2 books; On the Therapy of Chronic Diseases, 2 books. The convention of 8 books, but without the use of the cumbersome titles, has been adopted following Hude's edition (cf. p. 27, n. 2) whose page and line follow the book and chapter references. [3] Cf. Wellmann, Pneumatische Schule, pp. 23—64.

[4] Some years before Wellmann, it was argued that Archigenes borrowed from Aretaeus: C. W. Klose, Über das Leben des Aretaeus und seine auf uns gekommenen Schriften. *Janus, Central-Magazin* 1 (1851) 105—126, 217—246.

[5] Aretaeus himself mentions the titles of four other works no longer extant: *On Fevers* (V Proem. p. 91.7); *On Diseases of Women* (V 3 p. 102.19); *On Drugs* (V 4 p. 104.24; VI 2 p. 124.11—14) and *On Surgery* (VII 2 p. 145.21). One other title, *On Preservatives* is mentioned by Alexander of Aphrodisias *De Febribus* 24 ed. J. L. Ideler, Physici et Medici Graeci Minores. Vol. I (Berlin 1841) p. 97.

[6] The first Greek text, edited by Jacobus Goupylus, Paris, ap. Adr. Turnebum 1554 was preceded by a Latin translation in 1552 by Junius

not only by the numerous editions, translations, and commentaries, many of which were prepared by physicians, but also by the unanimity of medical historians. "As a clinician", GARRISON wrote, "Aretaeus ranks next to the Father of Medicine in the graphic accuracy and fidelity of his pictures of disease . . .[1]." This judgment is typical of the praise bestowed upon ARETAEUS by medical historians[2].

The literature devoted to him has tended, as GARRISON's remark suggests, to concentrate on his semiotics and on the identification of the diseases described by him. As a consequence, other aspects of his writings have been neglected. This has led to some misunderstanding concerning his pharmacology and materia medica[3]. The fact that many of the diseases described by ARETAEUS can be identified supplies the historian with an opportunity of correlating a recognizable disease with the corresponding forms of therapy. Not only is this an important chapter in the history of pharmacology and therapeutics, but the fact that ARETAEUS' writings are the only

P. CRASSUS, Venice. A copy of the 1554 edition is in the National Library of Medicine. For other editions, translations and commentaries cf. LUDWIG CHOULANT, Handbuch der Bücherkunde für die ältere Medizin. Faksimile-Ausgabe, Graz 1956, pp. 84—88.

[1] The passage continues ". . . of which he has given the classic accounts of pneumonia, pleurisy with empyema, diabetes, tetanus, elephantiasis, diphtheria (*ulcera Syriaca*), the aura in epilepsy, the first clear differentiation between cerebral and spinal paralysis, indicating the decussation of the pyramids, and a very full account of the different kinds of insanity." FIELDING H. GARRISON, An Introduction to the History of Medicine. 4th ed., Philadelphia 1960, pp. 110—111.

[2] ACKERKNECHT stated that ARETAEUS' "extraordinary clinical descriptions of diabetes, tetanus, diphtheria, and leprosy show clearly that the clinical genius of the Greeks was still as alive and powerful in the second century A.D. as it had been in the sixth century B.C." ERWIN H. ACKERKNECHT, A Short History of Medicine. New York 1955, p. 65. In the same spirit OSLER said that "his descriptions of diseases are patterns of fidelity to nature." WILLIAM OSLER, Bibliotheca Osleriana, Oxford 1929, p. 31.

[3] E. g. NEUBURGER stated that his drug therapy was characterized by the "making use of only a small number of remedies, mostly mild in their action. Where energetic interference is indicated he makes use of opium . . ." History of Medicine. Translated by ERNEST PLAYFAIR, Vol. I (London 1910), p. 232. That this rests on a misunderstanding, cf. p. 53 n. 2. SARTON's comment is hardly less enlightening: Aretaeus' "therapeutics were very simple; his prescriptions were largely restricted to physical agents and to dietetic regulations (e.g., use of cow milk)." GEORGE SARTON, Introduction to the History of Science. Vol. I (Baltimore 1945), p. 308.

extant complete works produced by a member of the pneumatist sect make such a study a valuable reference point for a comparative study of the development of Greek medical botany. It would seem worthwhile, then, to examine the extent of his materia medica and to place his pharmacological practice within the context of the pneumatic theories by which it was philosophically supported.

Prior to examining the composition of his materia medica, it is necessary to consider some of the beliefs which supplied the philosophic framework for his general medical theory. It has often been claimed that Aretaeus, if not a full-fledged pneumatist, shows a close kinship with the pneumatist sect[1]. This sect, which may be said to date from Athenaeus of Attalia (second half of first century, A.D.), is based upon the philosophic theories of the Stoa. The most important of the doctrines borrowed from Stoicism were those connected with pneuma. Other beliefs, such as the use of *tonos* to explain the qualitative changes of the pneuma and a technical vocabulary to describe different states of the pneuma, are derived from the same sources[2]. The most significant feature of this pneumatically-based medical theory is that it led, in physiology as well as in pharmacology, to the substitution of a humoral theory for the solidist theories advocated by the methodist sect and others.

Health and the normal functions of the internal organs were explained as a harmonious balance of pneuma and the humors[3]. Through various causes, both those directly attributed to overt

[1] G. Verbeke, L'Evolution de la Doctrine du Pneuma. Paris 1945. Pp. 191—206 and Wellmann, Pneumatische Schule, *passim*.

[2] *Tonos* or tension characterizes the state of pneuma under normal conditions of health. When the pneuma becomes thin and diffuse it is said to be *ἄτονος* (without tension) and the body as a whole is characterized by the state of *ἀτονία*. "Hysteric suffocation" is described as "prostration, loss of tonos and strength of the knees," II 11, p. 33.16. In the same way, Chrysippus correlated health with the state of well-balanced *τόνος* or *εὐτονία* and disease with *ἀτονία*. Chrysippus Fragment 471, in Hans von Arnim, Stoicorum Veterum Fragmenta. 4 vols. 1905—24. Vol. III (Leipzig 1913), p. 120. A closely related concept, innate heat (*ἔμφυτος θερμή*) occurs in Aretaeus in connection with the etiology of arthritis (IV 12 p. 83, 9) and satyriasis (VI 11 p. 142.14). The Stoa held that the innate heat, properly balanced with the other humors, was required for normal physiological processes, cf. Chrysippus Fragm. 410, Vol. II p. 135 von Arnim.

[3] "For when nature is strong and well-blended, it dominates and rules all—the moisture, the pneuma and the solids which, by their good order and symmetry, regulate man's life." II 3, p. 23. 7—10.

pneumatic changes and to such indirect causes as dietary deficiencies, this equilibrium is upset. Depending upon the severity of the cause and on the pneumatic reactions thereto, various pathological conditions are produced.

Because pneuma is transported through the body by the vascular system, there is a close relation between pneuma and blood. The liver, the organ in which blood is produced and which is described as a "concretion of blood", is especially sensitive to any form of pneumatic imbalance[1]. Likewise, the heart, described as the origin of life, is given special consideration[2]. If, through a hemorrhage or a too-drastic purge, a large amount of pneuma escapes, the body is greatly weakened and is subject to further complications. Death, accordingly, is attributed to an excessive loss of pneuma[3]. The other most important organ is the brain and here again, a Stoic influence is evident. ARETAEUS, following the lead of ZENO and CHRYSIPPUS, placed the origin of *neura* and the seat of sensation in the head[4].

Except for his dependence on those doctrines associated with pneuma, ARETAEUS is remarkably free from the philosophic vagaries and rationalistic dogmas that transformed medicine from a

[1] "The origin of blood is in the liver as well as its discharge to all the inner parts. And the entire liver is like a concretion of blood. Its inflammations are the most acute for the seat of nutrition is located there," VI 6, p. 134. 18—21. Regarding a cure for apoplectics, excessive bleeding is dangerous "for to them even a little blood is most important for it is the heat of the life of the body and of food," V 4, p. 103. 1—2.

[2] The heart is the origin of life, V 1 p. 92.28 and hence responsible for the innate heat transmitted through the vascular system, VI 7 p. 136.12—13.

[3] V 1 p. 97.19—22. Cf. the Stoic belief, "On the departure of the spirit [*spiritus* = pneuma], which is generated with the body, the living being dies." ZENO, Fragm. 137, VON ARNIM, Vol. I (Leipzig 1905), p. 38.

[4] VI 1, p. 92.28; V 2 p. 100.29. The seat of sensation is also located in the head, V 6 p. 108.17—18. ARETAEUS' knowledge of the brain may have been based partially on clinical experience for he seems to allude to the fact that an injury on one side of the brain caused paralysis on the opposite side of the body, III 6 p. 45.7—14. It has been argued that this was borrowed from GALEN who actually performed several experiments on the spinal cord, H. M. THOMAS, Decussation of the Pyramids—An Historical Inquiry. *Bull. Johns Hopkins Hosp.* 21 (1910), pp. 304—05. According to GALEN, CHRYSIPPUS argued that since the *hegemonikon* or ruling principle is located in the head and because the *neura* proceed from it, therefore they originate in the head. CHRYSIPPUS Fragm. 898, VON ARNIM, Vol. II (1903), p. 247. Cf. FRIEDRICH SOLMSEN, Greek Philosophy and the Discovery of the Nerves. *Museum Helveticum* 18 (1961), pp. 181—84.

healing art into a scholastic dialectic. Occasionally there is to be seen beneath the surface some fanciful beliefs, but these were far from unusual in the medico-philosophical circles of the times[1]. Although his pneumatic theories, rather than epistemological theory or anatomy, govern his choice of therapy, he is never reluctant to admit that he has derived valuable experience from his medical practice[2]. This emphasis on personal experience, summed up in his advice that the succesful physician must experiment, has led some historians to place him among the eclectics[3].

In Greek medicine generally, drug therapy was only one of several types of therapy available. The etiology of the disease and the philosophic theories guiding physiology and pharmacology were as important in determining the proper form of therapy as was the availability of drugs. Hence, in Aretaeus as in other Greek medical writers, there is to be found side by side with drug therapy, various surgical techniques, physiotherapy and exercise, and above all, an emphasis on dietary regimen.

The relation between these different forms of therapy follows a rather uniform pattern in the twenty odd diseases described by

[1] The remarkable description of the uterus, and how its movement is the cause of "hysterical suffocation" concludes with these words: "it wanders erratically. And it delights in fragrant odors and moves toward them. But it has an aversion to unpleasant odors and flees them. On the whole, the human uterus is like an animal within an animal." II 11 p. 32.26—33.1. For an attempt to place this passage within the context of ancient gynaecology cf. Johann Lachs, Gynäkologisches von Aretaios. Leipzig 1904, p. 227. Other fancies of Aretaeus include a belief in the divine origin of madness (III 6, p. 44.2), that blood was formed from wine (VIII 2, p. 163.1—2) and that the milk of a woman who has just delivered a male child is particularly effective in a fever diet (VI 3, p. 130.26). For a possible reference to the "doctrine of signatures" cf. p. 44 n. 1.

[2] Cf. VII 5, p. 158.12—13 where his use of a decoction of mallow seeds is based on a previous successful cure. He concludes by saying "but it is necessary that the physician himself contribute towards these matters and not apply himself solely to the writings of others." VI 11, p. 143.9—10. In discussing the cause of arthritic pain, he states, "only the gods understand the true cause, but men may know a probable cause," IV 12, p. 83.4—5.

[3] In discussing a phlegmagogue for lethargy, Aretaeus states that he had good results when he administered it "to a man who drank it at my advice. For experience is a good teacher. It is necessary, then, to experiment, for caution is ignorance" V 2, p. 102.1—2. — For Aretaeus as an eclectic, cf. T. Clifford Allbutt, Greek Medicine in Rome. London 1921. Pp. 265—279 and J. L. Pagel, Geschichte der Medizin. 2. Aufl. Berlin 1915, p. 104.

ARETAEUS. Prior to any form of therapy, the patient was carefully examined and special attention was given to the condition of his skin, his color, posture and general appearance. Pulse and respiration were, because of their obvious importance to the pneumatist, observed as were various bodily secretions[1]. Tests were performed on the patient's sputa in diagnosing *phthisis* while instruments used for the purpose of examination are referred to several times[2]. During the course of the examination the patient was questioned and a rudimentary case history was obtained[3]. Following the examination the patient was ready to undergo the first in a series of routine therapeutic measures.

The first of these measures was bleeding, including scarification and the use of leeches. Owing to the transportation of pneuma by the blood, an alteration in the blood supply was thought to be a reasonable form of therapy. In those conditions characterized by a surfeit of pneuma, or of blood, bleeding is the most direct means of lessening their quantity[4]. Precise directions are given where bleed-

[1] Scrapings of flesh were noted in the feces in dysentery (IV 9, p. 76.30) and urine was examined for sediment in diseases of the liver and kidneys, VII 13, p. 160.18—19. The force and frequency of the pulse was noted in cardiac disorders (VI 3, p. 127.27) and the return of a strong pulse is included in a prognosis (V 4, p. 103.26). Renal calculi were retained for purposes of examination and their color, shape and consistency were noted (III 3, p. 68.14—18).

[2] The testing of fluids "by fire or water" (III 8, pp. 47. 29—48.2) probably referred to the same type of test as those in the Hippocratic writings, e.g. *De morbis* II 47 (VII 72 L); *De morbis mulierum* I 83 (VIII 204 L). Twice the colon is examined by means of a probe or rectal speculum, II 6, p. 26.10; II 8, p. 28.21—22. Examination of a dropsical patient included percussion of the abdomen "for if you strike with the hand, the abdominal cavity murmurs" (IV 1, p. 63.19). For percussion and its Hippocratic antecedents, cf. E. F, CORDELL, Aretaeus the Cappadocian. *Bull. Johns Hopkins Hosp.* 20 (1909) p. 374.

[3] In seeking to determine the nature of coughing up blood, ARETAEUS states, "if it is produced by an erosion, it is necessary to ask whether the man has previously coughed, or has experienced a difficulty in breathing or whether he has suffered earlier from nausea or vomiting," II 2, p. 19.2—4. It is also noted that those whose occupations are inside heated buildings, e.g., blacksmiths or bath attendants, are less likely to suffer from asthma, III 11, p. 52.18—19. Some half a century earlier, RUFUS OF EPHESUS wrote a short tract entitled *Medical Questions* which dealt with just such matters as these. Cf. RUFUS, pp. 195—218, ed. DAREMBERG et RUELLE.

[4] Satyriasis, caused by a surfeit of blood, hence bleeding was required (VI 11, p. 142.6—7) while lethargy was caused either by excessive blood or "by phlegm or other humors." If the former, bleeding was indicated, if the

ing is to be performed[1]. Kindred forms of therapy, cupping and the use of vesicants and rubefacients, were employed on the same routine basis in the belief that pneuma and the humors tend to be collected or localized in specific organs. The local inflammation produced by such means was regarded as evidence of withdrawing or expelling an unnatural collection of pneuma. A further aim which these techniques served was to divert the collection of pneuma from an organ or area inaccessible to surgery or to a direct application of a drug, to an organ or area where medication would be more effective. Thus on several occasions cupping is advised in order to localize the superfluous pneuma or fluid and render the subsequent therapy more effective[2].

The second of the routine measures was purging. As early as the Hippocratic writings a connection was established between the inspired atmospheric pneuma, necessary for life, and a superfluous residue, or flatus, which was to be eliminated[3]. The rationale for routine purging thus depends upon the pneumatic etiology to which Aretaeus often refers in describing diseases.

The routine therapy of bleeding and purging may extend for a period up to ten days. During this time the physician was concerned with promoting the general physical and mental health of the patient prior to the advent of more specific forms of therapy[4]. If the

latter, a purge was ordered (V 2, p. 98.21—25). A collection of pneuma localized in the thorax caused the pain associated with pleurisy (V 10, p. 118. 14—15).

[1] Bleeding at the vein "by the elbow" is mentioned several times (VII 2, p. 144.21; VII 3, p. 149.18) as are the veins "by the nose" (VII 3, p. 149. 23—24), on the forehead (VII 2, p. 145.8), "by the genitalia" (VI 10, p. 141.13) and behind the ears (VII 2, p. 145.16).

[2] VII 1, p. 145.10—14; VII 5, p. 157.21.

[3] "Wind (πνεῦμα) in bodies is called breath (φῦσα), but outside the bodies it is called air (ἀήρ)." Hippocrates, *De flatibus* 3 (VI, 94 Littré). All references to Hippocrates are to the edition of E. Littré. By the time of Aretaeus, and due largely to Stoic teachings, φῦσα came to be identified with flatus while πνεῦμα was regarded as the necessary "life force" within the body. The corruption of pneuma into flatus is mentioned by Aretaeus in his account of the etiology of elephantiasis, IV 13, p. 88.2—4.

[4] A form of psychotherapy is mentioned in regard to the therapy of phrenitis. The household is to preserve silence and a plain room with a minimum of clutter and decoration is required in order to avoid exciting the patient. Moreover, the bed must be soft and the bed clothes without fringe or nap (V 1, p. 91.14—25). For a study of this and related passages, cf. Georg Ilberg, Das neurologisch-psychiatrische Wissen und Können des Aretäus von Kappadokien. *Zschr. f. d. ges. Neurol.* 86 (1923), 227—46.

patient's health permitted it, he was encouraged to exercise or at least submit to various forms of physiotherapy. But by far the most important part of what might be called the interim therapy was the institution of a dietary regimen.

Such a combination — physiotherapy and dietary regimen — served two functions. In some cases, this program was merely supplementary and was designed to make the patient as comfortable as possible, to build up his strength in case a further course of bleeding or purging might be required, and to guard against any lowering of personal hygiene[1]. In other cases, however, it served as the immediate precursor of drug therapy. This was true especially when surgery was considered to be ineffective, impossible or too dangerous[2]. But, neither surgery nor drug therapy was undertaken in isolation from the all-important dietary regimen.

One of the most pervasive and characteristic features of Greek medicine was the intimate relation between dietary regimen and drug therapy. The normal Greek diet included as staples, condiments and even confections, many of the same articles which were administered for medicinal purposes. The choice of a particular drug, consequently, was often motivated by the same considerations as those governing the choice of an article of food. Personal preference and ready availability probably determined the choice of a drug fully as much as medical theory in those cases where the drug was *also* an article of the normal diet.

As I have indicated above, a dietary regimen was carried on in conjunction with other forms of therapy. The close connection between the several forms of therapy available required the physician to choose what he believed to be the proper therapy for the case at

[1] For elephantiasis, Gallic soap was used in a bath (VIII 13, p. 169.10). A little later, it is stated that "let the clothing be clean, not only because it is seen, but because dirty things cause irritation to the skin (p. 170.2—4). For the relation between cleanliness and what may have been a venereal disease, cf. IWAN BLOCH, Aufgaben und Ziele einer Geschichte der Hautkrankheiten im Altertum. *Mhefte f. prakt. Dermatol.* 27 (1898), pp. 615—16.

[2] ARETAEUS opposed tracheotomy for quinsy on the grounds that "the heat of the inflammation becomes greater due to the trauma and promotes suffocation and coughing", V 7, p. 111.19—20. Trephining for *kephalaia* was likewise a dangerous procedure and permitted only "if the headache continues after all forms (sc. of therapy)," VII 2, p. 147.16. Despite ARETAEUS' reluctance to operate, this procedure was praised by G. DIDSBURY, Considérations sur la migraine d'après Arétée de Cappadoce. *Bull. Soc. Franç. d'Hist. Med.* 30 (1936), 266—67.

hand. The choice, and a justification for this choice in pneumatic terms, is illustrated by considering ARETAEUS' discussion of the treatment for acute diseases of the *Bladder*.

The bladder is a cold organ for it lacks any innate heat of its own; it is, moreover, prone to communicate its disorders and pains to adjoining organs via the *neura* by some sympathetic process[1]. Its proper function, that of storing and passing the urine, is threatened by any of a number of causes which lead ultimately to the production of calculi.

Surgery, states ARETAEUS, may be advisable because prompt and decisive action is required in a disease which so rapidly terminates in death[2]. Due to the lack of aseptic technique, death by secondary infection following surgery was probably not uncommon and ARETAEUS openly admits this danger in connection with surgical drainage of the bladder and the use of catheters[3]. These facts, coupled with the likelihood of hemorrhage probably motivated the search for an effective non-surgical form of therapy[4]. It is at this point that the relation between dietary regimen and drug therapy becomes most evident. After noting that astringent epithemes were counterindicated by reason of their weight on an already-distended bladder, ARETAEUS immediately turns to the proper diet. This is to consist essentially of foods possessing diuretic properties and those which are easily digested and nourishing, such as milk or wine[5]. Then he states: The best thing for the bladder is cicadas, in season and roasted as an article of food; out of season, dried and moistened in water[6]. Cicadas may also be boiled with nard for internal use while a sitz bath prepared from the same ingredients is also of benefit[7].

[1] "It is dangerous when the bladder suffers in acute diseases, even if it is the result of having suffered together (συμπεπόνϑῃ) with other diseases. It is more dangerous and more likely to be fatal if the disease originates in the bladder for it is especially able to make other organs sympathize (συμπαϑέα) with it, such as the *neura* and the understanding (γνώμη)" II 10, p. 31.14—17.

[2] ARETAEUS seems to have thought death was due to pain and suffering: to bolster his opinion he cites a verse of HOMER (*Il.* XIII 568—69) VI 9, p. 138.22—26.

[3] VI 8, p. 139.18—19.

[4] Pharyngeal ulcers ought not to be treated by cautery, rather rubefacients and "drugs resembling fire" are preferred, V 8, p. 113.8—9. Inflammation, *nekrosis* and gangrene follow hemorrhage from the bladder, II 10, p. 32.15—16. Cf. also VI 8, p. 139.3 for another reference to hemorrhage of the bladder.

[5] VI 8, p. 139.7—13.

[6] VI 8, p. 139.14—16.

[7] VI 8, p. 139.16—17. Cicadas were commonly used for genito-urinary disorders, cf. GALEN XII 360 (All references to GALEN are to the edition of

The use of the same item, now as a foodstuff, now as a drug, appears in the accounts of other diseases as well[1]. Thus, in treating internal hemorrhage, he is careful to specify the properties which both the food and the drugs must possess. According to the ancient conceptions of therapy which focused almost exclusively on overt symptoms, the primary aim was to arrest the flow of blood. Therapy was thought successful if this was accomplished and there was little further effort to determine the cause of the hemorrhage. To control internal hemorrhage, the first line of defense was the administration of astringents or hemostatics. These were to be taken internally for "things drunk or swallowed come nearest the injured parts[2]." Among the remedies mentioned for hemorrhage, coagulants were thought to be of especial value. Several such drugs are mentioned by Aretaeus including the juice of coriander, vinegar, rennet and several of the medicinal earths[3]. All of these were to be administered in a decoction of dates, carob or bramble[4]. Starch is singled out for special mention for not only does it aid in the coagulation, it also lubricates an irritated trachea[5]. If the hemorrhage is not excessive, daily administration is sufficient while a greater loss of blood may require medication three times a day. Thick syrups with powdered gall to which sumach or dock might be added were also thought beneficial by reason of the astringent ingredients[6]. The discussion concludes in these words:

C. G. Kühn). Nard, from *Nardostachys Jatamansi* DC possessed diuretic properties (Galen XII 84) and was used for genito-urinary disorders, Dioscorides I 7. (All references to Dioscorides are to the edition of Max Wellmann, 3 vols, Berlin 1907—1914.

[1] Epilepsy was treated by a similar combination of foods and drugs (VII 4, p. 152.13—155.30). Cf. Owsei Temkin, The Falling Sickness. Baltimore 1945, pp. 67—68.

[2] VI 2, p. 122.23—24.

[3] Coriander, *Coriandrum sativum* L.; said to be slightly styptic by Galen XII 36. For rennet as a hemostatic, cf. Dioscorides II 75.

[4] VI 2, p. 124.1—2. Carobs, the pods of *Ceratonia Siliqua* L. The styptic properties were commonly recognized, Diosc. I 114; Galen XII 23. The root of bramble, *Rubus fruticosus* L. also classed as a styptic, Galen XI 848. The date, *Phoenix dactylifera* L. was explicitly mentioned for internal hemorrhage, Diosc. I 109.

[5] VI 2, p. 124.3—5.

[6] VI 2, p. 124.6—9. Sumach, *Rhus coriarius* L. was frequently used as an astringent (Diosc. I 108, Galen XII 115) as was dock, *Rumex crispus* L. (Diosc. II 114, Galen XII 56). The hemostatic action of galls, from *Quercus infectoria* Oliv. was known already to Hipp., *De morbis mulierum* 192 (VIII 372 L.).

Food, like the drugs, are of various kinds, yet the drugs are in the food For neither all the good properties of food are found in a single item nor, even if a single item suffice for healing, should it be used alone for it easily leads to satiety. For we must admit variety, especially so in prolonged cases... But if death by hemorrhage appears evident, even unpleasant and undigestible items are to be administered: let even harsh things be given to save the patient's life[1].

Perhaps the most interesting example of a combined dietary and drug therapy occurs in connection with stomachic disease. This disease, unlike the others so far considered, is caused by an improper diet. Successful therapy thus depends completely upon a regulated diet. As Aretaeus notes, this disease poses problems not encountered in those having an external cause:

Passive exercise, walking, gymnastics, voice exercises and readily-digested food are sufficient to counteract the inability of the stomach to digest. But it is impossible that such things as these could remove chronic indigestion and create plump flesh from a withered body[2].

In this difficult situation, the physician must ignore theory and do everything in his power to promote the comfort of the patient, even to the point of gratifying his deranged appetite, so long as the foods are not demonstrably injurious[3].

Although Aretaeus' dietary program belongs to the tradition of Greek medical writing going back to the Hippocratic dietary treatises, it can, nevertheless, be justified on theoretical grounds. The basic idea is one of avoiding foods which were known to produce deleterious effects. Since the same plant or plant product may occur both as a foodstuff or as a drug, indiscriminate or excessive use may lead to serious disorders and even small quantities of some plants were considered dangerous[4].

[1] VI 2, pp. 124.19—125.7.

[2] VIII 6, p. 165.8—11.

[3] Cf. VIII 6, p. 165.13. In the treatment of cardiac disease, Aretaeus states "Food every day, light, easily digestible and mainly from grain. Let it be pleasant even though it may be less suitable. For in these cases, more than in any other, the patient ought to be gratified." VI 3, p. 127.21—23. Elsewhere, in connection with a restorative diet he cites the Delphic maxim *μηδὲν ἄγαν*, "nothing in excess" (VI 2, p. 125.23) and notes that apoplexy is caused by an immoderate use of food and wine (V 4, p. 103.16—17).

[4] Cf. III 6, p. 41.16 where the eating of hyoscyamus and mandragora is counterindicated. Both *Hyoscyamus reticulatus* L. and *Mandragora autumnalis* Spreng. were considered dangerous in antiquity, Galen XII 147 and Dioscorides IV 68 mention that mental derangement follows the eating of hyoscyamus while Dioscorides IV 75 states that mandragora produces a loss of speech. Already in Hipp., *De locis in homine* (VI 328 L.) mandragora

Running parallel to the advisability of certain foods for specified purposes and the prohibitions placed upon others, the administration of drugs depends both upon the availability of medicinal substances and a rationale for their administration.

Another instructive parallel between ARETAEUS' dietary regimen and his drug therapy is the extent to which he relies on his own experience. Despite the fact that he often alludes to highly regarded, but anonymous physicians who can testify to a drug's effectiveness, there is an equal number of cases where his therapy is supported by his personal experience. He mentions, e.g. that he has successfully used a certain drug in a former case while in another instance, notable for its honesty, he admits that he used cold water with no success and on that basis he now questions its effectiveness[1].

Because of the closely-knit texture of dietary regimen and drug therapy, a sharp distinction between foodstuffs and drugs is difficult to maintain. Nonetheless, the Greek physician placed considerable reliance on a classification which has come to apply to drugs to the exclusion of foods. The classification of substances in terms of their physiological action provides the basis for distinguishing such classes of drugs as diuretics, phlegmagogues, sternutatories, purgatives, and the like.

Among the beliefs adopted by ARETAEUS in order to explain physiological action, none was more important than the old belief that "opposites are cured by opposites[2]." The fact that drugs are conveniently distinguished from foodstuffs on this basis does not rule out another traditional classification of drugs. The form in which a drug is administered, if not philosophically important, is nevertheless a matter of practical concern to the physician[3].

root was said to produce a suicidal form of melancholy. Dysentery, ARETAEUS states, was caused by sharp, acrid or highly seasoned foods as fish paste, garlic, etc. and by alcoholic beverages IV 9, p. 76.11—14. Another interesting example concerns the cuttlefish, *Sepia officinalis* L. Although it was eaten as a delicacy (GALEN VI 736—39) and used for medicinal purposes (GALEN XII 347), ARETAEUS warns that its ink caused colic, II 6, p. 25.20.

[1] V 10, p. 118.8—11. Cf., p. 32, notes 2 and 3.

[2] The object in treating syncope produced by phrenitis is "to add *tonos* to *tonos*, to rouse the deadened pneuma, to warm what is cold, to constrict what is relaxed", V 1, p. 97.25—26.

[3] Cf. ARETAEUS' preference for ointments on the grounds they are more powerful than fomentations and more readily tolerated "for it [sc. an ointment] does not run off and stain the bed clothes which is unpleasant. It remains on the body until melted by the heat and absorbed" V 4, p. 104.16—19.

With but few exceptions, ARETAEUS follows the conventional pattern in administering drugs in the form of aqueous or vinous decoctions, cataplasms, clysters, etc. all of which go back ultimately to the Hippocratic writings[1]. The particular drugs which fall into these two related classifications must now be examined in greater detail. For reasons of convenience, they will be considered in terms of their origin—mineral, animal and vegetable.

About a dozen drugs of mineral origin are mentioned by ARETAEUS. It is difficult to be more precise because the distinction between varieties of the same mineral was probably based on differing amounts of impurities or a different crystalline structure[2]. On the basis of their most probable identifications, it appears that most of these minerals occured locally in natural deposits in a sufficiently purified state that little additional preparation was required[3]. Owing to their relatively pure state and the effects due to their concentrated form, minerals were sometimes preferred to drugs of animal and vegetable origin. This may account for the numerous uses found for nitron which occurs in 25 prescriptions. Because of the diuretic properties attributed to nitron, it was often used to rid the body of flatus[4]. Other uses include a mixture with sulphur to be used as an expectorant and as a seasoning agent in various dietary articles[5].

Other mineral drugs included sulphur, alum, copper, calamine, verdigris, bitumen, and several salts and medicinal earths. Sulphur occurs in a cerate applied to the site of scarification for pleurisy[6]. Another form of therapy for pleurisy was to wrap the patient in

[1] Cf. JERRY STANNARD, Hippocratic Pharmacology. *Bull. Hist. Med.* 35 (1961), 497—518.

[2] An Egyptian nitron is mentioned as an ingredient in a rubefacient, V 4, p. 104.23. It is clear from DIOSCORIDES V 113 that this was one of several varieties distinguished by color or by point of shipment. In the same way, HIPP., *De morbis mulierum* 78 (VIII 192 L.) speaks of a red nitron while a nitron from Berenike is mentioned by AETIUS OF AMIDA, VIII 47, ed. A. OLIVIERI, Leipzig 1950 (*Corp. Med. Graec.* VIII/2) Vol. II, p. 468.28.

[3] Only the simplest procedures are mentioned, e.g. reducing salt and nitron to a powder and sifting them (V 4, p. 104.11). Elsewhere, in describing pastils, it is stated that their composition can be varied according to the strength or odor desired, VI 2, p. 122.20—21. A double-vessel is referred to once for purposes of boiling an oil preparation (V 6, p. 108.14). ADAMS, op. cit. p. 402n. likens it to a *balneum mariae* and gives references to other occurrences of similar apparatus.

[4] V 5, p. 106.28—29; VI 1, p. 119.14.

[5] V 10, p. 115.20, 23.

[6] V 10, p. 117.10.

wool which had previously been impregnated with sulphur fumes[1]. Hot sulphureous baths were advised for melancholy and elephantiasis[2]. Alum occurs under two names: as *stypteria*, the usual name for any of a group of natural crystalline substances composed mainly of potassium alum and ferrous sulphate, it was used primarily as a styptic[3]. As *chalchitis*, or rock alum, it was powdered and blown into the throat to arrest ulceration of the pharynx[4]. Copper salts occur in a strong emetic containing also narcissus bulb, mustard, pepper, and hyssop. This emetic was believed to be effective in controlling an epileptic paroxysm, though for what reason he does not state[5]. Calcined copper also was included as an ingredient in a plaster containing bitumen, verdigris, alum, vinegar and a vegetable gum. Not without reason this plaster was stated to be astringent, and desiccative and hence useful as a hemostatic[6].

Salts of various sorts were used for a variety of different purposes. Taken internally with a gruel, salt was believed to contribute to the astringent and refrigerating effects necessary for the control of internal bleeding[7]. As an ingredient in a spelt gruel, salt was thought useful for liver complaints because of its detersive and diuretic properties: for it withdraws through the canals which lead from the liver to the kidneys, "and this is the best outlet for things passing from the liver because of the width of the canals and the straightness of the passage[8]." Other uses for salt include fomenta-

[1] V 10, p. 116.1.

[2] VII 5, p. 158.19—21 (melancholy); VIII 13, p. 170.11—12 (elephantiasis).

[3] VI 2, p. 122.9. Powdered oak galls were added to enhance the astringency, VII 2, p. 146.30.

[4] V 8, p. 113.13—12. Included in this was powdered *kadmia*, probably an impure zinc carbonate.

[5] V 5, p. 107.8—10. Narcissus, probably *Narcissus poeticus* L. and *N. serotinus* L. described as an emetic by Dioscorides IV 158. There are two species of mustard, *Brassica nigra* L. and *Sinapis alba* L.; it cannot be determined to which Aretaeus refers but both held to be valuable for epilepsy (Dioscorides II 154). Hyssop is difficult to identify. Alfred Andrews, Hyssop in the Classical Era. *Class. Philol.* 56 (1961), 230—248 has examined the many proposed identifications. Whatever its identity, a plant called *hyssopus* was listed as a phlegmagogue in Hipp., *De diaeta* 54 (VI 560 L.). Pepper, from *Piper nigrum* L. was already used in the Hippocratic writings for a wide variety of purposes, cf. *De morbis mulierum* 84 (VIII 206 L) where it occurs twice in purgative pessaries.

[6] VI 2, p. 122.10.

[7] VI 2, p. 124.23—27.

[8] VI 6, p. 135.23—25.

tions, in a cataplasm for eschars produced by surgery for the relief of headache, and as a purgative clyster[1]. Salt was considered to be less effective in the form of brine than in the crystalline state[2].

This preference for the crystallized form may explain ARETAEUS' use of *limnestis*, a substance rarely encountered in Greek medicine[3]. As a saline substance containing animal and vegetable impurities, it is understandable why *limnestis* was employed as a rubefacient and in heating ointments. The concentration of salts, plus the decayed organic remains would almost certainly lead to a localized inflammation, especially if it were combined in a greasy salve with such rubefacients as spurge or bay[4]. In a cerate with dried mustard it was applied to the chest in peripneumonia in order to withdraw fluid, heat, and pneuma[5]. One especially powerful vesicant included castoreum, turpentine resin, spurge, *limnestis*, pepper, pellitory, galbanum and nitron[6]. It can readily be appreciated why ARETAEUS advised the physician to raise the plasters often in order to prevent the formation of blisters[7].

There remain the so-called medicinal earths, the preparation of which GALEN has left an informative account[8]. These substances

[1] V 6, p. 108.14 (fomentation); VII 2, p. 147.7—10 (cataplasm); V 2, p. 98.19 (clyster). [2] V 10, p. 117.2.

[3] By comparing the accounts of GALEN XII 370 and DIOSCORIDES V 119, it appears that this was a saline efflorescence collecting about the stems of various marsh plants. GALEN states that it is very sharp and heating and hence useful where heat is required, but, he adds, it is too powerful to be given internally.

[4] VI 3, p. 130.7. Spurge, any of a number of closely related plants of the genus *Euphorbia*, all of which contain a milky sap, usually mildly toxic and capable of producing local irritations; for this reason it was classed by GALEN XI 879 and DIOSC. III 82 as violently heating. An undetermined *Euphorbia* was used in the Hippocratic *De superfetatione* 32 (VIII 500 L.) as an ingredient in an ointment. The volatile oils in the leafs and fruit of bay, *Laurus nobilis* L., made it useful for local applications (DIOSC. I 78; GALEN XI 863; CELSUS V 19.13.). [5] VI 1, p. 120.20.

[6] V 4, p. 104.20—22. Turpentine resin, from *Pistacia Terebinthus* L. was occasionally used as a vesicant (CELSUS V 3 and V 5). Pellitory, *Anacyclus Pyrethrum* DC because of its extreme heating properties (DIOSC. III 73) was used as a counter-irritant in febrile conditions (GALEN XII 110). Galban, the resin from *Ferula galbaniflua* Boiss et Buhse possessed a warming faculty and was useful for asthma and dyspnoia (DIOSC. III 83).

[7] V 2, p. 101.19—20. For a very powerful plaster, it is advisable to raise it every hour, VI 3, p. 130.9—10.

[8] GALEN, XII 168—178 KÜHN.

which were much used in antiquity occur but twice and since ARETAEUS does not explain their physiological action it may be supposed that he is following the conventions of current medical practice. For an inflamed uvula, an astringent is recommended containing the juice of sumach, acacia, hypocistis and unripe grapes, vinegar and Samian, Lemnian and Sinopic earths[1]. Somewhat later, a hemostatic decoction is mentioned containing dates, carobs, bramble roots and four earths—the white *aster*, Eretrian and Sinopic earths and the Lemnian seal[2]. The effectiveness of these remedies probably lay as much in the vegetable products as in the minerals for, with the possible exception of Samian earth, they were practically inert[3].

Drugs of animal origin exhibit the same wide array of uses as the preceding class and, like drugs of vegetable origin, there is the same close relation between their alimentary and medicinal uses.

This relation is nowhere more evident than in ARETAEUS' use of vipers. As a remedy for elephantiasis, pastilles made from their flesh are to be taken in a draught like squills. "The vipers themselves", he adds, "are used as a seasoning at dinner, and should be prepared like fish. But if the compound medicine "from vipers" is available, it should be drunk in place of all others for it contains [the properties] of all the others together. It cleans the body and makes smooth

[1] V 8, p. 112.19—21. The astringent action of acacia, including *Acacia vera* Willd. was well-known in antiquity; DIOSCORIDES I 101 recommends it for ulcers of the mouth. Hypocistis, *Cytinus hypocistis* L. was also used as a styptic, especially for running sores, GALEN XII 27.

[2] VI 2, pp. 123.28—124.2.

[3] Samian earth was a clay rich in kaolin (hydrated aluminum silicate) found in natural deposits on Samos. cf. THEOPHRASTUS, On Stones, ed. with translation and commentary by E. R. CALEY and J. F. C. RICHARDS, Columbus, Ohio. 1956, p. 209. The absorbent properties of kaolinite, the chief constituent of kaolin, determined its use as an astringent, cf. DIOSC. V 153. Lemnian earth, also called Lemnian seal because it was packaged and sealed for commercial purposes on the Island of Lemnos. Cf. GALEN XII 169 and DIOSCORIDES V 97. It was a reddish clay containing ferric oxide. Sinopic earth, a red ochre, produced, according to DIOSC. V 96, in Cappadocia and shipped to Sinope. Eretrian earth was less frequently used but it appears from DIOSC. V 159 that it was a siliceous earth. White ἀστήρ (lit. star) was a white form of Samian earth (DIOSC. V 153, GALEN XII 181). The name derives from the star-shaped packages in which form it was shipped, cf. CALEY and RICHARDS, op. cit., p. 212.

the tubercles"[1]. Previously in discussing the remedies for epilepsy, he advised that both the food and the drugs should be digestive, heating, drying and diuretic[2]. Of special mention are the compound "from vipers" and the even stronger drugs of Mithridates and of Vestinus[3]. Elsewhere, the two former composita are described as possessing a drying faculty and hence useful in melancholy and diabetes, while the digestive properties of the compound "from vipers" make it suitable for coeliac disease[4].

Other animal products taken internally include flesh and certain by-products. Melancholy, characterized by a dryness and density of the body is corrected by eating rare and soft flesh, such as the head and feet of swine, hares, goats or deer or the wings of fowls[5]. The limbs of unspecified animals, made into a soup, and roasted swine brain were thought useful for pleurisy. For, like ptisan, a common remedy for this disease, they disperse the heat, concoct coughs and soothe the lungs[6]. Apples roasted in suet were included in a restorative diet for a febrile condition in which dryness is corrected by moistening foods[7].

More exotic animal products are combined into an ointment for elephantiasis: For the tubercles of the face, anoint with the ashes of

[1] VIII 13, p. 169.6—9. Behind this usage may lie a belief in the "doctrine of signatures": the use of the smooth and sleek snake to make the body smooth.

[2] VII 3, p. 153.20—23.

[3] VII 3, p. 153.25—26. Other compounds mentioned by Aretaeus include the drug of Philon (VIII 5, p. 164.25), several cataplasms and a variety of drugs whose chief ingredient is indicated by its name, e.g. the one "from swallows" (*διὰ τῶν χελιδονίων*, V 7, p. 111.6) or the one "from besasa" (ib., cf. p. 53, n. 1) or the one "from amber" (*δι' ἠλέκτρου* VI 2, p. 124.13). These compounds tended to become increasingly complex in later times, reaching the peak in Servilius Damocrates' recipe for theriac containing over 150 ingredients (U.C. Bussemaker, ed. Fragmenta poematium rem naturalem vel medicinam spectantium, Paris 1862, pp. 118—127). In general, cf. Owsei Temkin, Galenicals and Galenism in the History of Medicine, in: Iago Goldston, ed. The Impact of the Antibiotics on Medicine and Society, New York 1958, pp. 18—37.

[4] VII 5, p. 159.8—9 (melancholy); VIII 2, pp. 162.27—163.2 (diabetes); VIII 7, p. 166.21 (coeliac disease).

[5] VII 5, p. 158.29.

[6] V 10, p. 115.1—3.

[7] V 1, p. 93.26. There are many species closely related to the domestic apple. *Pirus Malus* L. and it is difficult to distinguish which of several varieties or species is referred to in the ancient writers who mention "sweet," "sharp," "wild" varieties, (cf. Hipp., *De diaeta* 55, VI 562 L) and Galen XII 75—76 who mention several therapeutic uses.

grape vine mixed with the suet of some wild animal such as the lion, panther, bear or, if these are not available, of the Egyptian goose, for like in the unlike, as the ape to man, is best[1].

But there is a limit to the drugs ARETAEUS is willing to admit. Some physicians, he states, have used the brain of a vulture, the heart of a raw *aithyia*, and the common weasel but he admits to no experience with these and is opposed to the use of blood drawn from a man recently slain. "I have described these things," he concludes, "for those who might dare to use them[2]."

No drug of animal origin was more extensively used in antiquity than *castoreum*. ARETAEUS included it in ointments because of its heating and alleged narcotic properties[3]. It was, for this reason, useful in lethargy where a combination of coldness and neck pain contribute to the patient's discomfort[4]. The internal uses of castoreum center around its diuretic and purgative properties as well as in the fact that "it alters the habits of the body to the hot and dry and changes the constitutions of diseases[5]." Again in the case of lethargy, it was administered as a sternutatory in order to expel pneuma[6]. In the same fashion, castoreum with pepper and soapwort is given to women suffering from "hysteric suffocation" in the belief that by sneezing the uterus will forcibly return to its normal position[7].

Of the many drugs used for elephantiasis, two further drugs may be mentioned. *Alcyonium*, dried and powdered, is an ingredient in a detergent ointment along with nitron, alum, sulphur, burnt wine lees, costus, iris and pepper[8]. Shortly before this passage, ivory

[1] VIII 13, p. 169.17—20. The last clause is explained by ADAMS, op. cit., p. 496n as an analogy—the Egyptian goose is to quadrupeds as ape is to man—although I have not been able to locate on what evidence this is based. The Egyptian goose has been identified as the *Chenalopex aegyptiacus* Steph. by D'ARCY THOMPSON, A Glossary of Greek Birds, Oxford 1936, p. 330.

[2] VII 4, p. 154.1—9. D'A. THOMPSON, op. cit. p. 27, identified the *aithyia* as probably a shearwater. It was used by GALEN XII 336 as a digestive and stomachic drug.

[3] It is specified to be of value in cases where the tendons "suffer dreadful pains and tension," V 6, p. 108.6.

[4] V 1, p. 99.14. [5] V 2, p. 100.22—23. [6] V 2, p. 100.23—29.

[7] VI 10, p. 141.8—9. Soapwort, *Saponaria officinalis* L. Frequently mentioned as a sternutatory (DIOSC. II 163 and GALEN XII 131) and used for gynaecological disorders by HIPP., *De morbis mulierum* 78 (VIII 174 L.).

[8] VIII 13, p. 169.14—16. *Alcyonium* was a generic term covering several genera of marine zoophytes, including sponges and holothurians, cf. DIOSC. V 118 and GALEN XII 370. For an attempt to identify some of them, cf.

shavings in wine are recommended, but with no explanation of why they might be useful[1]. These complicated recipes, a reminder of the polypharmacy prevalent in some of the medical sects, and the futility of such measures, were probably determined by the lack of any better means to arrest such a hideous disease. This seems to be reflected in ARETAEUS' remark that "it is good to give drugs frequently for a beneficial purpose[2]." Presumably, frequent and involved medication gave the patients, as well as the physicians, renewed hope[3].

As external applications, spider webs were employed as a hemostatic as was unwashed wool[4]. The latter was often used to wrap a patient in, not only to conserve the bodily heat but also in the belief that wool possessed certain intrinsic healing properties[5]. This same belief occurs also in regard to eggs. Freshly laid eggs are recommended for pleurisy on the grounds that the heat of the egg is more humid than articificial heat and more natural proceeding, as it does, from one animal to another[6].

In common with other Greek medical writers, the majority of ARETAEUS' drugs are of vegetable origin. Some 177 different plants are mentioned of which only a selection can be discussed here. Most of these drugs were herbs used as simples. Some, however, were included in composita along with mineral and animal products.

One of the interesting features of his medical botany is the frequent occurrence of plants indigenous to the Near East. An examination of the plants and vegetable products of an Eastern origin will set the stage for considering other portions of his medical botany.

Alexander von Tralles, ed. THEODOR PUSCHMANN, Vol. I (Wien 1878), p. 442 n. They were used for medicinal purposes as early as *De morbis mulierum* 106 (VIII 230 L). Iris, including *Iris Florentina* L., *I. pallida* Lam. etc. was said to be of value for ulcers and skin conditions, DISCORIDES I 1.

[1] VIII 13, p. 168.28. According to DIOSCORIDES II 57 W., the shavings are astringent.

[2] VIII 13, p. 168.25.

[3] With reference to the treatment for cardiac disease, ARETAEUS states "the patient must be both courageous and cheerful and the physician must bid him to be hopeful and aid him with various foods and wines." VI 3, p. 128. 27—30.

[4] VI 5, p. 135.13 (spider webs); VI 1, p. 121.30 (unwashed wool).

[5] This was attributed to *oisypos* the grease from unwashed sheep's wool. ARETAEUS refers to it several times, e.g. VI 2, p. 121.30 where it is used as a hemostatic. Cf. DIOSC. II 74 W for the preparation and medicinal uses, and N. WULFSBERG, Geschichtliche Notizen über Oesypum und therapeutische Versuche mit dem reinen wasserfreien Lanolin. *Therapeutische Mhefte* 1 (1887). 92—97.

[6] V 10, p. 115.21—22.

In describing the remedies for epilepsy, several composita are recommended on the grounds that they contain all of the virtues of cinnamon, cassia, malabathron, pepper and seseli combined[1]. All of the plants mentioned here are of an Eastern origin though most of them were available in the West as spices and condiments[2]. Owing to their demand as spices, various aromatic substances were frequently incorporated into ointments and other forms of medication. One such ointment for renal calculi contained cassia, cinnamon, nard and myrrh[3]. Another ointment, administered as a pessary for "hysteric suffocation" contained nard, Egyptian bachar, malabathron or the "Indian tree" and cinnamon[4]. It is not clear what value these preparations possessed except for their soothing properties and fragrance. Elsewhere, cinnamon is said to be a restorative while seseli, along with ginger and pepper, were regarded as condiments and as digestive and heating drugs[5].

[1] VII 3, p. 153.28—29. The identification of these three plant products is difficult for it can no longer be determined which of several related species of *Cinnamomum* furnished the bark for the ancients. Probably *C. Cassia* Bl. and *C. Ceylanicum* Nees were among the sources. Products from these trees were much used in antiquity, cf. Diosc. I 12—14; Galen XII 13, 26, 66 and Hipp., *De morbis mulierum*, 51 (VIII 110) and 195 (VIII 378), although there was confusion over their sources (Pliny XII 42, 89). For an attempt to identify some of the species involved, cf. B. Laufer, Malabathron. *Journal Asiatique* 12 (1918), 5—49.

[2] Cf. Jacques André, L'Alimentation et la Cuisine à Rome. Paris 1961, pp. 201—212.

[3] VIII 3, p. 164.2—3. Myrrh, the resin from several species of *Commiphora* including *C. Schimperi* Engl. and *C. simplicifolia* Engl. was used in many ointments. I have found no instance of its being used for calculi but its reputed diuretic properties are mentioned often, Hipp., *De morbis mulierum* 209 (VIII 404); Diosc. I 74; Celsus V 25, 16, Pliny XXV 18, 41, etc.

[4] VI 10, p. 140.22—24. Egyptian bachar (*βάκχαρι τὸ Αἰγύπτιον*) if it is identical to *baccar* (Pliny XXI 16, 29) and to *bakkaris* (Hipp., *De morbis mulierum* 182 (VIII 364 L.) may be *Helichrysum sanguineum* Boiss. Cf. J. André, Lexique des Termes de Botanique en Latin. Paris 1956, p. 49. The later writers who mention it follow Dioscorides III 44 who notes its usefulness in gynaecological disorders. Galen does not mention it, but it occurs in Ps-Galen XIX 87 in a gloss on the above passage in Hippocrates.

[5] VI 5, p. 134.14—15 (restorative); VII 13, p. 161.10 (condiment); VIII 6, p. 165.16—17 (digestive); VIII 7, p. 166.19—20 (heating drug). Ginger, *Zingiber officinale* Rosc. "imported into Italy" (Diosc. II 160, cf. Galen XI 880) is said to be used as a spice with meat and to possess a warming property, beneficial to the stomach, while seseli, *Tordylium officinale* L. heats (Galen XII 120) and is good for the digestion (Diosc. III 53).

Another Eastern plant which was later to achieve importance in Arabic medicine, is *myrobalanos* whose medicinal uses are mentioned here for the first time[1]. It occurs as a cerate applied to the hypochondrium following deep scarification and, in the form of a powder it was thought to be of value for inflammations of the spleen[2].

There occur several other plants of Eastern origin which though used but sparingly by Aretaeus are not uncommon in other Greek medical writings. For an ulcerated uvula an application of gum, starch, and the decoction of various plants is recommended. One of the ingredients of this decoction is *costus* which is described as one of the stronger and more acrid drugs like myrrh and cyperus[3]. The extracted gum or resin of styrax is mentioned as an ingredient in a semi-liquid sternutatory for relief of headache. This substance was injected into the nostrils by means of a Y shaped pipe[4].

The most frequently-used plants of Eastern origin are pepper and nard. These were used for so many different purposes that they must be examined in more detail.

Pepper is listed as a sternutatory, emetic, expectorant, diaphoretic, phlegmagogue, carminative, diuretic, purgative and heating drug. Corresponding to these many uses, it is dispensed in the form of a drink, clyster, linctus, ointment, cataplasm and powder. For this wide array of uses there was a simple but wide-sweeping ratio-

[1] *Myrobalanos* is mentioned by Dioscorides I 109 (in a chapter on the palm tree) and by Ps. Aristotle, De Plantis II 10, 829 b 33 ed. O. Apelt (Leipzig 1888), but neither state any medicinal uses. The identity of this tree is doubly-difficult for it was sometimes confused with ben, *Moringa aptera* Gaertn. (e.g. Pliny XXIII 52, 98) while there were four different trees supplying what come to be known in medieval medicine as *myrobalanus*: *Terminalia Chebula* Retz; *T. citrina* Roxb.; *T. belerica* Roxb. (= *Myrobalanus belerica* Gaertn.) and *Emblica officinalis* Gaertn. Cf. J. Lindley, Flora Medica (London 1838) pp. 67—69, 176. For the medieval uses, cf. P. Dorveaux, ed. L'Antidotaire Nicolas (Paris 1896), pp. 75—76. For the Arabic uses, cf. the references in M. Levey, Mediaeval Arabic Bookmaking and its Relation to Early Chemistry and Pharmacology. *Trans. Amer. Philos. Soc.* n. s. 52/4 (1962) p. 44, n. 301.

[2] VI 6, p. 135.15; VII 14, p. 161.22.

[3] V 8, p. 112.22—24. Cyperus, including *Cyperus rotundus* L. and *C. longus* L. was described as heating and drying (Galen XII 54) and possessing a drawing faculty (Diosc. I 4). Costus, the root of *Saussurea Lappa* Clarke was used for various eruptions of the skin according to Galen XII 40.

[4] VII 2, p. 146.3—5. *Styrax officinalis* L. yielded a gum which was used for upper respiratory diseases (Galen XII 131) and in fumigation for gynaecological disorders (Hipp. *De superfetatione* 32, VIII 500 L.). Dioscorides (I 66) however, states that it causes a headache!

nale resting upon pneumatic grounds. In virtue of its penetrating and heating properties pepper is the ideal drug to dissolve and dissipate cold, viscid collections of pneuma. Nearly all of its uses can be traced back to the expulsion of pneuma which, as in lethargy, results from the congealing of the innate heat[1]. In pleurisy, where collections of heavy, viscid phlegm were thought to be responsible for the difficult and painful breathing, pepper was regarded to be of benefit[2]. The use of pepper for peripneumonia is based upon the same belief: "Drugs are to be given such as thin the fluids so that they may be removed by perspiration and which will make them slippery and thin so that they may be expelled as sputa[3]." Coeliac disease likewise is remedied by such drugs as pepper and ginger which warm the viscera and thus dissipate the cold, viscous deposits of undigested food[4].

The many different uses of nard are equally-well explained in pneumatic terms. Whereas pepper was stated to be heating and attenuating, nard exhibits the opposite properties of cooling and congealing. This is clearly brought out in the case of the profuse and debilitating sweats characterizing cardiac disease. First an epitheme of nard, dates, aloes and mastich is applied to the left breast in order to restrain the copious sweating[5]. Should further medication prove necessary, astringent and cooling drugs such as nard and juice of unripe grapes are to be given[6]. The etiology of diabetes—the melting of the bodily humors and their transformation into water—though naive today, did account for the marked thirst and frequent urination. A drug is thus indicated which would congeal the humors and preserve the body from an excessive loss of fluids. For this reason an astringent cataplasm containing nard, raw quinces, mastich and dates was ordered or a lotion containing both astringent and cooling substances such as nard, acacia and hypocistis[7].

The pneumatic framework by which the physiological action of pepper and nard was explained applies equally well to other vege-

[1] V 2, p. 98.10. For the expelling of pneuma by sternutatories, cf. V 2, p. 100.24. [2] V 10, p. 115.23—25. [3] VI 1, p. 119.27—28.

[4] VIII 7, p. 166.20; cf. IV 7, p. 74.7—10.

[5] VI 3, p. 129.1—3. Aloes from *Aloe vera* L. and mastich, the gum from *Pistacia Lenticeus* L. were both regarded as drying and styptic (Galen XI 821; XII 68; Diosc. III 22, I 70). [6] VI 3, p. 129.11.

[7] VIII 2, p. 162.18—23. The astringent property of quince, *Pirus Cydonia* L. was noted by Dioscorides I 115.

table products. Within this framework, ARETAEUS made the fullest use of the drugs available to him[1].

Since many of the diseases result from an unnatural balance in the pneumatic constitution, those drugs which act directly on the pneuma will be of especial value, e.g. purgatives, diuretics and carminatives. Other diseases were explained in humoral terms, e.g. asthma which is characterized by thick and viscous humors collected in the lungs and chest, especially in persons of a cold temperament[2]. Hence, other classes of drugs will be useful, including phlegmagogues, cholagogues, astringents, diaphoretics and those drugs which cool or heat. In fact, the only portion of ARETAEUS' pharmacodynamics not directly explained in pneumatic terms is his frequent employment of hemostatics. But, as I have shown above, these are explained indirectly in pneumatic terms owing to the belief that pneuma was carried in the blood and the consequent routine practice of bleeding.

The administration of drugs for specific diseases is thus fully justified on pneumatic grounds. Renal calculi, e.g. were stated to be produced by a condensation of the bodily fluids[3]. Because the aged were popularly believed to be of a cold temperament, the age factor must be considered in determining the proper form of therapy[4]. Therapy begins with warm baths and fomentations with which are blended herbs possessing diuretic properties—dill, rose-

[1] Several references indicate that certain drugs were not always available and that supplies depended upon imports. In specifying the ingredients of a carminative, he states that the best type contains the juice of the *silphium* from Cyrene "if it is readily available" (V 6, p. 109.14—15). Elsewhere, gourds and cinnamon are preferable to other drugs "if some are available" (VI 7, p. 136.28—29). That ARETAEUS expected the physician to prepare his own composition seems to be indicated by a reference to the availability of *chamaeleon*, cf. p. 51, n. 6.

[2] III 11, p. 52.12—14.

[3] The etiology of several diseases was described in terms of the refrigeration and the consequent condensation of the innate heat. The formation of renal calculi is explained thus: "Cold more quickly than heat solidifies the thick humors. The proof of this is supplied by thermal springs for when the warm waters become congealed they are compacted into petrified stone" IV 3, p. 68.20—22. A similar explanation is given of the tubercles produced by elephantiasis and the knobby joints of arthritis; both are likened to hard, white hailstones caused by the congealing of water, IV 13, p. 87.19—21; IV 12, p. 84.18—20.

[4] IV 3, p. 68.19—20.

mary and marjoram[1]. Next, these herbs and others, including valerian, spignel, hazelwort, *prionitis*, parsley, and water parsnip are to be given as a drink[2]. Finally, a "hot-water bottle", in reality an ox-bladder filled with warm oil of chamomille, is placed upon the abdomen[3]. This combination of a relaxing, gentle heat and the administration of diuretics remained the standard form of non-surgical therapy for calculi for centuries.

Coeliac disease, of which mention has been made above, is produced by a lack of tonos of the warm pneuma in the digestive tract. As a consequence, food remained undigested and flatulence is increased[4]. Successful therapy requires astringents, purgatives, emetics, and heating drugs. The cold state of the stomach, resulting from a diffusion or dissipation of the innate heat required for digestion, is corrected by astringents for they "mollify, irritate, rouse the heat and reduce the flatulency of the stomach[5]." Several such epithemes are advised but the only vegetable products mentioned are bay berries and the root of the *chamaeleon*[6]. However, heating

[1] VIII 3, p. 163.20. Dill, *Anethum graveolens* L., rosemary, *Rosmarinus officinalis* L. and marjoram, *Maiorana hortensis* Moench. were all frequently used in antiquity, but only dill is mentioned elsewhere as possessing diuretic properties (Diosc. III 58).

[2] VIII 3, p. 164.1—2. All these plants are mentioned elsewhere as possessing diuretic or lithontriptic properties. Valerian, *Valeriana Phu* L. (Diosc. I 11; Galen XII 152); spignel, *Meum Athamanticum* Jacq. (Diosc. I 3; Galen XII 78); hazelwort, *Asarum Europaeum* L. (Diosc. I 10); *prionitis*, perhaps *Betonica officinalis* L. This last plant does not occur in Theophrastus, Hippocrates, Dioscorides or Galen. Its lithontriptic properties are mentioned by Alexander Trallianus II 465 and Rufus Ephesius p. 393 and its diuretic properties by Alexander Trall. II 489 and Rufus, p. 419. Parsley, *Petroselinum hortense* Hoffm. (Diosc. III 66; Galen XII 99) and water-parsnip, *Sium angustifolium* L. (Diosc. II 127; Galen XX 124). With the exception of *prionitis*, all are mentioned in a list of diuretics by Galen XI 747.

[3] VIII 3, p. 163.21—22. A similar form of treatment consisted in putting heated salt crystals or millet seeds in sacks which were then placed upon the affected part, cf. V 6, p. 108.14. Chamomille, from *Matricaria Chamomilla* L. was stated to have a warming and drying effect and useful for diseases of the bladder, Diosc. III 137. [4] IV 7, p. 74.10—12. [5] VIII 7, p. 166.12—13.

[6] After leeches "then apply digestive epithemes, such as 'that from seeds' (τὸ διὰ τῶν σπερμάτων), if the root of *chamaeleon* is available. For this purpose, the best ones are 'the fruit of bay', the one called *malagma*, 'the Green', or mine, called 'Mystery'." VIII 7, p. 166.9—12. Although two plants named *chamaeleon* were known to the medical writers, the black and the white, this is probably the former or *Cardopatium Corymbosum* Pers. The black variety is

epithemes of mustard and spurge are specifically mentioned on the grounds that they "prevent refrigeration . . . and cause a restoration of heat[1]." The entire course of medication for coeliac disease includes, in addition to the foregoing, astringent drinks containing plantain, myrtle and quince and another drink designed to warm, composed of ginger, pepper and wild parsley[2]. Should all else fail, there was recourse to a radish emetic and a purge with white hellebore by which "there will be a purification and elimination of the cold fruits and a kindling of the heat[3]."

No disease of antiquity was more celebrated than phrenitis[4]. This disease is caused by thick, hot vapors and humors which collected in the chest, the seat of the φρήν, *phren*, and in the head. On pneumatic grounds, two types of therapy were required—the dispersal of the thick humors and a cooling of the hot vapors. A cooling fomentation of rose-oil to the head is to be followed by draughts containing sulphurwort and cow-parsnip for these, along with vinegar dispersed the vapors and the heat[5]. If, because of the febrile

described as drying, warming, emollient and detersive (Galen XII 154). The epitheme Aretaeus refers to is mentioned also by Alexander of Tralles, II 347 Puschm. and Paul of Aegina VII 18, ed. J. L. Heiberg (Leipzig 1924: Corpus Med. Graec. IX/2) II, p. 372, but neither include *chamaeleon* as an ingredient.

[1] VIII 7, p. 166.15.

[2] VIII 7, p. 166.19—21. Plantain, from *Plantago* spp. was regarded as an astringent (Galen XI 838) and specifically mentioned as useful for coeliac disease (Diosc. II 126). Myrtle, *Myrtus communis* L., was also regarded as an astringent and as good for the stomach (Diosc. I 112; Galen XII 82). Wild parsley, *Apium graveolens* L. var. *silvestre* was used for many purposes but there is no record of its use for coeliac disease (cf. Diosc. III 64; Galen XII 118).

[3] VIII 7, p. 166.24—26. Radish, *Raphanus sativus* L. used as an emetic (Diosc. II 112) and as a phlegmagogue [Hipp., *De diaeta* 54 (VI 558 L.)]. White hellebore, *Veratrum album* L., was a well-known purge (Diosc. IV 148). Probably many of the references to *elleboros* as a purge in Hippocrates refer to this species rather than to black hellebore, *Helleborus niger* L., which in Hippocrates is always called ἐλλέβορος ὁ μέλας [cf. *Praenotiones Coacae* 556 (V 710 L.)]; *De affectionibus* 25 (VI 236 L.); *De morbis mulierum* 16 (VIII 54 L.) etc.

[4] The febrile condition often associated with phrenitis, e.g. V 1, p. 92.10—13 has been identified both as pleurisy, Cecilia Mettler, History of Medicine, Philadelphia 1947, p. 348 and as pernicious malaria, Pliny, Natural History ed. W. H. S. Jones, Vol. VI (London 1951) p. 8n.

[5] V 1, pp. 93.32—94.5. Sulphur-wort, *Peucedanum officinale* L. mentioned as useful for phrenitis by Dioscorides III 78 and for "hysteric suffocation", possibly a related psycho-somatic disorder in Hipp., *De morbis mulierum* 201 (VIII 384 L.). Cow-parsnip, *Heracleum sphondylium* L. regarded as useful for phrenitis (Diosc. III 76) and for epilepsy (Galen XII 135). According to

nature of this disease, the patient was unable to sleep, soporifics were indicated. These include marjoram, ivy, thyme, melilot and poppy[1]. It is curious that ARETAEUS does not refer to the well-known narcotic properties of poppy taken internally. Instead, he orders that the head and face be fomented with a decoction of poppies and dried poppies placed under the patient's head[2]. As curious as this may seem, it was justified on pneumatic grounds. Sleeplessness, characterized by a dry and rarefied pneuma, is corrected by a drug which will thicken and moisten the pneuma[3]. It is improbable that the dried poppies would accomplish this; a more effective method of expelling the vapors by purges is followed by "stroking lightly the feet with oil and caressing the head. But especially effective is a gentle patting of the temples and ears. For by this means, wild beasts are restrained from their anger and fierceness. Each person, moreover, is lulled to sleep by things familiar to him: to a sailor lying in a boat, feeling the motion of the sea, the sounds of the beach, the crash of the waves, the rustling of the wind and the swell of sea and ship[4]."

Acknowledgement. Assistance received from the National Institutes of Health (RG 9361) is gratefully acknowledged.

CAELIUS AURELIANUS, ASCLEPIADES used *spondylion* for phrenitis and lethargy (On Acute Diseases and On Chronic Diseases, II 37, ed. I. E. DRABKIN, Chicago 1955, p. 144).

[1] V 1, p. 94.14—20. There is no evidence that ivy, *Hedera Helix* L., thyme, *Satureia Thymbra* L., or melilot, *Trigonella officinale* L. possess any soporific properties. Opium, from *Papaver somniferum* L. et varr D C was known as a soporific (cf. DIOSC. IV 64) but the confusion between the juice of the leaves, nearly inert, and that from the unripened seed capsule long continued, cf. SCRIBONIUS LARGUS, Conpositiones 180 ed. G. HELMREICH, Leipzig 1887, pp. 73—74. ARETAEUS, in another passage, mentioned a plant *βησασᾶ* (*Withania somnifera* Dun.), known to possess several narcotic alkaloids, cf. JERRY STANNARD, The Plant Called Moly. *Osiris* 14 (1962), esp. pp. 267—284.

[2] V 1, p. 94.20—22. A few lines later, an application of poppy juice directly into the ears is advised "if more effective treatment be required" p. 94.24—26. Neither of these forms of treatment, however, is evidence of NEUBURGER's assertion (cf. p. 29 n. 3) that ARETAEUS used opium.

[3] V 1, p. 94.22—24.

[4] V 1, p. 94.26—32; for the purgative clysters cf. p. 96.2—7.

VI

Marcellus of Bordeaux *and* The Beginnings of Medieval Materia Medica*

MEDIEVAL materia medica began, in one sense, in Graeco-Roman antiquity.[1] Except for some Arabic introductions, the majority of the simples used for therapeutic purposes in the Middle Ages were identical to those of earlier times. Moreover, their names as well as their uses frequently can be traced back to Graeco-Roman texts, especially to Dioscorides' Περὶ ὕλης ἰατρικῆς (*De materia medica*) written *circa* A.D. 50.

In another sense, however, medieval materia medica began several centuries later with the replacement of new and different intellectual forms for those of antiquity. The establishment of a distinctive medieval materia medica did not take place overnight and the speed by which it was accomplished varied from region to region. Generally speaking, much of the content of Graeco-Roman materia medica was retained but it was modified in accordance with the values, needs, and institutions of an incipient culture.[2]

One stage of the transition from ancient to medieval materia medica is represented by Marcellus of Bordeaux. Much of the content of his book—the simples, their names, uses, and methods of preparation and administration—goes back to earlier Graeco-Roman texts. On the other hand, his book, the *Liber de medicamentis*, points ahead to doctrines and uses which are characteristic of materia medica for centuries to come.

The *Liber de medicamentis* was written sometime between the years A.D. 395-410.[3] Exact dates are lacking and must be approximated on the basis of the scanty biographical data supported by some external evidence.[4] The most

---

*Presented before the Section on Contributed Papers, American Institute of the History of Pharmacy, 1972 meeting, Houston.

important datum is the dedication, to the Emperor Theodosius, (*ob.* A.D. 395) whom Marcellus *vir illustris* served as *magister officiorum.*[5] The purpose of the *Liber*, Marcellus continues, was to provide medical counsel for his sons in the absence of trained physicians. Whether Marcellus accompanied the Emperor on his travels to and from the capital in Augusta Treverorum, now Trier, cannot be determined, though a number of references to Rome is suggestive. Equally uncertain are his birthplace and the city where he obtained his education. While he may have been born and educated in Bordeaux, a Gallic background seems reasonably certain. And that he was well trained is clear, as the classical allusions and his knowledge of Greek testify. Although it is known that a classical education was obtainable in the larger cities of Gaul,[6] that in itself does not prove a Bordelais ancestry. There is, significantly, no conclusive evidence that Marcellus was a physician or that his training in any way specifically equipped him for his self-appointed task.

Formally, the text, first edited by Janus Cornarius (1500-1558) in 1536, is divided into thirty-six chapters of varying length.[7] They follow the well known arrangement *a capite ad calcem* of his major sources, Scribonius Largus and the so-called *Medicina Plinii.*[8] Preceding the text proper, are a short, bilingual tract on metrology and six *epistolae*, purporting to have been written by famous physicians of earlier times.[9] A short poem concludes a text that has been extensively studied by philologists,[10] and folklorists,[11] for quite different reasons, but that has been somewhat neglected by historians of pharmacy.[12]

Given Marcellus' background, his choice of the simples mentioned in the *Liber*, as well as many of the recipes themselves, can readily be explained. For with respect to content, he drew heavily upon earlier writings, in some cases, doing little more than rearranging the order or varying the ingredients or their amounts in the complex recipes.[13] But unlike the earlier texts on which he depended or the Byzantine compilations which, beginning in the next century preserved ancient Greek medicine in a relatively pure form, the *Liber* is essentially a collection of recipes. Like the many medieval texts representative of *Rezeptliteratur*, some of which depend on Marcellus,[14] the *Liber* pays little attention to the philosophic rationale of therapeutics.[15] There is no indication, for example, that Marcellus appreciated the issues dividing the humoralists and the solidists or that he was aware of the internecine quarrels of the medical sects. Indeed, the epithet *Empiricus*, sometimes attached to his name, does not imply an adherence to the empirical school, but rather derives from a passage in the *Preface* in which he describes the purpose of the *Liber.*[16] He is likewise innocent of physiological theory and demonstrates little concern in explaining how a particular drug acts. It was sufficient for him to state, usually at the conclusion of a recipe, a formulaic advertisement of its wonderous curative properties.[17] The numerous references to his own successes in prescribing a remedy may, of course, rest on an historical basis, but that they produced miraculous cures, may be doubted.[18]

But what Marcellus lacked in the way of medical sophistication or logical rigor is more than balanced by an impressive number of recipes, boundless confidence, some clever name-dropping, and an immediacy of language. The latter, though crude when compared to the best models of latinity, was apparently the local *patois*. There are different opinions concerning the origin, nature, and extent of Marcellus' bizarre latin, but that the syntax, orthography, and vocabulary contain many gallicisms cannot be denied.[19] But what is more germane to our purpose here is his explicit admission in the *Preface* that his remedies were collected, not only from older medical writers, "but also from country and common people" (*sed etiam ab agrestibus et plebeis*). The strange, sometimes revolting, remedies which Marcellus advocated, as well as their magical and paramedical reinforcements, may thus rest on folk uses;[20] likewise, they may underlie later forms of folk therapy, however difficult that may be of conclusive demonstration.[21]

As an extension of classical materia medica, plants and plant products naturally predominate, though substances of mineral and animal origin are well represented.[22] The pronounced emphasis on human and animal substances, especially ordure, probably reflects the folk uses noted above. The number of botanicals—exotica, indigenes, and domesticates alike—has not been calculated with exactness.[23] The reason is not because of the occurrence of names of unknown, and hence unidentifiable plants, though a few unusual names have created both puzzles and polemics.[24] The lack of precision is due rather to the inconsistent fashion by which synonyms are applied.[25] On the basis of a few which are demonstrably incorrect, at least by classical standards, it appears that Marcellus himself was uncertain of the identity of some of the plants whose usefulness, however, he never questioned.[26] Whether it was very helpful to add to Latin plant names their Greek and vernacular, i.e., their Celtic synonyms,[27] that very preoccupation with such lexical matters as synonomies and etymologies, indicates a significant change of emphasis.

The change from describing medicinal plants to discussing their names is intimately connected with the development of medieval materia medica. The northward and westward expansion of European civilization away from its earlier Mediterranean homeland to the south, meant an encounter, not always quickly recognized, with a new and different flora.[28] As a consequence, many of the locally abundant plants were unrecognized or unknown except insofar as they could be identified with plants of the *flora classica*.[29] This in turn led to the adoption of vernacular names, in latinized form, or those plants were given post classical latin names on the basis of some supposed similarity to those of the *flora classica*.[30] Because of the uncertainty regarding the identity, and hence the properties, of those plants, names and synonyms assumed the role formerly played by descriptions. A series of synonyms, perhaps with some clue as to its economic[31] or alimentary[32] use, thus became the normal means of referring to a plant such that it could be recognized and collected.

It is instructive to observe how Marcellus faced that problem. Although there are a few passages which charitably might be regarded as descriptions,[33] it may be doubted whether, in the absence of some antecedent information, especially the various synonyms, a particular plant could be differentiated from other plants of a similar appearance.[34] Moreover, some of the descriptive passages in the *Liber* were simply borrowed by Marcellus from his sources with little indication that he understood the technical details of the original description.

The simples mentioned in the *Liber* include many exotica, with some 25 botanicals heading the list. However, references, for example, to *galbanum*, *sagepenum*, or *zingiber* do not necessarily mean that they were available in the West.[35] If such exotica, including animal and mineral products were available,[36] either through commercial channels or from craftsmen such as unguent makers, tanners, or dyers, their cost would have put them beyond reach of all but the wealthy.[37] Whether the mention of cinnamon and cloves, candied tragacanth, Falernan wine, imported cheese, Alexandrian niter, African snails, etc., accurately reflect the uses of Marcellus' clientele, cannot be determined. It is, however, at least open to speculation that he copied such recipes from older texts without regard to the special socio-economic conditions of early fifth century Gaul.

But Marcellus did not rely exclusively on imported luxury items. On several occasions, he specified the location where various plants[38] and small animals[39] might be collected. This reliance on indigenes is one more sign of things to come for in some early medieval texts, the ratio of indigenes to exotica is significantly higher than in texts from antiquity.[40]

Turning from the simples to composita, yet another indication of the future is Marcellus' commitment to polypharmacy. Few of his recipes contain less than ten ingredients, and often they include 20 or more. In one notable instance, a recipe for an *antidotus Cosmiana*, said to be effective against all diseases and dangerous inflammations of the intestines, contains 73 ingredients.[41]

Another aspect of polypharmacy, the instructions for the preparations of such heroic medicaments, is no less filled with involved and complicated details. Special attention is given to the repetition and proper sequence of certain operations as well as to the external conditions without which a successful outcome cannot be assured.

In regard to Marcellus' belief in superstition and the efficacy of magic, the *Liber* once again anticipates medieval practices. Whether he was a Christian cannot be answered definitely for the few references to Christianity were commonplace and would have been appreciated by the educated classes whatever may have been their religious persuasion.[42] By the same token, however, the pagan or semi-pagan charms described in the *Liber* do not necessarily mean that Marcellus was not a Christian for the occurrence of pagan charms in Christian texts of the Middle Ages is well known.[43] But, Christian or not, some 65 incantations, magical formulas, and charms, are included as intrinsic parts of the therapy.[44] They range from simple commands, prayers, and nonsense rhymes[45] to involved ceremonies accompanied by recitation, spitting, or physical action.[46] Especially noteworthy is an example of a charm well represented by the familiar childhood jingle of "Ten Little Indians." Marcellus' example, *Novem glandulae sorores*, presumably for inflamed tonsils, may have been the source for a similar charm in Anglo-Saxon.[47]

In addition to the *incantamenta*, Marcellus also provided directions for a variety of amulets, phylacteries, and rings, all of which were regarded as beneficial either in averting or in curing a disease.[48] Underlying such uses was an implicit belief in sympathetic magic as the agency by means of which a disease was transferred from a sick person to an animal or to an inanimate object. In one passage, a series of names of widows is to be recited as the knots of a string are fingered.[49] Since this is a charm to prevent swelling in the groin, the widows, presumably no longer capable of bearing children, became a magical prophylaxis against swelling.[50]

Closely related to such apotropaic devices were numerous superstitious practices, some harmless forms of which still survive. Thus leaves, or the root, of *cicuta* (*Conium maculatum* L.) are worn about the neck for headache,[51] a root is bound to the right arm for nasal complaints,[52] or lucky pebbles are bound to the head for headache.[53] Elsewhere, he advises the root of wild sorrel (*Rumex* sp.) to be collected during the waning of the moon. "So long as you keep this with you," he concludes, "you will not suffer from *lippitudo* (inflammation of the eyes)."[54]

It is in keeping with the magical arts, in which nothing is left to chance, that rules for the collection and preparation of simples must be scrupulously followed.[55] Thus the use of iron was forbidden in the cutting or digging of cer- plants or in the preparation of certain composita.[56] Similar in nature are the ritualistic instructions for collecting simples in which the correct hand, time, and color are specifically indicated.[57] There may have been some hygenic value in the injunction that neither simple nor amulet touch the ground.[58] It is equally likely, however, that such advice rested on a ritual similar to other astrological and numerological fancies with which the text abounds. These, it may be suspected, guaranteed it some measure of success in subsequent centuries, both directly[59] and indirectly.[60]

References

1. See, Jerry Stannard, "Materia Medica and Philosophic Theory in Aretaeus." *Sudhoffs Archiv* 48 (1964) 27-53.
2. See, Jerry Stannard, "Benedictus Crispus, An Eighth Century Medical Poet." *Jrnl. Hist. Med.* 21 (1966) 24-46.
3. All references to Marcellus are to Marcelli *De Medicamentis Liber.* Herausgegeben von Max Niedermann. 2. Aufl. besorgt von Eduard Liechtenhan. Uebersetzt von Jutta Kollesch und Diethard Nickel. Berlin: Akademie-Verlag, 1968. 2 vols. (*Corpus Medicorum Latinorum*, 5).
4. Among the cives ac maiores nostri (*Praef.* 2) Marcellus mentions Ausonius (cf. Ausonius medicus, XXV, 21), the father of the poet Decimus Magnus Ausonius of Bordeaux (*ob. ca.* A.D. 395). A reference to Gamaliel, Jewish Patriarch of Constantinople (*ca.* 395-415) helps to fix the date of composition (XXIII, 77). Two passages in Libanius mentioning a physician Μάρκελλος have been rejected as referring to a different person, Eduard Liechtenhan, *Sprachliche Bemerkungen zu Marcellus Empiricus.* Inaug.-Diss. Basel 1917 p. 6.
5. Marcellus is twice referred to in the *Codex Theodosianus*; see the transcriptions in Liechtenhan, *op. cit.* p. 5.
6. As evidence of the level of education see Ausonius, *Commemoratio professorum Burdigalensium, Opuscula* ed. Rud. Peiper. Leipzig, 1886, pp. 48-71. For the background, see Camille Jullian, "Ausone et son temps. II: La vie dans une cité Gallo-Romaine à la veille des invasions." *Rev. Historique* 48 (1892) 1-38.
7. On Cornarius and the out-dated Teubner edition of Georg Helmreich (Leipzig, 1889), see Max Niedermann, "Hyperkritische Fehlgriffe." *Berl. Philol. Wochschr.* 34 (1914) cols. 92-95.
8. Scribonius Largus (*ca.* A.D. 1-50), author of the *Compositiones* (ed. G. Helmreich, Leipzig, 1887) is probably identical to the Designatianus mentioned by Marcellus, *Praef.* 2, as one of his sources. Parallels between Scribonius, the *Medicina Plinii* (ed. A. Oennerfors, Berlin, 1964 [*Corp. Med. Lat.*, 31]) and Marcellus have been assembled by Niedermann, *ed. cit.* pp. 350-366. See also, G. Helmreich, "Zu Scribonius Largus and Marcellus Empiricus." *Blätter f. das bayer. Gymnasialschulwesen* 18 (1882) 394-95. For Marcellus' relation to Celsus, also named as a source (*loc. cit.*), see Joh. Ilberg, "A. Cornelius Celsus und die Medizin in Rom." *Neue Jahrb. f. d. klass. Altert.* 19 (1907) 377-412.
9. On their fictitious nature, see Hermann Stadler, "Epistola Pseudohippocratis." *Arch. f. lat. Lexikogr.* 12 (1902) 21-25.
10. Paulus Geyer, "Spuren gallischen Lateins bei Marcellus Empiricus." *Arch. f. lat. Lexikogr.* 8 (1893) 469-481; Max Niedermann, "Sprachliche Bermerkungen zu Marcellus Empiricus de medicamentis" (in) *Festagabe Hugo Blümner* überreicht zum 9. August 1914 von Freunden und Schülern. Zürich, 1914 pp. 328-339; Benvenuto Terracini, "Gallico e Latino." *Riv. di Filol.* 49 (1921) 401-430, and Samuel Chabert, *Marcellus de Bordeaux et la Syntaxe Française.* Paris, 1901.
11. Armand Delatte, *Herbarius.* 3e éd. Bruxelles, 1961; Jacob Grimm, "Ueber Marcellus Burdigalensis." *Abh. Akad. Wiss. Berlin, Phil.-hist. Kl.* 1849 pp. 429-460; *ibid*, "Ueber die Marcellischen Formeln." *Abh. Akad. Wiss. Berlin, Phil.-hist. Kl.* 1855 pp. 51-68; Eugene Tavenner, *Studies in Magic from Latin Literature.* New York, 1916. See below notes 20, 47, 50.

12. For a brief note on Marcellus, see Hermann Schelenz, *Geschichte der Pharmazie*. Berlin, 1904, pp. 185, 213; Alexander Tschirch, *Handbuch der Pharmakognosie*. 2. Aufl. I/3 (Leipzig, 1933) p. 1344, and Glenn Sonnedecker, *Kremers and Urdang's History of Pharmacy*. 3rd ed. Philadelphia, 1963, p. 332, n. 37. Much material on Marcellus is contained in Max Höfler, "Volksmedizinische Botanik der Kelten." *Sudhoffs Archiv* 5 (1911) 1-35, 241-279. For references to Marcellus' collyria, see C. L. Grotefend, "Drei und Siebenzig Stempel römischer Augenärzte." *Philologus* 13 (1858) 122-164.
13. For examples, see Liechtenhan, *op. cit.* pp. 13-38.
14. For example, the *Antidotarium Bruxellense* (*saec.* XII) (in) Theodorus Priscianus, *Euporista*, ed. Val. Rose. Leipzig, 1894, pp. 363-396. See below notes 59 and 60.
15. For the level of pharmacological and medical knowledge in the early medieval antidotaria, see Julius Jörimann, *Frühmittelalterliche Rezeptarien*. Zürich, 1925, pp. 78 sqq. and Henry E. Sigerist, *Studien und Texte zur frühmittelalterlichen Rezeptliteratur*. Leipzig, 1923 especially Section 8 "Form und Inhalt der Antidotarien" (pp. 168 sqq.).
16. . . . libellum hunc de empiricis quanta potui sollertia diligentiaque conscripsi. . . . *Praef.* 1.
17. . . . mirabiliter . . . sedat (III. 10), . . . mirum est (XX. 46): . . . certum et praesans remedium sentiet (XV. 10). . . . mira celeritate sanabis (XV, 47). Sometimes the curative properties are announced at the beginning of a recipe: antidotum incomperabile (XVI. 93); certissimum remedium (XXXVI, III), medicamentum mirum (XXVII 8).
18. See, for example, his claim to have restored a patient's sight after twelve years of blindness (VIII, 126: overlooked by Alb. Esser, *Das Antlitz der Blindheit in der Antike*. Leiden, 1961) and the claim for an unguent, beginning, Harum conpositionum, si qua fides est, et ipse ego utilia remedia semper expertus sum et alios idem sensisse confirmo. . . . (XXXV, 9).
19. For examples, see Carl Hamp, "Die zusammengesetzten Präpositionen im Lateinischen." *Arch. f. lat. Lexikogr.* 5 (1888) 321-368; Max Niedermann, "Ueber einige Quellen unserer Kenntnis des späteren Vulgärlateinischen." *Neue Jhrb. f. d. klass. Altert.* 29 (1912) 313-342; A. Sonny, "Demonstrativa als Indefinita." *Glotta* 6 (1915) 61-70, and Antoine Thomas, "Une nouvelle édition de Marcellus de Bordeaux." *Journal des Savants*. Année 1920, pp. 15-21.
20. Eduard Liechtenhan, "Beiträge zur Erklärung und Emendation von Benennungen von Heilkräutern u. ä. des ersten Glossars des Codex Vaticanus Reginae Christinae 1260 saec. X." *Sudhoffs Archiv* 13 (1921) 123 n. 1 suggests that Marcellus' references to *fungus lucernae* (lampblack from the wicks) (cf. XVIII. 31; XXXIII, 68) as ingredients in medicaments rests on folk use. Cf. also Pliny XXVIII xlvi, 163; which may have been Marcellus' source.
21. Max Höfler, *Mitt. z. Gesch. d. Med. u. Naturw.* 13 (1914) 439 in his review of Clauss' study, suggests a parallel between Marcellus and a passage in Clauss' text (anno 1826), but prints neither passage for comparison. The relevant passages are: Limaci calculum, quem in capite habet, tolle.. . . . (I, 41) and Eine Schnecke soll ein Sternlein in ihrem Kopf haben.. . . . H. Clauss, "Volksmedizinisches und Abergläubisches aus Franken und Schwaben." *Blätter d. Ver. f. bayer, Volksk. u. Mundartf.* 2, (1913) p. 22 sect. 23.
22. For animal substances employed by Marcellus, see Max Höfler, "Organotherapie bei Gallo-Kelten und Germanen." *Janus* 17 (1912) 3-19, 76-92, 191-216. I know of no study devoted to Marcellus' use of substances of mineral origin, but since many derive ultimately from Pliny, D. E. Eichholz' "Index of Minerals" (in) Pliny, *Natural History* (Cambridge, Mass., Loeb Classical Library, 1962) vol. X, pp. 341-344 may be consulted.
23. Meyer lists 262 different names, which reduce to 131 when synonyms are omitted, E.H.F. Meyer, *Geschichte der Botanik*, II (Königsberg, 1855) pp. 305-315. My own studies indicate that the total number of different plant names is closer to 350.
24. On the poorly understood *(h)al(us)* or *alum gallicum* see Jerry Stannard, "*Classici* and *Rustici* in Clusius' *Stirp. pannon. hist.*" (to appear in *Clusius-Festschrift*, Vienna, 1973). On the debate concerning the identity of Marcellus' *Saliunca* (*Valeriana celtica* L.) see J.-A. Guillaud, "Le nom de plante 'saliunca.'" *Rev. Etudes Anc.* II (1909) 246-252, 364-365; A. Cuny, "Note sur saliunca." *Rev. Etudes Anc.* II (1909) 252, 365-366, and Guillaud, "La 'saliunca' dans Pline le Naturaliste et dans Marcelle de Bourdeaux." *Rev. Etudes Anc.* 12 (1910) 183-185. The final word, however, belongs to Vittorio Bertoldi, "Droghe orientali e surrogati alpini." *Archivum Romanicum* 10 (1926) 201-220.
25. Compare the following series: Polygonos quoque herba quae Latine sanguinaria [*Polygonon aviculare* L.] dicitur (I, 2); millefolium [*Achillea millefolium* L.], quod Graeci polygonon appellant (I. 28); herbam polygonon, quam rubiam [*Rubia tinctorum* L., perhaps also *R. peregrina* L.] vocamus (I, 44); Herbam, quam Graeci polygonon, nos aliquot vocabulis, sed praecipue sanguinalem [*var. lect.* for *sanguinaria*] dicimus (IX, 81 cf. IX, 21); Verbenam herbam [a collective term], quae, quia multa est et ubique nascitur, polygonos appellatur (X, 5).
26. The statement, Veretrum, id est bulbus, quod est genus carpae (XII, 62) confuses *veratrum*, as it is usually spelled (*Veratrum album* L.) with the common onion (*Allium Cepa* L.), but the former possesses a rhizome quite unlike the fleshy, tunicate bulb of the latter. Equally revealing is his identification of *sarcocolla* [*Pennea (Astragalus) Sarcocolla* L.] with *argimdnia* [*Papaver Argemone* L.] (VIII, 163). The origin of Marcellus' error doubtless lay in the fact that *sarcocolla* was an import from the Near East. It may be noted that Marcellus' synonomy was repeated by an anonymous glossator of the early eighth century, cf. J.-L. Heiberg, "Glossae medicinales." *Medd. kgl. Danske Videnskab. Selskab., Hist.-filol.* IX/I (1924) p. 73: Sartocolla: argemonia.
27. Twelve Celtic plant names, ten of which are also provided with Greek or Latin synonyms, and one bird name, are mentioned by Marcellus. The two plant names without synonyms—and yet to be conclusively identified—are: herbam, quae Gallice vernetus dicitur (IX, 131) and herbam, quae Gallice dicitur blutthagio, nascitur locis umidis (IX, 132). Höfler (note 12) p. 17 connects *vernetus* with Celt. *verno* (cf. germ. *Erle*) i.e. the alder (*Alnus*). The same etymology, but without mention of Marcellus' *vernetus*, occurs in Evald Lidén, "Baumnamen und Verwandtes." *Indogerm. Forschg.* 18 (1905/06) 485. Cf. Gustav Körting, *Etymologisches Wörterbuch der französischen Sprache*. Paderborn, 1908, p. 407b: *vergne*, Erle: lat.-kelt. vernium. Meyer (n. 20) p. 315 does not propose an identification. As regards *blutthagio*, Höfler, *op. cit.* 262 suggests it represents *Ranunculus* while Dottin, in his list of celtic plant names, does not offer an identification. Georges Dottin, "La langue des anciens Celtes." *Rev. Etudes Anc.* 7 (1905) 46.
28. See Auguste Chevalier, "Les transformations survenues dans le paysage rural et dans les cultures du N.-W. de la France, depuis le début du Néolithique jusqu' au début de l'ère chrétienne." *Compte Rendus, Soc. de Biogéographie* 22 (1945) 2-8.
29. For examples, see Ernst H. L. Krause, "Zur Vorgeschichte von Kresse und Waid." *Naturw. Wchschr.* N. F. 10 (1911) 453-454.
30. A case in point is herba gladiatoricia (XVI, 104) which has not been satisfactorily identified, see Jacques André.*Lexique des Termes de Botanique en Latin*. Paris, 1956, p. 149. Other examples are tiniatica (XVII, 12) and herba ianthidis (see below n. 38). Despite the explicit synonomy papaver silvestre [*Papaver rhoeas* L.], quod Gallice calocatanos dicitur (XX, 68), the latter may be of Greek origin, see Otmar Schroefl, *Die Ausdrücke für den Mohn im Galloromanischen*. Inaug.-Diss. Zürich, 1915, p. 35 and note 2.
31. A good example of Marcellus' technique is his reference to *agnus castus* (*Vitex agnus castus* L.): . . . hagni folia, quae nascuntur in virgultis, quibus Graeci pro viminibus utuntur eademque lygus appellant. . . . (I. 14). See also Cedria [the resin of *Juniperus excelsa* MB.] qua librarii utuntur (XXXI, 21) and the reference to a lampwick made of papyrus (Lucernam fictilem de papyro, VIII 179).
32. Rafani [*Raphanus sativa* L.] quoque, id est radicis quae manducatur (IX. 13, cf. IX 30), . . . herba

Brittanice [*Rumex* sp.] viridis sumpta in cibo lactucae [*Lactuca sativa* L.] modo sanat (XI, 16), pastinacae [either *Daucus carota* L. or *Pastinaca sativa* L.], hoc est holeris quod manducatur summitatem . . . (XXVI, 16).

33. Herba est vel frutectum, quod maxime circa sepes vinearum nascitur; Latine cytisus [? *Medicago arborea* L.] dicitur. Virgultis est subtortis vel curvis, foliis oblongis bene virentibus, floribus conclusis oblongis albis vel etiam galbineis bene olentibus, id est mellioto odore (XXV, 11); . . . accipies herbam tribolum [*Trapa natans L.*], cuius herbae ipsum tribolum, hoc est semen spinosum (XXVI, 95), Rosae silvaticae [*Rosa sempervirens* L.] bacas oblongas, rubicundas, maturas. . . . (XXVII, 42).

34. Ros marinum [*Rosmarinus officinalis* L.] sive rhus orientale [*Rhus coriarius* L.], quod est minutum, robium, simile lenticulae [*Lens culinaris* Medikus] (XII, 40).

35. Some evidence of imported drugs still remains, cf. B. Reber, "Pharmacie de poche d'un médecin romain." *Bull. Soc. franç. Hist. Med.* 2 (1903) 369-375; F. A. Flückiger, "Zur Geschichte der Gewürznelken." *Journ. de Pharm. d' Alsace-Lorraine* 12 (1885) 343-345; and Marcel Baudouin et Louis Reutter, "Analyse de résidus de vases gallo-romains et d'un vase à parfums, trouvés dans un puits funéraire de la Vendée." *Comptes Rendus, Acad. des Sciences* 157 (1913) 57-58.

36. That the required ingredients were not always available is evident from the following: Quod si non habueris cinnamomum, duplicem cassiam coicies (XVI, 8); Lacertam viridem excaecatam acu cuprea in vas vitreum mittes cum anulis aureis, argenteis, ferreis et electrinis, si fuerint. . . . (VIII, 49); . . . picem Bruttiam mitti debere [*sc.* in ollam novam], quia ges ampelites vera difficile invenitur (VII, 19). I leave open the question of how readily available were fel cameli (VIII, 83) and crocodili stercus (VIII, 123) despite Marcellus' advice custodire [*sc.* stercor lupi] debes (XXIX, 35).

37. There are several references to unguent makers (IX, 59; XXIII, 3), tanners (IX, 35; XXVII, 25; XXVIII, 4), and dyers (XXIII, 42 cf. XXV, 42) as sources of various raw products.

38. Herba senicio [*Senecio vulgaris* L.], quae in tegulis nascitur (XXXVI, 9); . . . ianthidis herbae [unidentified *fide* André, *op. cit.* 168], quae humidis locis nascitur (XVII, 14); the reference to caulibus non plantatis (XXX, 50) may refer to any of a number of alimentary plants. The garden also provided supplies: rutae hortensis [*Ruta graveolens* L.] (XX, 64) betas [*Beta* sp.] de horto (XXX, 53).

39. Lacertas pissinnas, quae in segetibus morantur (VIII, 90); Muris agrestis, id est qui in campis invenitur (VIII, 90); Bestiolae multipedes, quae in locis umidis et sordidis nascuntur (XXXII, 43 cf. XII, 64 Bestiola, quae brassicae innascitur).

40. Cf. Jerry Stannard, "Graeco-Roman Materia Medica in Medieval Germany." *Bull. Hist. Med.* 46 (1972), 455-468.

41. XXIX, 11. A confection, called *dictamnos* (VII, 18) contains 38 ingredients, and allows for substitutions if certain ingredients are not available.

42. Herba salutaris, id est spina alba [? *Crataegus oxycantha* L.], qua Christus coronatus est (XXIII, 29). Many guesses have been made as to the identity of "Christ's-thorn," but H. Wehrhahn examined some holy relics (unfortunately not dated) in Trier and identified the plant material as *Zizyphus Spina-Christi* L., "Was ist Christusdorn? *Mitt d. deutsch. dendrol. Ges.* Jhrg. 1920, p. 335). Another reference to Christ occurs in a magical incantation, XXV, 13.

43. Out of a vast literature, see J. Klapper, "Das Gebet im Zauberglauben des Mittelalters." *Mitt. d. schles. Ges. f. Volksk.* 18 (1907) 5-41; Reinhold Köhler, "Segensprüche." *Germania* 13 (1868) 178-188 [cf. p. 185 for references to Marcellus], and W. Drexler, "Alte Beschwörungsformeln." *Philologus* 58 (1899) 594-616 (NB: Drexler's discussion of Marcellus (pp. 608-09) refers to XXIX, 23 not XXVIII 23).

44. For a thorough analysis of such material, and a classification into distinct categories, see Ricardus Heim, "Incantamenta Magica Graeca Latina." *Jahrbücher f. class. Philol.* Supplm.-Bd. XIX/2 (1893) 465-576.

45. Commands: VIII, 93; Prayers: XXV, 13; Nonsense rhymes: XXVIII 72 et 73. A charm, to be written on unused parchment, against inflammation of the eyes:
ρουβρς ρποπειρασρ 'Ηέλιος, ὃς πάντ' ἐφορᾷ καὶ πάντ' ἐπακούει
(VIII 58) is composed of two words of gibberish, followed by a verse from Homer, *Iliad* III, 277.

46. Recitation: XII, 24; Spitting: VIII, 171; Physical action: VIII, 191.

47. XV, 102. The Anglo-Saxon charm derives from Marcellus is asserted by Wilfrid Bonser, *The Medical Background of Anglo-Saxon England*. London, 1963, p. 252. It is denied by G. Storms, *Anglo-Saxon Magic*. The Hague, 1948, p. 152.

48. To avert a disease: I, 41; VIII, 27, 45, XV, 67; To cure a disease: I, 85; II, 7; XXVIII, 21; XXIX, 35.

49. XXXII, 19. Cf. Pliny, *Historia Naturalis* XXVIII xii, 48 for a similar charm. Niedermann, in his edition of Marcellus, does not mention this passage of Pliny's as Marcellus' source.

50. Many examples from Marcellus, but not XXXII, 19, are analyzed by Eduard Stemplinger, "Die Transplantation in der antiken Medizin." *Sudhoffs Archiv* 12 (1920) 33-49.

51. Cicutae folia aut, si folia non fuerint, radicem eius teres et fronti inlines aut in saccello lineo ad frontem suspendes. I 38.

52. X, 81. The text, *radicem conupem* is meaningless; Niedermann notes in the apparatus criticus Heraeus' emendation of *conuzae*; Helmreich, *ad loc.*, writes *cannabis*. For a similar use see XXIII, 26.

53. I, 54. See Heim, *op. cit.* p. 474 for parallels.

54. VIII, 41. See Tavenner, *op. cit.* p. 86 for parallels.

55. This seems to be the reason behind the complicated instructions governing the writing of a charm. For example, Ad coli dolerem scribere debes in lamina aurea de grafio aureo infra scriptos characteres luna prima vigesima et laminam ipsam mittere intra tubulum aureum. . . . (XXIX, 26). The instructions continue for another 7 lines before the actual charm is revealed. Cf. also XXVI, 43.

56. The use of iron is forbidden in the collection of simples: XVI, 19; XXIII, 35; XXV, 14; ii the preparation of composita: XXV, 13; XXVI, 10.

57. Left hand: VIII, 191; XXVI, 41; Right hand: XXIX, 43; Time: XIV, 65; XV, 101; XXVI, 41; Color: XXIX, 55.

58. XXIII, 22; XXVIII, 71; XXIX, 35.

59. Traces of a knowledge of Marcellus in the Middle Ages are sporadic, but enough direct evidence remains to hypothesize an established, if not continuous, tradition and to suggest, moreover, that the indirect evidence, such as parallels, rests on an historical basis (see next note). Beginning with a list of physicians, preserved in Cod. Laurentianus 73. 1 (*saec.* IX/XI) fol. 143[r], in which Marcellus' name appears (printed by Max Wellmann, "Zur Geschichte der Medicin im Alterthum." *Hermes* 35 (1900) 369), he is explicitly cited in the Reichenau Antidotarium (*saec.* IX/X, cf. Sigerist *op. cit.* p. 41) and extracts from lib. XII appear in Cod. Bamberg L. III. 9 (*saec.* XIV) fol. 140[a] (cf. Elias Steinmeyer und Eduard Sievers, *Die Althochdeutschen Glossen* IV (Berlin, 1898) 380). Finally, in a 15th century MS, there occurs the statement, Von demselbm kraut [sc. verbena] schreybt uns Marco der hoch arzt, Anton Birlinger, "Bairische Besegnungen." *Germania* 24 (1879) 75.

60. Parallels, some of which may indicate Marcellus as an indirect or intermediate source, have been noted by Christoph Ferckel, "Medizinische Marginalien aus dem Cod. Trevirens Nr. 40." *Sudhoffs Archiv* 7 (1913) 130 n. 2, 137 n. 1; Hermann Fischer, "Mittelhochdeutsche Receptare aus bayerischen Klöstern und ihre Heilpflanzen." *Mitt. d. bayer. bot. Ges.* 4 (1925) 71, 72, 74; Friedrich Wilhelm, *Denkmäler deutscher Prosa des 11. und 12. Jahrhunderts.* München, 1960, 2. Abt. pp. 95, 96; W. M. Lindsay, *The Corpus Glossary*. Cambridge, 1921, p. 211, and Sven Norrbom, *Das Gothaer mittelniederdeutsche Arzneibuch und seine Sippe*. Hamburg, 1921, p. 205 s.v. *athanasia magna*. Finally, Ludwig Choulant, in his edition of Macer Floridus, *De viribus herbarum* (Leipzig, 1832) suggests (*ad versum* 1389) that Macer's reference to Justus medicus derives from Marcellus XXV, 32.

VII

LOST BOTANICAL WRITINGS OF ANTIQUITY

The recovery of the Dead Sea Scrolls and Menander's long lost *Dyskolos* remind us that only a portion of ancient literature has survived the accidents of time. In the past half-century new material relating to ancient botany has also come to light. Our over-all picture of ancient botany has not, perhaps, been fundamentally altered, yet our appreciation of ancient science has been enriched. Although the four known papyrus fragments of lost redactions of Dioscorides have not enabled us to reconstruct a complex text tradition, they have added data which must be considered in future studies. If one wishes, therefore, to obtain a comprehensive survey of ancient botany, the older material as well as the new must be brought together and both must be related to the remains of what did not survive intact.

Ancient botany was strongly bound by tradition. Endless copying, reediting, and abridging disguised the lack of original thought while the marshalling of authorities *pro* and *contra* often took the place of empirical data. In virtue of this traditionalism, ancient botanical writings are filled with the remnants of earlier writings which have since perished. It is this record, preserved by the surviving fragments and in the references to nearly-forgotten authors and titles which I propose to examine. In doing so, I shall call attention to a few examples representative of the major divisions of ancient botany.

It is well known that systematic, philosophic studies of plant life in general were rarely undertaken in antiquity. Save for Theophrastus works and the pseudo-aristotelian *De plantis*, now ascribed to Nicolaus of Damascus, no texts have survived and there are no fragments which unequivocally point to other such works as having been written in antiquity. It is thus a matter of regret that Aristotle's *Περὶ φυτῶν* has disappeared but for a few probable fragments and some testimonia in Syriac[1].

---

[1] H. J. DROSSAART LULOFS, *Aristotle's* Περὶ φυτῶν, „Jrnl. of Hellenic Studies" 77/1, 1957, 75—80.

Owing to the high regard for him in the Islamic world and the spurious botanical writings transmitted in his name, it is possible that someday a literal Arabic translation will partially restore our greatest loss.

The paucity of texts on general botany is somewhat offset by the extant pharmacological and agricultural texts, and the numerous references to similar writings which have vanished. In the Preface to his *De materia medica*, Dioscorides mentions by name nine writers whose inadequacies prompted him to write a more comprehensive work. The titles of their writings are not stated and their dates have proven difficult to determine. Yet, by collating these names with similar references in later writers, especially Pliny and Galen, enough can be learned of Krateuas, Sextius Niger and others to trace the development of medical botany to within a generation or two of Hippocrates.

In this regard, Caelius Aurelianus has preserved an interesting fragment from Praxagoras' lost *Θεραπευτικά*. Among the simples listed for liver complaints, the name *eryngion* stands out. The plant in question is a species of *Eryngium*, probably *E. maritimum* L. or *E. campestre* L. Yet, among the 225 plants mentioned by Hippocrates, *ἠρύγγιον* does not occur, despite the fact that he and Praxagoras were natives of Cos, and lived within a century of one another.

Ancient pharmacology, like ancient agriculture, survived into modern times, though often in strange forms. As an example, the Latin translation of Nicolaus Myrepsus' *Δυνάμερον* (ca. 1280) was the basis of the pharmacopea of the Paris Faculty of Medicine until the seventeenth century. Interest in Nicolaus is heightened by the fact that texts, since lost, were perhaps available to him, e. g. the *Νάρθηξ* of Heras of Cappadocia, known to us otherwise chiefly through Galen. That other classical texts survived into the Middle Ages is indicated by Simon Cordo's citations from Demosthenes Philalethes, especially in regard to drugs for various ophtalmological disorders [2].

With the advent of poisoning as a social institution, the demand for antidotaria increased. Mithridates Eupator, King of Pontus, was a devotee of the art and an experimental toxicologist. His writings, translated into Latin by Pompeius Lenaeus, were still extant in Pliny's time. The popularity of writings on antidotes for venemous beasts likewise called forth many texts. Among the most popular were Nicander's *Alexipharmaca* and *Theriaca*. From the anonymous scholia to the latter, we learn that Alexander, Apollodorus, Mikion, Numenius, and Petrichos wrote on similar subjects.

---

[2] Simon Januensis, *Synonyma medicinae*, Venice, G. Anima Mia, 1486. Fol. 21r s. vv. calasma, calaza.

There is no better illustration of ancient literary tradition than the corpus of agricultural writings. From Cato to Palladius the precepts as well as the errors were repeated, despite some fundamental changes in agricultural practice and dietary habits. Fragments relating to the agricultural tracts up to the time of Columella were collected by Reitzenstein but the later period has been rather neglected[3]. From the *Geoponica*, edited in its final form by Cassianus Bassus (ca. 950 A. D.), 21 authors excerpted for the botanical sections alone. Their dates are conjectural but they testify to the number of agricultural treatises once available. If Diophanes, one of Cassianus' sources, is the same person who was responsible for the six-book abridgment of Mago's *De agricultura*, then *Geoponica* covers a period stretching back over a millenium. Anonymous vernacular tracts served to keep the agricultural tradition alive in Western Europe also. In a Middle English manuscript on arboriculture a passage is introduced by the phrase "as aristotil seyth in his boke of plantes"[4]. The passage following is neither a translation nor a close paraphrase of anything in the spurious *De plantis*. Although its provenance is undetermined, it is not altogether impossible that by coping from a lost Latin compendium, material survived which has yet to be properly interpreted.

In addition to pharmacology and agriculture, other forms of botanical writings existed in greater number than is evident today. Horticulture is represented by several authors to whom Pliny attributed a *Κηπουρικά*. It is difficult to form an adequate idea of their contents for Gargilius Martialis *De arboribus pomifers* provides a slender basis of camparision. On the strength of fragmentary remains, information is obtained regarding the introduction into Rome of exotic ornamental and fruit trees. Such evidence is often open to question, but when combined with archeological findings, there is little doubt that gardening was as much a part of the daily life as it is for some of us today.

Space permits us to indicate only a few additional literary genres. Because the historian of botany is limited by his evidence, he must resort to writings whose titles seem strange to modern ears. Book XV of Athenaeus' *Deipnosophistae* has preserved a score of fragment dealing with floral wreaths. These miscellaneous passages are among our most valuable sources for information regarding the purely aesthetic approach to plants. From fragments deriving from writings on agriculture, often cited by Pliny, information is obtained regarding cultivated plants and their economic uses. Economic considerations led to writings on perfumery

---

[3] R. Reitzenstein, *De Scriptorum Rei Rusticae... Libris Deperditis*. Diss. Berolini, 1884.

[4] Cambridge University Library Ee I. 13. Fol. 110r.

and scents, of which the sole survivor, Theophrastus' *De odoribus*, may also be the earliest classical text on the subject and one of his few writings in which he does not mention previous authors. Passages from such diverse writings often are of aid in the identification of ancient plants and in the interpretation of color terms. Ancient references to the color of a flower have been utilized by phytogeographers intent upon tracing the evolution and distribution of sub-specific taxa[5].

Gastronomy and the alimentary uses of plants were represented by a large body of literature. From writings such as Chrysippus' *Π. λαχάνων* or Antilochus' *Π. κυτίσου* we can form a better picture of those pot herbs which, having fallen into disuse, are now ranked as weeds.

We cannot conclude without mentioning one further genre. Today, dictionaries of plant names play an important role. In antiquity, with many less avenues of scientific communication, lexica played an even greater role and as a source of basic information, they continued to circulate, in manuscript, into the age of printing[6]. Numerous botanical lexica survive and we possess the raw materials for others. I refer to the Alexandrine commentaries in which the learned etymological notes often contain passages of genuine botanical interest[7]. As might be expected, Dioscorides became the major source for the Byzantine multi-lingual glossaries. For this reason, a gloss which cannot be traced back to him deserves close scrutiny. It may be the clue to the detection of one of the many missing links some of whose cacatenations I have tried to indicate here.

---

[5] W. B. Turril, *A Study of Variation in Glaucium Flavum*, "Kew Bulletin" **4**, 1933, 174—184.

[6] J. Stannard, *A Fifteenth-century Botanical Glossary*, "Isis" **55**, 1964, 353—367.

[7] M. Wellmann, *Hippokratesglossare*, "Quell. u. Stud. z. Gesch. d. Naturw. u. d. Med." **2**, 1931, 1—88.

VIII

BYZANTINE BOTANICAL LEXICOGRAPHY

To Professor Alexander Turyn
on his seventieth birthday

The botanical sections or lists of medicinal plants in Byzantine medical texts may appear distant to us today. But for nearly one millenium they were carefully studied, as references to them in Renaissance herbals amply testify [1]. Of greater significance is the fact that such material supplied continuity to a branch of science in which nomenclature, synonomies, and even orthography play an unusually important role, viz., botanical terminology.

In the days before antibiotics and chemotherapy, anaesthesia and the germ theory of disease, plants and plant products dominated the physician's armamentarium. Nearly every common or conspicuous plant was invested with one or more, specific, therapeutic properties (δυνάμεις). Many of those plants, which also served alimentary and other economic needs, were already known in earlier periods of Greek literature. They appeared in the writings of physicians such as Hippocrates, of scientists and philosophers such as Aristotle and Theophrastus, while a considerable number of them appeared even in literary texts, for example, in Homer, Herodotus, and others.

Greek medicine entered into a long, gradual decline after Galen, relieved only by the voluminous encyclopedias of Oribasius (325-403), Aetius of Amida (*fl.* 540-550), and Paul of Aegina (*ca.* 625-690). Thereafter it was retrospective and Byzantine physicians, for the most part, were content either to read the earlier, and by then, thoroughly-acknowledged medical classics, or to edit and abridge them in accordance with contemporary standards of literary scholarship [2].

Our concern here, however, is with one aspect only of this larger picture, viz., the lists of medicinal plants which formed a characteristic feature of the Byzantine encyclopedias mentioned above and of the lexica which derived therefrom. It was by means of such lists and the information contained in the glosses that a large por-

1. Paul of Aegina was cited by OTTO BRUNFELS, *Contrafayt Kreüterbuch*. Strassburg: Hans Schott, 1532, fol. CCXXIX as one of his sources. Much later, ADAM LONICERUS, *Kreüterbuch*, Frankfort: M. Wagner, 1674, p. 544 cited both Paul and Aetius. In general, see JERRY STANNARD, «The Graeco-Roman Background of the Renaissance Herbal». *Organon* 4 (1967) 141.

2. Cf. L. LABOWSKY, «Bessarion Studies, II». *Medieval and Renaissance Studies* 5 (1961) 132-154. In 1467, Michael Apostolis was engaged in copying Galen's «On Simple Drugs» according to D.J. GEANAKOPLOS, *Greek Scholars in Venice*. Cambridge, Mass., 1962, P. 90.

tion of the knowledge of Greek medical botany was preserved, later to be transmitted to the Latin West as well as to Islam[3].

There were two practical reasons why the compilation of plant lists was undertaken. We have referred above to the fact that the physician's main line of defense was drugs of plant origin. To that extent, a list of medicinal plants correlated with the diseases or complaints which they alleviated, would be of benefit to the practicing physician. Closely related is a second reason, and one which reveals a characteristic side of Byzantine biology and medicine. Lists, and especially those containing descriptive data about those plants and not merely their δυνάμεις, obviously were useful for purposes of recognizing and collecting medicinal plants and of avoiding noxious or poisonous plants. The value of such lists, however, depended primarily upon the extent and accuracy of the description and only secondarily upon the name of the plant, its synonyms, and its medical application. Beginning with Galen, and increasing with the passage of time, there was a tendency to go back to Dioscorides' Περὶ ὕλης ἰατρικῆς. Usually his nomenclature was adopted by the Byzantine physicians, though not always his synonyms. His descriptions of the plants, however, were, in almost all cases, abridged by omitting some or all of the pertinent information by means of which a plant could be recognized. The listing of medicinal plants thus came to be an implicit commentary on Dioscorides. And after Galen, compiling plant lists became a literary activity in which scholarship and editorial expertise supplanted a personal knowledge of the plant in its living state[4].

The final stage of this medico-botanical scholasticism was reached in late Byzantine times (from the thirteenth century onwards) with the rise of multilingual lexica. Often they recorded little more than synonyms and even then some of the synonomies were as questionable as their orthography. These lists, whatever their medical value may have been, are, nonetheless, important documents for classicists, Byzantinists, and historians of science insofar as they represent a stage in the evolution of Greek botanical scholarship. Byzantine botany, in fact, cannot be understood apart from its scholarly, lexicographical orientation[5].

There were, in addition to the practical medical reasons for the compilation of plant lists, another set of influences which helped to shape the course of Byzantine

3. As examples, cf. H. Mörland, ed. *Oribasius Latinus*. Oslo, 1940, pp. 11 sq. and Jerry Stannard, «Benedictus Crispus, An Eighth-Century Medical Poet» *Journal of the History of Medicine* 21 (1966) 24-46. For the Islamic tradition of Paul, see, G. Dugat, «Etudes sur le traité de médecine d'Abou Djàfar Ahmad, intitulé: Zad al-Moçafir». *Journal Asiatique* (sér. v) 1 (1853) 326 and M. Steinschneider, «Die toxicologischen Schriften der Araber bis Ende xii. Jahrhundert». *Virchows Archiv* 52 (1871) 361-62; for Aetius, see H. Lehmann, «Zu Aetius Amidenus». *Sudhoffs Archiv* 23 (1930) 205-206. In general, see F. Brunet, «Contribution des médecins byzantins à l'histoire des plantes et à la botanique médicale en France». *Hippocrate* 5 (1937) 524-531 and Cesare Brescia, *Prodotti Italiani nella Farmacopea dei Medici Greci*. Naples, 1956.

4. Cf. Jerry Stannard, «Dioscorides and Renaissance Materia Medica». *Analecta Medico-Historica* 1 (1966) 1-21.

5. Cf. M.H. Thomson, *Textes grecs inédits relatifs aux plantes* (Paris, 1955) pp. 127-130.

botanical lexicography. We have noted above that the names of a number of medicinal plants occured in even the earliest, non-medical writings. Already by Hellenistic times, a large body of exegetical material had developed whose primary function was to preserve the ancient wisdom of the classics and to transmit it to later generations. In the process, difficult or unusual words were explained, and particularly those words which were no longer a part of ordinary speech. Among the specialized vocabularies, the names of plants and animals bulked large in the subsequent commentaries and lexica [6]. The preservation of plant names, particularly those which had been superseded by synonyms or which were threatened by regional or dialectal forms, thus came to be a part of antiquarian scholarship.

Although the plant lists of the Byzantine physicians and the lexica of literary scholars were designed for different purposes and originally had little in common save the names of certain plants, together they helped to keep alive a knowledge of their synonyms, uses, and appearance.

The historical development of Greek botanical lexicography up to and including the specialized medico-botanical lexica of late Byzantine times, is closely connected with a parallel development in plant names. A few guidelines will indicate their quantitative increase and will suggest some of the problems faced by the compilers of lists and lexica and, later, by students of ancient and medieval botany. (In the following, chronologically-arranged list, the figures must be regarded as provisional for reasons that will become apparent in later portions of this paper. Moreover, each of the following writers presents some special problem – textual, lexical, or botanical – some of which will be mentioned in the notes thereto. Finally, even approximate figures are unavailable save for a few non-medical writers who have been explicitly studied for this purpose and who are included here for the sake of comparison only).

Homer: 63 different species [7]
Herodotus: 63 different species [8]
Hippocrates: *ca.* 225 different species [9]
Theophrastus: 500-550 different species [10]

6. Cf. M. Wellmann, «Hippokratesglossare». *Quell. u. Stud. z. Gesch. d. Naturw. u.d. Med.* 2 (1931) 1-88.

7. F.A.W. Miquel en W.H. de Vriese, «Tentamen Florae Homericae». *Tijdschrift voor Natuurlijke Geschiedenis en Physiologie* 2 (1835) 111-165.

8. F. Kannngiesser, «Die Flora des Herodot». *Archiv. f. d. Gesch. d. Naturw. u. d. Techn.* 3 (1910) 81-102. It is a mere coincidence that Kanngiesser should reach the same figure as Miquel did for Homer, for other historians have counted Herodotus' plants differently, e.g. A. Tschirch, *Allgemeine Pharmakognosie* (Leipzig, 1933), p. 1264 enumerates 65 species.

9. Jerry Stannard, «Hippocratic Pharmacology». *Bulletin of the History of Medicine* 35 (1961) 497-518.

10. K. Regenbogen, «Theophrastos von Eresos» *Pauly-Wissowa* Supplm. Bd. vii (1950) col. 1467. and E.L. Greene, *Landmarks of Botanical History* (Washington, D.C. 1909) P. 69.

Theocritus: 87 different species [11]
Nicander: *ca.* 299 different plant names [12]
Erotian: *ca.* 52 different species [13]
Dioscorides: 550-600 different species [14]
Galen: *ca.* 438 different species [15]
Oribasius: *ca.* 439 different species [16]
Aetius of Amida: *ca.* 356 different species [17]
Alexander of Tralles: *ca.* 370 different species [18]
Paul of Aegina: *ca.* 489 different species [19]
Michael Psellos: 59 alimentary plants [20]
Simeon Seth: 140 chapters devoted to plants [21]
Nicolaos Hieropais of Agrapha: 553 drug names [22]

11. A. LINDSELL, «Was Theocritus a Botanist?» *Greece and Rome* 6 (1936/37) 78-93.

12. This figure is based on Index I (pp. 228-237) in NICANDER, *The Poems and Poetical Fragments*. Edited by A.S.F. Gow and A.F. Scholfield, Cambridge, 1953. I have eliminated, from the above figure, orthographical variants and obvious synonyms.

13. *Vocum Hippocraticarum Collectio* ed. E. Nachmanson. Goteburg, 1918.

14. JERRY STANNARD, «P.A. Mattioli: A Sixteenth Century Commentator on Dioscorides». *Bibliographical Contributions, University of Kansas Library* I (1969) 59-81.

15. This figure is based on the alphabetically-arranged list in Books VI-VIII of *De simplicium medicamentorum temperamentis ac facultatibus* (in) Galeni *Opera Omnia* ed. C.G. Kühn. Vol. XI, 789-892 and vol. XII, 1-158 (Leipzig, 1826). It does not include the plants listed in *De alimentorum facultatibus* (vol. VI, 453-748) nor those in the pharmacological writings of vols. XIII-XIV K. For a sample of the plant-names listed in the latter, see G. HELMREICH, «Zu Galen». *Philologus* 75 (1919) 77-94.

16. Based on Books XI-XII, *Collectionum Medicarum Reliquiae*, ed. I. Raeder, Vol. II pp. 80-159 (*Corp. Med. Graec.* VI, 1, 2). Leipzig/Berlin, 1929. (Cited below by book and page).

17. Based on Book I, *Libri Medicinales* ed. Aless. Olivieri, Vol. I pp. 30-146 (*Corp. Med. Graec.* VIII, 1). Leipzig/Berlin, 1935 (Cited below by book and page).

18. *Alexander von Tralles*. Original-Text und Übersetzung von Theodor Puschmann 2 vols. Vienna, 1878-79. Unlike the writings of Oribasius, Aetius, and Paul, there is no chapter or book devoted solely to plants. (Cited below by volume and page).

19. Based on Book VII, cap. 3 ed. J.L. Heiberg, Vol. II pp. 186-274 (*Corp. Med. Graec.* IX, 2) Leipzig/Berlin, 1924. (Cited below by book and page).

20. Based on Psellos' Πόνημα ἰατρικόν (sections Περὶ λαχάνων, Περὶ ὀσπρίων and Περὶ κολοκύνθης, πεπόνων κτλ. (in) J.L. IDELER, ed. *Physici et Medici Grceci Minores*. Vol. I (Berlin, 1841) pp. 206-208.

21. *Syntagma de alimentorum facultatibus* ed. B. Langkavel. Leipzig, 1868. For a discussion of some of the more noteworthy plants in Seth, see E.H.F. MEYER, *Geschichte der Botanik* III (Königsberg, 1856) pp. 361-365.

22. Λέξικον τῶν Βοτάνων συλλεχθὲν παρὰ τοῦ ἐν ἰατροῖς λογιωτάτου Νικολάου τοῦ Ἀγραφιώτου κατὰ διαλέκτους... (in) A. DELATTE, *Anecdota Atheniensia et Alia* vol. II (Liège, 1939) pp. 394-417. (This is number XII of the lexica edited by Delatte, thus cited below as Delatte XII plus page).

It was thus for a two-fold purpose – medical pragmatism and antiquarian scholarship – that plant lists and lexica were compiled. Although there is no such thing as a typical plant list or botanical gloss, certain types of information are commonly encountered therein:

I. *Synonyms, including new names and foreign names*

II. *Medical Uses*

III. *Descriptive Data*

IV. *Etymologies*

V. *Economic Uses*

Each of these will be discussed below.

I. *Synonyms*

One of the two most frequently found types of information in both lists and lexica is the synonym or synonyms of the plant in question. Obviously, its medical uses and physiological properties were the most important data. Nonetheless, the recording of a synonym, or perhaps two synonyms, served a practical purpose by calling to the attention of the physician that the same plant was known under different names.

Already in Dioscorides and Galen, a plant may be recorded under a heading composed of two names. The form varies somewhat, but often it is a simple disjunction of the form typified by Galen's ἀνθημὶς ἢ χαμαίμηλον (XI, 833 K.). Entries of this form presumably meant that the plant was known equally well under either of the two names[23]. This entry is distinguished from a second type in which the synonym, while well known, had not become universally established. The form of this entry is typified by the phrase following the lemma, for example, λωτὸς ὁ ἥμερος, ὃν ἔνιοι τρίφυλλον ὀνομάζουσι (Galen XII 65) or Ῥῆον ποντικόν... ἔνιοι δὲ ῥᾶ προσαγορεύουσι (Aetius I, 127). Finally, there is a third class of entries containing synonyms which exhibit so many variations that no formula has been sought. The synonomies falling within this class suggest, however, that sometimes, and in some places a plant was known under one synomym while in other places it was known under another synonym. For example, Πολεμώνιον, οἱ δὲ φιλεταίριον, οἱ δὲ χιλιοδύναμον (Aetius, I, 124); Βότρυς, οἱ δὲ ἀμβροσίαν, ἄλλοι ἀρτεμισίαν (Paul VII,

23. As an illustration of how names changed and the problems they caused cf. Galen Κορίανον, ἢ κόριον, ἤ, ὅπως ἂν ἐθέλῃς ὀνομάζειν, οἱ μὲν γὰρ Ἕλληνες οἱ παλαιοῖ χορίανον χαλοῦσιν, οἱ νεώτεροι δ' ἰατροι πάντες ὀνομάζουσιν κόριον... (XII, 36 K.). And to illustrate the acceptance of an exact synonomy, cf. Aetius I, 111: Μορέας ὁ καρπὸς συκάμινα καὶ συκόμορα καλεῖται συνήθως ὑπὸ πολλῶν.

201); ῎Εχιον, οἱ δὲ Δωριάδα, ἔνιοι δὲ καὶ ταύτην ᾿Αλκιβιάδειον (Paul VII, 212). In this class, the synonym or synonyms following the main entry have apparently not become sufficiently established with the result that a simple disjunction, had it been employed, would have led to more confusion than clarity.

In Dioscorides, and in many of the entries of other writers derived from Dioscorides, the recording of a synonym is followed by a description of the plant or some remark which characterized that plant in such a way that it could be distinguished from other plants and/or collected. The one great exception to this is Dioscorides' habit of occasionally referring to a plant as «known to all» or «well-known» (γνώριμος). Such a technique obviated the necessity of providing a description, though not necessarily a synonym[24]. The use of such Dioscoridean phraseology occurs sporadically in Byzantine writers, for instance, Νήριον ἢ ῥοδοδάφνη γνώριμος ἅπασι θάμνος (Aetius I, 114) or Γλήχων πόα γνώριμος (Oribas. XI, 97).

As the recording of synonyms (and other non-medical data to be discussed below) took up more and more space, the descriptive data were curtailed. In extreme cases, there are entries which contain no descriptive data at all. Some contain synonyms and medical uses (for example, ἀπαρίνη, Aetius I, 41; ἔμπετρον Paul VII, 210) others contain the medical uses of the plant without synonyms (for example, ἕρπυλλος, Paul VII, 211; λεπίδιον, Aetius I, 102).

One other technique requires mention, if only to indicate the practical needs served by recording synonyms. This is the technique of cross-reference, employed consistently, though not frequently, by Aetius and Paul. It is not clear what criterion was used in deciding under which of the two names a plant should be listed, but common usage probably determined the listing as it did in the case of synonyms followed by ἔνιοι κτλ. mentioned above. For example, Aetius described χονδρίλλη thus: Περὶ χονδρίλλης προείρηται ἐν τῷ περὶ σέρεως τόπῳ, ἐστὶ γὰρ σέρεως εἶδος (I, 146) and s.v. σέρις he writes: Σέρις ἢ κιχόριον ἢ πικρὶς ὑπόπικρόν ἐστι λάχανον καὶ μᾶλλον τὸ ἄγριον· ὅπερ καὶ διὰ τοῦτο πικρίδα προσαγορεύουσιν ἔνιοι, τινὲς δὲ κιχόριον ὀνομάζουσιν... ὥσπερ καὶ ἡ χονδρίλλη· καὶ γὰρ καὶ αὐτὴ σέρεώς ἐστιν εἶδος. (I, 130). Common usage also seems to have dictated Paul's entry - Κάσαμον. τοῦτό φασιν εἶναι τὸν καρπὸν τοῦ βαλσάμου, περὶ οὗ κατ' ἐκεῖνο λέλεκται (VII, 222) though s.v. βάλσαμον (VII, 199) no mention is made of κάσαμον. This last remark calls attention to the lack of consistency in cross references. Paul, like Aetius, states that χονδρίλη (sic!) is a σέρεως εἶδος (VII, 273) but, unlike Aetius, he does not mention χονδρίλη in his discussion of σέρις (VII, 258).

In keeping with the practical function served by recording synonyms of medically-useful plants, it may be noted that new names entered the plant lists and lexica.

24. Dioscorides used γνώριμος in this sense approximately thirty times. As typical examples, cf. II, 109: θέρμος ὁ ἥμερος γνώριμος· (the remainder is concerned with its medical uses) and III, 34: ἡδύοσμον, οἱ δὲ μίνθην· γνώριμον βοτάνιον, δύναμιν ἔχον κτλ. All references to Dioscorides are to Wellmann's edition, 3 vols. Berlin, 1906-1914.

Some of these names became established while others survived only in dialectal form as folk names or book names (see below p. 9).

These new names fall rather generally into two classes, each of which will be briefly examined.

(i) *Adaptation of Non-Greek Names*

Beginning with the alphabetical recensions of Dioscorides, which Wellmann dated as early as the first century A.D.[25], there occurs a profusion of synonyms for many of even the most common plant names. Ostensibly, they are of Egyptian, African (? Punic), Tuscan, Dacian, and Latin origin, to name just a few of the «dialects» mentioned by Pamphilus, who apparently compiled lists of such synonyms in his now lost Περὶ βοτανῶν or Περὶ γλωσσῶν ἤτοι λέξεων[26]. Despite much effort and ingenuity on the part of comparative philologists to corroborate the exotic names of Pamphilus and Pseudo-Dioscorides, no general agreement has yet been reached as regards either the languages involved or the identification of the plants[27].

There were, of course, plant names of non-Greek origin long before Dioscorides, but such names as πέπερι, βδέλλιον, κιννάμωμον, μαλάχη, νάρδος, etc. had become so thoroughly naturalized by the first century that they were no longer felt to be foreign[28]. A few new, but also non-Greek words were introduced by the Byzantine physicians and, significantly enough, some uncertainty still attaches to such names as ἀρναβώ and μαμιράς. On the other hand, γάλαγγα, which first appears in Aetius (I, 131) is now quite conclusively identified as the root of *Alpinia officinarum* Hance.

There also occur a few synonyms of older plant names which are explicitly stated to be of foreign origin. Thus, both Aetius (I, 113) and Paul (VII, 244) record βήσασα as the Syrian name of μῶλυ, an old Homeric name (*Od.* X, 305). Alexander also mentions βήσασα but in the one occurrance of that term he merely remarks, ὅ τινες ἀρμαλᾶ προσαγορεύουσιν, οἱ δὲ ἄγριον πήγανον· (II, 135), thus suggesting that, like Oribasius (XI, 124), he had not learned of βήσασα as a synonym of μῶλυ[29].

25. MAX WELLMANN, «Die Pflanzennamen des Dioskurides». *Hermes* 33 (1898) 369.

26. Cf. MAX WELLMANN, «Pamphilos». *Hermes* 51 (1916) 1-64.

27. Out of the large literature on this subject, see V.L. BOLOGA, «I Sinonimi 'Daci' delle piante descritte da Dioscoride possono servire alla ricostruzione della lingua Daca?» *Archeion* 12 (1930) 166-170; V. BERTOLDI, «'Nomina Tusca' in Dioscoride». *Studi Etruschi* 10 (1963) 295-320; G. ALESSIO, «Phytonymata». *Riv. di Filol.* 66 (1938) 377-394 and JOH. HUBSCHMID, «Ein etruskisch-iberischer Pflanzenname». *Museum Helveticum* 7 (1950) 221-226.

28. Cf. P.AALTO, «Wo der Pfeffer wächst». *Neuphilogische Mitteilungen* 50 (1949) 13-21; I. LEWY, «Sur quelques noms sémitiques de plantes». *Revue Archéologique* (III) 36 (1900) 336-344, and R.C. THOMPSON, «The Migration of Assyrian Plant-Names into the West». *Classical Review* 38 (1924) 148-149.

29. Cf. JERRY STANNARD, «The Plant Called Moly». *Osiris* 14 (1962) 254-307.

Simeon Seth provides three examples of a slightly different case by his use of ζίνζιφα, καφουρά, and σάχαρ. The plant or plant-product denoted by each of the three terms was known at a much earlier period. However, the orthography was different and, more significantly, the terms did not occur in the plant-lists of the Byzantine encyclopedists [30].

Synonyms of latin origin pose less a problem in identification owing to Pliny's industriousness as much as to later Greco-Roman glossaries [31]. Latin synonyms occur as early as Galen, but are rare in the Byzantine physicians except in those passage that derive from Galen and Pseudo-Dioscorides [32]. For example, Aetius' synonomy (I, 87) Κέστρον τὸ ψυχρότροφον λεγόμενον, ῥωμαιστὶ δὲ βεττονίκη καλεῖται paraphrases Galen's Κέστρον ἢ ψυχότροφον (!) (XII, 23 K) just as Paul's Ἐπιμηλίς, ἥτις καὶ οὐνέδων παρὰ τοῖς Ἰταλικοῖς ἀγροίκοις καλεῖται (VII, 210) is derived from Galen's account of ἐπιμηλίς (XI, 876) [33]. Oribasius' synonymy, Ἱεράκιον τὸ μέγα (Ῥωμαῖοι λαπτούκα ἠρράτικα) (XI, 106) stems not from Galen but from Pseudo-Dioscorides (III, 64 R.V.) as does his Ἱπποσέλινον (Ῥωμαῖοι δ' ὀλέσαθρον), mentioned a few lines below (cf. Ps-Diosc. III 67 R.V. where Oribasius' bizarre orthography can be corrected to ὀλυσάθρουμ, i.e. *(h)olusatrum*, Pliny, *Hist. nat.* XIX, 48, 162). Latin names become more frequent in the late Byzantine lexica and only two examples need be cited here: λευκόϊον· βίολα λευκή (Delatte, X, 383); κασσία· κασσία φίστουλα (Delatte XI, 389).

(ii) *Folk names.*

Opinions vary as to what constitutes a folk name of a plant in Antiquity and in the Middle Ages [34]. Moreover, no hard and fast rules are applicable to pre-linnaean

30. Ζίνζιφα p. 40 (codd. ζίτζινφα, ζίζυφα) the fruits of the jujube tree (*Zizyphus Jujuba* Mill), known to Columella IX 4, 3, *Geoponica* X 3, 4 p. 267 Beckh. and Alex. Trall. II, 219 Puschm. Καφουρά is described as δένδρου κόμμι ἤτοι δάκρυον φυομένου ἐν Ἰνδιᾷ... φύεται δ' ἐν ὄρεσι πλησίον τῆς ἐκεῖσε θαλάσσης (p. 58). It is not certain whether Seth's καφουρά derives from *Cinnamomum Camphora* Nees et Eberm. (so LSJ s.v.) or from *Dryobalanops aromatica* Gärtn.) the source of Borneo camphor. Cf. F.A. FLÜCKIGER and D. HANBURY, *Pharmacographia* (London, 1874) 458-464 for further details. Σάχαρ p. 96 (codd. σάχαριν) or sugar (from *Saccharum officinarum* L.) occurred in Galen XII, 71 and Dioscorides, *Euporista* I, 40; II 107, 112 Wellm. Cf. E.O. VON LIPPMANN, *Geschichte des Zuckers.* (Leipzig, 1890) 59-90 for classical references.

31. For example, J. SCHMIDT, «Das medicinisch-botanische Glossar von Siena». *Hermes* 18 (1883) 521-545 and the anonymous *Alphita* (in) S. de Renzi, *Collectio Salernitana* vol. III (Napoli, 1854) pp. 271-322. The glossaries in Goetz, *Corpus Glossariorum Latinorum* have been analyzed by J. ANDRÉ, *Notes de Lexicographie botanique grecque* (Paris, 1958).

32. Cf. H. STADLER, «Lateinische Pflanzennamen im Dioskorides». *Arch. für lateinische Lexikographie* 10 (1898) 83-115 and H. KÄSTNER, «Pseudo-Dioscoridis de herbis femininis». *Hermes* 31 (1896) 578-636.

33. For the origin of *unedo* < **uledone*, cf. G. ALESSIO, «Fitonimi Mediterranei». *Studi Etruschi* 15 (1941) 188.

34. Much valuable material has been assembled by R. STRÖMBERG, *Griechische Pflanzennamen* (Göteborg, 1940), pp. 152-161. But neither he nor G. ROHLFS, «Kretische Pflanzennamen». ΚΡΗΤΙΚΑ ΧΡΟΝΙΚΑ 6 (1953) 315-318 define what they mean by a folkname.

and post-linnaean folk names. By international consent, every plant name since the publication of Linnaeus' *Species Plantarum* (1753) possesses a unique binominal name composed of genus and species. In Antiquity, where no such rules obtained, the nearest approximation to an «official» or scientific name was the name generally adopted by physicians, hence the majority of those appearing in the plant lists or in a medical context. It must suffice here, in the absence of a definition of an ancient folk name of a plant, to provisionally characterize the term negatively - the name of a plant which does not appear in the plant lists or in a medical, scientific or agricultural treatise before the Byzantine period. Modern studies of Greek folk names must be used with caution in this respect because what is recorded as a folk name may be nothing more than a corrupt form of an old «official» or «officinal» name. For instance τὸ κυπαρέσσ' [35], like the demotic κυπαρίσσι are nothing but debased forms of κυπάρισσος, already known to Homer (*Od.* v, 64) for its scented wood.

Within this vague but broad area of folk names, two sub-classes can be distinguished, though not in all instances is it clear in which of the two a given name should be placed.

There are, first of all, dialectal variants, noted by the Byzantine glossators. For example, for the well known food and crop plant ἐρεβίνθος (*Cicer arietinum* L.), Hesychius records four synonyms: γάλινθοι, γέλινθοι, γέρινθοι and λέβινθοι, each of which is glossed simply ἐρέβινθοι. The similarity of these four forms, plus the fact that another dozen variants can be added from modern studies, lead to the conclusion that these forms occured in actual speech [36]. Unfortunately, Hesychius neglects to state where or among whom these forms were used. Elsewhere, however, he is more specific, for example, βουκανῆ· ἀνεμώνη τὸ ἄνθος. Κύπριοι or λάφνη· δάφνη. Περγαῖοι. Whether these latter names, and many more hapax legomena like them, were actually used, we have only Hesychius' testimony.

The same lack of external evidence concerning usage appears in the second of the two sub-classes noted earlier. This class is composed of what might be called «book names», that is, names invented for a specific occasion but lacking any evidence of usage. Names belonging to this class often appear in a series of so-called synonyms. For example, in the chapter *Herba Verbenaca*, of Pseudo-Apuleius' *Herbarius*, the section entitled *Nomina herbae* contains the following synonyms: *Dios elacate* (= ? διοσηλακάτη), *Persefonion* (= Περσεφόνιον), and *panchroma* (= πάγχρωμα), plus other synonyms allegedly of Egyptian and Italic origin and one,

35. E. Κούσης, «Λεξιλογιον φυτολογικὸν Τραπεζοῦντος». Ἀρχεῖον Πόντου 1 (1928) p. 109.

36. Cf. C. Fraas, *Synopsis Plantarum Florae Classicae*. 2. Aufl. (Berlin, 1870) p. 56; T. Heldreich, *Die Nutzpflanzen Griechenlands*. (Athens, 1862) p. 71; B. Langkavel, *Botanik der späteren Griechen* (Berlin, 1866) p. 3, and H.O. Lenz, *Botanik der alten Griechen und Römer* (Gotha, 1859) p. 723.

erisisceptron, ascribed to that shadowy figure Pythagoras[37]. What lies in back of these strange words in difficult to determine. And in the absence of any evidence other than that afforded by the probable Greek form appearing within parentheses, it would appear that these synonyms were coined for the occasion and possessed no life of their own outside their present context.

It is not surprising that a certain amount of confusion attended the proliferation of new names and synonyms[38]. A good example of this is the transfer of a name from one plant to another. Usually, common and widely used plants (as opposed to widely distributed plants) preserved a simple name from Hippocrates onwards and in some cases going back to Homer, e.g., ἴον, κίσσος, μήκων and φοῖνιξ. Other plants, however, were not so fortunate, for their name was later transferred to another, often unrelated plant. A good example is provided by the late synonymy ἄμωμον ἤτοι ξυλοκινάμωμον (Delatte XI, 387). The former name is old and well-attested, even though its identity is not absolutely certain[39]. The latter name, however, does not appear until Dioscorides (I, 14.3) where it is clear that ξυλοκινάμωμον is the inner woody part (ἔστι... ξυλῶδες κινάμωμον, cf. Pliny XII, 42,91: *ipsum vero lignum... xylocinnamomum vocatur*), not the outer scented bark of *Cinnamomum Cassia* Bl. which was used as a spice. The identity of κινάμωμον has never been in doubt, except perhaps to those who took seriously Herodotus' (III, 111) fable about its collection[40]. Given the alleged identity of the two products, which have nothing in common save an Eastern origin, it is difficult to resist the conclusion that the author of the synonymy was more adept at scholarship than at botany. It was, apparently, for much the same reason that another glossator equated two unrelated Eastern plants: ζαδοάρ˙ τὸ ἀνακάρδιον[41].

An analogous case, but one in which the reason therefor is now unknown, concerns the replacement of one name for another in a close paraphrase. For example, Dioscorides' description of ἄλισμα˙ οἱ δὲ δαμασώνιον καλοῦσιν, οἱ δὲ λύρον (III, 152) is repeated except for word order, by Oribasius: Ἄλκμαρ (οἱ δὲ δαμασώνιον, οἱ δὲ λύραν καλοῦσι) (XI, 85). Ἄλκμαρ is unknown outside of this passage and one can only speculate why it replaced the better known ἄλισμα. A clue, perhaps, is provided by Oribasius' unusual omission of the medical uses reported by Dioscorides under ἄλισμα (III, 152.2), but beyond that his reasons are lost.

37. ANTONII MUSAE, *De herba vettonica liber*, PSEUDO APULEI, *Herbarius...* edd. E. Howald et H.E. Sigerist (= *Corpus Medicorum Latinorum* IV). Leipzig/Berlin, 1927 p. 31.

38. JERRY STANNARD, «The Medieval Reception of Classical Plant Names» *Revue de Synthèse* (III) Nos. 49-52 (1968) 153-162.

39. Identified as *Amomum Cardamomum* L. by J. ANDRÉ, *Lexique des Termes de Botanique en Latin* (Paris, 1956) p. 28.

40. For the later influence of Herodotus' account, see J. HUBAUX et M. LEROY, «Vulgo nascetur amomum», *Mélanges Bidez* (Bruxelles, 1934) I, 514.

41. V. LUNDSTRÖM, «Ett 'persiskt' - grekiskt medico-botaniskt lexikonfragment». *Eranos* 12 (1912) p. 173.

Library of the Seminario Vescovile of Padua, cod. 194, fol. 10v: Ἀλικάκαβος (*Parietaria officinalis* L.). The codex is strictly connected with the Cod. Vindob. Med. Gr. I (*Constantinopolitanus*, so-called of Juliana Anicia), from which it was transcribed by the monk Neophytus about 1350 in Constantinople; it belonged to the danish Johannes Rhodius (1587-1659), graduated in Medicine at the University of Marburg, who in 1631 was prefect of the Botanical Garden of Padua and professor of Materia Medica in the same University. The Paduan codex (*Patavinus*) is particularly important because it completes its own archetype, as it contains the miniatures that were lost in the Juliana Anicia codex. On the other hand, it is the intermediate codex from which the Cod. Vat. Chig. 53 (XV century) comes from (See: E. Mioni, *Un ignoto Dioscoride miniato - Il codice greco 194 del Seminario di Padova*, in *Libri e stampatori in Padova - Miscellanea di studi storici in onore di mons. G. Bellini*, Padova 1959, pp. 345-376; Id., *Un nuovo erbario greco di Dioscoride*, in «Rass. Med.», 36, 169-184, 1959.

Still another case concerns the replacement of one name by another where the new name coexists with the older name but does not entirely displace it. An example is provided by μαρούλιον (and cognate forms.) Apparently it first appeared in Alexander (II, 61 Puschm.) and was soon sufficiently established so that it could be used in the *Geoponica* (I, 12,1 p. 347 Beckh) without explanation. As for its coexistence with the older name, Seth (p. 64 Langk.) notes τὰ μαρούλλια, ἃ καὶ θριδακίναι καλοῦνται. Yet it did not entirely replace θριδακίνη, for in his list of λάχανα Psellus still preferred to write θριδακίνη (p. 206 Ideler).

LIBRARY OF THE SEMINARIO VESCOVILE OF PADUA, cod. 194, fol. 98^{r}: Λιβυστικόν (*Ligusticum* L.). Is missing in the Juliana Anicia codex.

The final stage in the evolution of Byzantine synonyms was reached in a series of inter-related lexica, dating from the thirteenth century on, which were devoted to materia medica. From an examination of eleven such texts, certain features stand out [42]. The single, most outstanding characteristic is that lexicographical considerations have virtually displaced the botanical data. Most of the glosses are bare syno-

42. In addition to the lexica in DELATTE, *op. cit.*, pp. 381-454, see also, A. DELATTE, «Le lexique de botanique du Parisinus graecus 2419». *Bibl. de la Fac. de Philosophie et Lettres de l'Université de Liège* 44 (1930) 59-101; M.H. THOMSON, *Textes grecs*, pp. 125-177; and V. LUNDSTRÖM, *op. cit.*, pp. 170-174.

nyms and only infrequently do they contain descriptive data pertaining to the plant name of the lemma. Moreover, the preoccupation with lexical matters is carried to the point of recording synonyms from several languages, each preceded by an abbreviation: ε (ἑλληνικόν); φ (φραγγικόν or φράγκικα i.e., Italian); τ (τούρκικον or τουρκικα); κ (κοινόν); Λ (λατινικά or λατίνα); and Α (ἀράβικα). In one of the lexica, a few synonyms in Romaïc have been added, probably by a later hand[43].

LIBRARY OF THE SEMINARIO VESCOVILE OF PADUA, cod. 194, fol. 132v: Ποταμογείτων ἕτερος (*Potamogeton rufescens*). Is missing in the Juliana Anicia codex.

With the exception of one of the glossaries, drawn from Galen, and an expanded plant list in alphabetical form based on Aetius, they represent a broad spectrum of the sources available to a Byzantine physician[44]. Although many of the lemma-names can be traced back to earlier writers, the sources for some of them, and for many of the synonyms mentioned in the glosses, have not been located. The botanical value of those synonyms is somewhat diminished by the high num-

43. DELATTE, *op. cit.*, p. 427. The multilingual lexica are Delatte XII (pp. 393-417), XIII (pp. 418-427), and XIV (pp. 427-450).

44. Λέξεις βοτανῶν ἑρμηνεῖαι κατὰ ἀλφάβητον τοῦ σοφωτάτου Γαληνοῦ (= Delatte XI pp. 385-393); Ἕτερον περὶ βοτανῶν κατὰ στοιχεῖον ἐκ τοῦ Ἀετίου (= Thomson VI [*Textes grecs*, pp. 90-107]).

ber of transfers (see above p. 10) and the lack of a consistent transcription of Arabic names. Despite these shortcomings, the lexica are important documents for the history of later Greek and deserve to be more closely studied, not only by historians of botany, for whom they may hold some unexpected surprises, but by students of Byzantine culture in general[45].

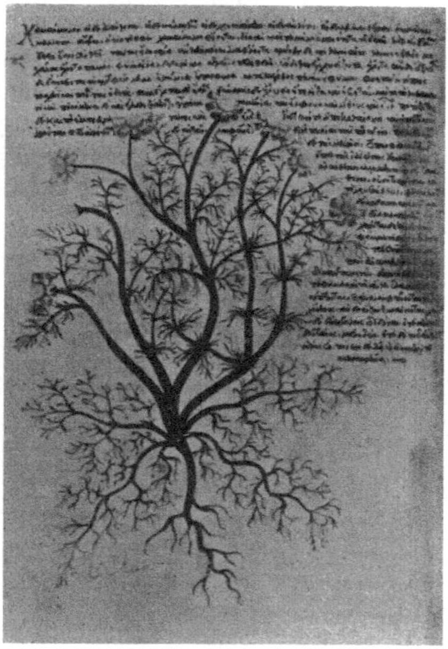

LIBRARY OF THE SEMINARIO VESCOVILE OF PADUA, cod. 194, fol. 177^{v}: Χαμαίμηλον (*Anthemis tinctoria* L.). Is missing in the Juliana Anicia codex.

II. *Medical uses*

In studying the development of Byzantine botanical lexicography, it is essential to bear in mind that the plant lists and lexica originally were designed to fulfill a medical need. What appears to us as a disproportionate amount of space devoted to non-medical matters reflects our modern notion of medical training and practice[46]. For the ancient physician, at least as regards therapy, most plants had some

45. A good starting point is provided by O. SCHISSEL, «Der byzantinische Garten». *Stzb. Wien. Akad., Ph-Hist Kl.* CCXXI/2 (1942) and M.H. THOMSON, ed. *Le Jardin Symbolique: Texte Grec tiré du Clarkianus XI*. Paris, 1960.

46. See JERRY STANNARD, «Materia Medica and Philosophic Theory in Aretaeus». *Sudhoffs Archiv f. Gesch. d. Med.* 48 (1964) 27-53.

healing property. With this as a premise, his most important task was to prescribe the proper plant for each ailment. Because there was no such thing as clinical or laboratory testing, effort was expended on obtaining the correct plant and on preparing and administering it in the most effective manner. This is, of course, an oversimplified account, but it will serve as a background for the following brief remarks on the medical matters insofar as they pertain to our present subject.

Essentially, four different kinds of medical information are contained in the plant lists, though usually the longer passages contain several or all of the following:

The simplest sort of entry is the statement that the plant in question possesses therapeutic properties similar to those of another plant. For example, τραγάκανθα παραπλησίαν τῷ κομμὶ δύναμιν ἔχει (Paul VII, 266) or Aetius' statement: Περὶ ἀράκων. Παραπλήσιόν ἐστι τὸ σπέρμα τοῦτο τῷ τῶν λαθύρων κατά τε τὴν χρῆσιν καὶ τὴν δύναμιν (I, 42).

LIBRARY OF THE SEMINARIO VESCOVILE OF PADUA, cod. 194, fol. 193v: Παιονία ἄρρην ε θήλεια (*Paeonia officinalis* L.). Is missing in the Juliana Anicia codex.

Not much more revealing, except to the physician versed in the Galenic theory of grades, is the statement that a plant possesses certain properties to a specified degree. Thus, Νᾶπυ θερμαίνει καὶ ξηραίνει κατὰ τὴν τετάρτην ἀπόστασιν (Paul, VII, 244) or Βλίτον ἐδώδιμόν ἐστι λάχανον ὑγρὸν καὶ ψυχρὸν τῇ κράσει ἐν τῇ δευτέρᾳ μάλιστα ἀποστάσει τεταγμένον (Aetius I, 47).

Of more immediate use is the listing of the diseases or complaints alleviated by the plant. E.g., μελάνθιον is indicated because ἀναιρεῖ... ἕλμινθας. λέπρας καὶ ἥλους καὶ μυρμηκίας ἐκβάλλει καὶ καταμήνια προτρέπει κτλ. (Aetius I, 108) while of κίσσος Paul states ὁ δὲ χυλὸς ἐρρινόν ἐστι φάρμακον καὶ τὰ χρόνια τῶν ἐν ὠσὶν ἰᾶται ῥεύματα· τὸ δὲ δάκρυον αὐτοῦ δριμύτερον ὑπάρχον φθεῖρας κτείνει καὶ τρίχας ψιλοῖ (VII, 225).

Most frequent, however, is a listing of the physiological properties of the plant: Καυκαλίς, οἱ δὲ δαῦκον ἄγριον· θερμαίνει τε γὰρ ὡς ὁ δαῦκος καὶ ξηραίνει καὶ διουρεῖται καὶ ταριχεύται εἰς ἀπόθεσιν (Paul VII, 222) or Ἀτρακτυλὶς ἢ κνῆκος ἄγριον, οἱ δὲ ἀνδρόσαιμον καλοῦσιν, ξηραντικῆς καὶ μετρίως θερμῆς ὑπάρχει δυνάμεως (Aetius I, 44).

It will be noticed that synonyms are recorded for the last two plants, for, as we have suggested above, the medical uses of the plant were never divorced from its name. And its name, except in the case of new names introduced in Byzantine times, was the means of locating its medical uses and its description in Dioscorides and Galen – the two major sources for later writers.

III. *Descriptive data*

We have seen in several of the glosses above that no descriptive data is contained such that one would be enabled to recognize the plant on the basis of that passage alone. If one did not know by some other means the name of a particular plant, the medical uses and synonyms contained in the gloss would be useless.

Accordingly, there are some glosses that contain descriptive data such that the recognition of a certain plant would be facilitated or that the plant being sought might be distinguished from other plants of a similar appearance.

Very rarely in ancient botany is there a description of a plant such that in the absence of other data (including linguistic), could that plant be identified. Ancient descriptions, which emphasized vegetative characters and gross morphology, cannot, of course, be compared with those of a modern flora in which the most important data concern the reproductive organs. On the other hand, the ancient physician possessed a store of information which, with but the slightest clue, could be brought to bear upon the plant in question.

One such clue was provided by the simple technique of comparing or likening one plant to another. For example, Aetius' account of σκάνδιξ begins with a comparison: τῶν ἀγρίων ἐστὶ λαχάνων παραπλήσιον τῷ γιγγιδίῳ (I, 132). Even in non-medico-botanical glossaries the same technique was employed as a convenient, if somewhat imprecise method of describing a plant. Thus Hesychius' gloss on μυόσωτον· πόα, ἐμφερὴς μυὸς ὠσί. Other kinds of information could be added to increase the possibility of recognition. Thus Aetius states Ὑδροπέπει· ἐν ὕδασι φύεται τὸ βοτάνιον ἐοικὸς τοῖς φύλλοις καὶ τοῖς κλωνίοις ἡδυόσμῳ (I, 142).

On the other hand, if the comparandum be uncommon or unknown, this technique loses its value. Judging from the rarity of silphium in Pliny's time (*Hist. nat.* XIX, 15,39-40). Oribasius' description seems of questionable value: ἀγαρικόν· ῥίζα φέρεται σιλφίῳ ἐμφερής... (XI, 80)[47]. Moreover, if the comparison is not obvious, the information serves little purpose. It is difficult, for example, to grasp the similarity adduced in a fourteenth century lexicon: γεράνιον ἔχει φύλλα ὅμοια ἀνεμώνης (the complete entry!, Delatte XI, 388).

Another kind of descriptive evidence is indicated by a determinative following or closely connected to the lemma. The habit of the plant is thus specified by being labeled πόα, θάμνος, δένδρον etc., for example, ὕσσωπον ἡ πόα· κτλ. (Aetius I, 142) or φοῖνιξ τὸ δένδρον as opposed to φοῖνιξ, οἱ δὲ ῥοῦν, πόα ἐστιν κτλ. (Paul VII, 270). The last example points up the fact that there were names which were ambiguous. Thus, Paul's lemma: Ὀθόνναν οἱ μέν φασι πόας εἶναι χύλον ἐν τῇ κατ' Αἴγυπτον Ἀραβίᾳ φυομένης, οἱ δὲ λίθον Αἰγύπτιον (VII, 246). A post-classical usage is indicated in a gloss on ὕσσωπον. It is, states the anonymous compiler ἡ βοτάνη καὶ ὁ ῥύπος τῶν ἐρίων τῶν προβάτων (Delatte, X, 384). Earlier, Hesychius used a similar technique in glossing words that denoted both a plant and some other object possessing the same name, cf. ἀρκτοῦρος· βοτάνης εἶδος. καὶ ἄστρον; βήρυλλος· λίθος. ἢ βοτάνης εἶδος and κορίς· ἰχθῦς ποιός. ἢ πόα, ἣν ἔνιοι χαμαίπιτυν. The use of a determinative did not always solve a problem of ambiguous usage, however, as the lemma λεύκη τὸ δένδρον (Aetius I, 103; Paul VII, 235) illustrates. Although there may have been two plants each named ἡ λεύκη there is no evidence now of this term, without a modifier, except as the name of the white poplar (*Populus alba* L.)[48].

Further descriptive data are sometimes supplied by a statement concerning a sensible property of the plant. Although the possession of this property was easily verified, it was also subjective and was probably of use only in confirming a provisional identification that the plant to be examined was the plant being sought. Phrases such as the following indicate not only the sort of information that was included but also its limitations. Σίνων· θερμὸς καὶ ὑπόπικρός ἐστι τὴν γεῦσιν,... (Aetius I, 132); Καρδάμου τὸ σπέρμα καυστικόν ἐστιν, ὥσπερ τὸ νᾶπυ (Paul, VII, 220) or Ἡδύσαρον ἢ πελεκῖνος· Πυρρὸν μὲν ἢ ὠχρὸν εὑρίσκεται τὴν χρόαν τὸ σπέρμα, ἀμφίστομον δὲ καθάπερ οἱ πελέκεις τῷ εἴδει (Aetius, I, 74).

A fourth sort of information contained in the descriptive portion of the gloss concerns geographical origin and habitat. Many ancient plant names were formed

47. Although genuine silphium had become extinct (A.C. ANDREWS, «The Silphium of the Ancients: A Lesson in Crop Control». *Isis* 33 (1941) 232-236), it was still listed as an ingredient in composita in a fourteenth century MS, cf. E. JEANSELME, «Sels médicamenteux et aromatiques pris par les byzantins au cours des repas». *Bull. Soc. franç. d'hist. de la med* 16 (1922) p. 326.

48. For other examples, see, J. ANDRÉ, «Confusions botaniques dans les textes des médecins et vétérinaires latins». *Latomus* 17 (1958) 488-492.

in a manner superficially similar to Linnaean binominals: A «generic» name (e.g., ἄμπελος, δάφνη, κύαμος, λωτός) plus an adjectival modifier. Our concern here is with only one sub-class of ancient binomials, viz. those in which the adjective denoted a country or a province. There are numerous instances of this kind, e.g., ἀβρότονον Σικελικόν (Orib. XI, 80), σίσων συριακός (Alex. Trall. II, 273), κίτριον, ὃ καὶ μῆλον Μηδικόν (Paul VII, 225)[49], σχοῖνος ὁ μὲν ἀραβικός, ὁ δὲ μηδικὸς λεγόμενος (Delatte XI, 392), κύαμος αἰγύπτιος (Alex. Trall. I, 401), δάφνη 'Αλεξανδρεία (Orib. XI, 98) etc.

Needless to say, the adjectives must not be pressed too far. But as it is not our purpose here to explore the lands referred to, it will suffice to say that the adjective called attention to the fact that the plant was believed to be indigenous in the country so designated and that a search for it locally would be futile. This is explicitly stated in another class of plant names where the geographical reference is not a part of the name, but indicates rather where a specific plant or product was prepared. E.g., σαρκόκολλα... δάκρυον δένδρου περσικοῦ (Aetius I, 129); καρυόφυλλον... ἐκ τῆς 'Ινδίας κομιζόμενον (Delatte XI, 390); ἀνακάρδια· καρπὸς ἰνδικός (Thomson IX, 134); 'Ιασμή· 'Η δὲ καλουμένη ἰασμὴ παρὰ Πέρσαις σκευάζεται ἐκ τῶν ἀνθῶν τῶν λευκῶν τοῦ ἴου καὶ ἐλαίου σησαμίνου (Aetius I, 61).

A reference to the habitat of the plant, which indicated its ecological niche, meant that the plant was available locally and might be found in the designated station or stations. For instance, 'Ελξίνη, οἱ δὲ σιδηρῖτιν, οἱ δ' 'Ηρακλείαν, οἱ δ' ἡλῖτιν, φύεται ἐν φραγμοῖς καὶ ἀμπελῶσι καὶ σίτῳ (Orib. XI, 101) or Πέπλιον ὀπὸν ἔχει καὶ τοῦτο καὶ φύεται τὰ πολλὰ παρὰ θαλάσσῃ (Paul VII, 250).

Data of the kind mentioned above supplemented the information inherent in the synonym or synonyms and naturally varied from author to author. The framework of the gloss was sufficiently flexible to permit the inclusion of whatever information seemed useful. For example, Paul incorporates a surprisingly large amount of information in his short account of κάγχαμον. It is, he states, δάκρυον ξύλου 'Αραβικοῦ σμύρνῃ προσεοικὸς εὐῶδες, ᾧ καὶ θυμιάματι χρῶνται and then continues with its medical uses (VII, 219).

IV. *Etymology*

Remarks concerning the etymology of a plant name are not plentiful. They occur sufficiently often, however, as to suggest that such information was considered useful and worthy of communication[50].

49. Hesychius' gloss κιτρίον· τὸ 'Ινδικὸν μῆλον, which may reflect the close commercial contacts between Alexandria and India, may also mean nothing more than «a fruit from an unknown Eastern source»; either way, it illustrates the elusiveness of geographical epithets as indicators of a plant's nativity.

50. E.g., Photius' gloss on 'Αδίαντον· ξηρόν, ἄβροχον, ἢ φυτὸν παρὰ ὕδασι φυόμενον τὸ καλούμενον πολύτριχον.

The purpose of an etymology, like that of the descriptive data, was to facilitate the recognition of a plant regarded as therapeutically useful. Thus Aetius explains the name βούφθαλμον in such a fashion that the plant could be recognized and collected. It is, he states, ὠνόμασται μὲν οὕτως ἀπὸ τῶν ἀνθῶν ἐοικέναι δοκούντων κατὰ τὸ σχῆμα βοὸς ὀφθαλμῶν (I, 47). Perhaps not quite so obvious is the explanation found under ἡλιοτρόπιον τὸ μέγα, ὃ ἔνιοι σκορπίουρον ἀπὸ τοῦ περὶ τὸ ἄνθος σχήματος ἐκάλεσαν, ἡλιοτρόπιον δ' ἐκ τοῦ συμπεριτρέπεσθαι τὰ φύλλα τῇ τοῦ ἡλίου κλίσει (Orib. XI, 104).

An etymology also served a second purpose, viz., to indicate either its medical use or one of its properties for which it was valued by the physician. This is clearly brought out by Paul's account of παρωνυχία· ὠνόμασται... ἐκ τοῦ παρωνυχίας ἰᾶσθαι (VII, 250). The medical use of a plant and a descriptive datum sometimes are both included in an etymology, e.g., θηλύπτερον, ἣν ἔνιοι νυμφαίαν πτέριν ὀνομάζουσιν, παρεοικυῖα τῇ πτέριδι καὶ κατὰ τὴν δύναμιν (Paul VII, 216), or γλαῦξ ἡ πόα γάλακτός ἐστιν γεννητική (Delatte, X, 388) which seems to depend upon the false etymology γάλα < γλαύξ.

It is more difficult to see the practical usefulness of some etymological explanations, for instnace, Paul's account of κάπνιος· οἱ δὲ καπνός, ὠνομάσται δὲ ἐκ τοῦ τὸν χυλὸν αὐτοῦ πολὺ δάκρυον ἐπισπᾶσθαι (VII, 220). At best, we can only say that such information must have been found useful in Byzantine medical circles.

V. *Economic uses*

There is, finally, one other type of information that is occasionally encountered and for reasons similar to those given above. A statement regarding the economic use of a plant, in the absence of other information, may have aided the recognition, collection, and perhaps even the medical employment of the plant. It is conceivable, for example, that the foliage and blossoms of ἐρυθρόδανον (madder or *Rubia tinctorum* L.) might not be recognized on sight. But the statement that its root was used as a dye could be correlated with other information about that plant. Hence, Aetius begins his account of ἐρυθρόδανον by stating ταύτης ἡ ῥίζα ἐρυθρά, ᾗ χρῶνται οἱ βαφεῖς (I, 73). In a more abbreviated form the same information is transmitted in a late lexicon where the economic use of madder has almost assumed the status of a synonym: Ἐρυθρόδανον ἤτοι ῥιζάριν τῶν βαφέων (Delatte XI, 388)[51].

In a similar fashion, a practical use of σπάρτον (*Macrochloa tenacissima* (L.) Kunth.) precedes its medicinal uses: σπάρτου, ᾧ παρ' ἡμῖν τὰς ἀμπέλους δεσ-

51. Already in the sixth century, Alexander directed, as an ingredient in a recipe θαψίας (sc. δραχ. β'), ᾗτινι οἱ βαφεῖς χρῶνται, ἣν οἱ Ῥωμαῖοι ἐρβαρυβίαν καλοῦσι (I, 457 Puschm).

μεύουσιν... (Aetius I, 134)[52]. Quite a different use is described in Hesychius' gloss to βράθυ· πόα τις θεοῖς θυομένη. And while this may have been meaningful to some, πόα is scarcely the right determinative to use to describe *Juniperus Sabina* L.

It must be stressed, in conclusion, that the evidence put forward above represents only a small fraction of a long and complex tradition which continued even after the fall of Constantinople[53]. Much work remains to be done in this subject, especially by the incorporation of medico-botanical materials, previously ignored by most scholars, within more general studies on the history of Greek lexicography. It is hoped, therefore, that the present study will serve as a stepping stone for further work in an area where classical antiquity and medieval Byzantium overlapped.

52. A further example of how an economic use may help in the identification of a plant, cf. Eustathius, *Comm. in Iliadem* ed. Rom. 744 (vol. II p. 356). Commenting on Homer's δόναξ and μυρίκη (K 467), Eustathius adds: Δόνακες δὲ εἶδος καλάμου... δῆλον δὲ ὡς οἱ δόνακες καλάμων μὲν λεπτότεροί εἰσιν, ὀρόφων δὲ ἁδρότεροι, καὶ δοκοῦσιν ἐκ δονάκων μὲν σύριγγες γίνεσθαι, αὐλοὶ δὲ ἐκ καλάμων.

53. For example, Stephanos Magnetes' (*saec.* XI) *Alphabetum empiricum, sive, Dioscoridis et Stephani Atheniensis... De remediis expertis liber...* [Zürich:] Froschauer, 1581 was translated by Caspar Wolff on the basis of a Greek MS belonging to Conrad Gesner. The Greek text still remains unedited as does Nicolaus Myrepsos' Δυναμερόν, two Latin translations of which were printed in the mid-sixteenth century (cf. L. Choulant, *Handbuch der Bücherkunde für die ältere Medicin*, Leipzig, 1841 pp. 156-57). On Nicolaus' influence and a specimen of the Greek text, see H. Lehmann, «Zu Nicolaus Myrepsus». *Sudhoffs Arch. f. Gesch. d. Med.* 17 (1925) 299-306 and E. Jeanselme et L. Oeconomos, «Un Dispensaire médical à Byzance au temps des Paléologues d'après une peinture du MS Grec Paris. 2243» *Aesculape* 15 (1925) 26-30.

IX

ASPECTS OF BYZANTINE MATERIA MEDICA*

Byzantine materia medica, although not coterminous with Byzantine medicine in general, was certainly one of its more prominent subdivisions. Many different kinds of texts, varying greatly in size, date, style and credibility were concerned, wholly or partially, with medicaments as the primary means of promoting health. The heterogeneous nature of those texts is such that some have been ignored by students of Byzantine medicine while others would scarcely be recognized as practical medical texts by physicians and pharmacists today. For that reason, it may be convenient to examine some of the major components of Byzantine materia medica and to determine their relations, one with another, within the wider context of Byzantine medicine.

Prior to our discussion of some of the characteristic features of Byzantine materia medica, a few remarks on sources and their utilization are required. These remarks, however, are not those of the *Quellenforscher*, bent on establishing a textual tradition or the reconstruction of a lost *Urtext*. Rather, they are designed to call attention to one of the most pervasive characteristics of Byzantine materia medica, viz., the reliance of Byzantine writers on earlier written sources and the frequency with which the same authorities and their claims were copied, recopied, paraphrased and excerpted century after century.[1]

In Byzantine texts on materia medica a reference to a written source, especially in the form of a personal name, for example, Hippocrates or Galen, was not the act of historical scholarship that today is associated with a learned footnote. Such a reference, it is true, sometimes furnishes a useful clue to the sources used by late Byzantine writers. But sometimes it appears as if a citation to an earlier writer was a ritualistic performance reflecting, in part, the traditionalism inherent in Byzantine medicine.

Hippocrates and Homer, as one would expect, were almost venerated by medical writers and, in the case of Hippocrates, for good reason. Leaving aside the question concerning the authorship of the texts traditionally ascribed to Hippocrates, there is little doubt that the observations contained in some of those texts provided valuable insights into the nature of disease and hence guidelines for therapy. On the other hand, citing the equally magical name of Homer was of questionable value with respect to therapeutic practice. For similar reasons references to Orpheus, Democritus, Poseidonius, and others were more a show of learning than an indication of how to minister to the sick.

References to the earlier, well-known Greek and Greco-Roman physicians are more readily understood. This applies especially to Dioscorides and Galen, but to others as well. Some of their writings, not merely extracts available in a florilegium, must have been readily available for consultation judging by the frequency of citation and the fidelity with which those passages agree with the original.[2] Even earlier Byzantine medical writers are sometimes cited by their successors, for example, Oribasius, Theophilus, Aitios, John Myrepsus and Jacob Psychrestus are all mentioned, favorably one might add, by later compilers.

In addition to citing sources and authorities by name, there are also references to anonymous oral

*I am grateful to Miss Mary Kay, Head, Reference Department, University of Kansas Libraries, and her staff for assistance in obtaining some of the materials necessary for the completion of this study.

[1]As an example, see the parallel passages assembled by A. Sideras, "Aetius und Oribasius," *BZ*, 67 (1974), 110–30.

[2]See J. M. Riddle, "Pseudo-Dioscorides' *Ex herbis femininis* and Early Medieval Medical Botany," *JHB*, 14 (1981), 43–81.

sources,[3] rustics,[4] books whose titles can no longer be ascertained,[5] and the vague references to *hoi Indoi, hoi Persai, hoi palaioi*, and the like. Finally there is a large but elusive class of references, typified by such phrases as *hos tines phasi* and *hos legetai*. Whether the passages introduced by such phrases are genuine references to information current at the time or mere literary formulae requires further research.

Last is a group of pseudonymous texts. The attribution of a text, often rather small, to an identifiable figure was a well known literary device in both the East and the West. Hence there are several texts on materia medica that were attributed to Hippocrates, Galen, Dioscorides, Aitios, Simeon Seth and others.[6] None of these is now accepted as genuine.[7]

It is important, then, to recognize the dependence of Byzantine writers upon their predecessors. This applies particularly to descriptions of plants and plant substances which, in fact, make up the bulk of the composita. As the writings of Aitios and Paul indicate, they often repeated, sometimes with only the slightest verbal modifications, descriptions of simplicia drawn from Galen's abridgments of Dioscorides or from the latter's *De materia medica*.[8] There is little evidence in such descriptions that the Byzantine compilers possessed any additional information, with the result that the contemporary reader of Aitios or Paul was reading an account already over 500 years old. While plant species do not normally change very much in so short a period, their geographical distribution, and hence market availability, may have changed dramatically. Silphium is a good example.[9] According to Pliny the Elder, it was already rare in the fir century A.D.[10] Yet Leon and Alexander include as an ingredient in composita without any hint th it might be rare or even unprocurable.[11]

With this by way of background, it is time to ex amine five other prominent aspects of Byzantin materia medica.

A convenient starting point is the recipe litera ture. For this is, in terms of size, one of the large single components of the literature on materi medica.[12] By that very fact, it reveals the attitud towards drugs and the reliance placed upon them.[1

I. Recipes

Recipes played the same role in Byzantine med icine as they do in modern folk medicine.[14] The were, so to speak, the distillate of traditional wis dom. They were the tangible evidence that dis eases and other complaints were capable of bein cured, for, after all, what other meaning can b placed on such phrases, usually at the conclusio of a recipe, as: "this has been tried," "he will b cured," or "you will be amazed." Other phrases o similar intent perhaps exaggerated the efficacy o the remedy.

The sheer mass of recipes, whether or not ac companied by magical devices,[15] gave the impres

[3] ἐγὼ γοῦν οἶδά τινα, ὃς . . . ἔφασκε Alexander (ed. Puschmann), II, 485.

[4] Alexander identified his sources for useful medical information as παρ' ἀγροίκου (II, 563 Puschm.) and παρὰ Κερκυραίου ἀγεοίκου (II, 565 Puschm.) Dialectal forms of plant names re likewise credited παρὰ τοῖς ἰδιώταις (Delatte, *Anecdota*, II, 83.13; 300.14; cf. also 345.12 . . . οἱ ἰδιῶται καλοῦσι).

[5] The source for the Mousarion collyrium is described by Alxander (II, 15 Puschm.) ἐκ τοῦ ἱερατικοῦ τόμου.

[6] As an example of such pseudonymous texts, see H. Schöne, d., "Hippokrates Π. φαρμακῶν," *RhM*, 73 (1920), 434–48.

[7] See Delatte, *Anecdota*, II, pp. 339, 385, 456, 466.

[8] *De simplicium medicamentorum temperamentis ac facultatibus*, in Galen (ed. Kühn), XI, 379–892; XII, 1–377.

[9] Asafoetida, the congealed sap of silphium, was also known under other names, for example, λάσαρον καὶ λάσαρ ὁ ὀπὸς οὗ σιλφίου, Delatte, *Anecdota*, II, 290.7. It was probably obained from one or both of two closely related species, *Ferula narthex* Boiss. and *Scorodosma foetidum* Bunge. See, Vladimir Vientiev, "Le Silphium," *BIE*, 37 (1954), 123–50.

[10] *NH* 19.15.39.

[11] Leon Philosophus, *Conspectus medicinae* I, 8 (in Ermerins, 97); Alexander, I, 407 Puschm.

[12] E. Jeanselme, "Sels médicamenteux et aromatiques *BullSocFrançHistMédical*, 16 (1922), 324–34; *id.*, "Sur un aid mémoire de thérapeutique byzantine," *Mél. Ch. Diehl*, I (Pari 1930), 147–70; A. P. Kousis, "Quelques considérations sur l traductions en grec des oeuvres médicales . . . par Constant Melitiniotis," Πρακτ.Ἀκαδ.Ἀθ., 14 (1939), 205–20; *id.*, Ἐκ τῶ Μητροδώρας περὶ τῶν γυναικείων παθῶν τῆς μήτρας, Πρακτ Ἀκαδ.Ἀθ., 20 (1945), 46–68; Emile Legrand, "Formulair Médical de Jean Staphidas," in his *Bibliothèque Grecque Vulgair* II (Paris, 1881), 1–27. Miscellaneous recipes are noted in Ch Daremberg, "Notices et extraits des manuscrits médicaux grecs *AMSL*, 2 (1851), 24–37. Because of a Byzantine substratum i late Coptic *Rezeptliteratur*, cf. Walter Till, "Koptische Rezepte *BSACopt*, 12 (1949), 43–54 and *id.*, *Die Arzneikunde der Kopte* (Berlin, 1951). For Western medieval analogues, cf. J. Stannar "Rezeptliteratur als Fachliteratur," *Scripta* (Brussels), 6 (1982 59–73.

[13] For useful supplementary references to the use of and valu placed upon drugs and drug therapy at the popular level, se J. H. Magoulias, "The Lives of the Saints as Sources of Data fo the History of Byzantine Medicine in the Sixth and Sevent Centuries," *BZ*, 57 (1964), 127–50.

[14] J. Stannard, "Albertus Magnus and Medieval Herbalism, in James Weisheipl, ed., *Albertus Magnus and the Sciences* (To ronto, 1980) 355–77.

[15] Cf. the series of "Recettes magiques" edited by A. Delatte *Anecdota*, I, *passim*.

sion that, provided the enumerated ingredients were available, prepared and administered in the manner stipulated, there was, ready at hand, one or more means of combating a disease. The plurality of recipes for the same complaint, often in succession, seemed to demonstrate that there was a variety of procedures, all equally efficacious. The conclusion that *we* would reach, viz. that such a sequence was a virtual admission that none was predictably reliable, was apparently not drawn. The usefulness of these recipes, however, was not confined to curing diseases or healing wounds. They also provided one with the means of averting misfortunes, prognosticating the future, and a wide range of techniques applicable to daily life, either at home or when traveling abroad.[16]

Because of the wide range of processes and events that could be corrected or controlled by following a recipe, it is impossible to even guess at their number. They occur in profusion, as one might expect, in the general medical treatises, from Oribasius to Actuarios. In addition they occur, though with a lesser frequency, in lexica, herbals, and dietetic texts. Finally, of course, there were the recipe collections themselves. But even here, there is considerable variety in the individual recipes. Many were straightforward statements of what to prepare and how to administer it for a specific medical problem, for example, gout, fever, headache, and the like. Others, however, were more akin to what we would call *secreta*, that is, pieces of useful information, not generally known, relating to the solution of mundane problems, for example, how to rid the house of mice,[17] how to disguise gray hair,[18] or the means of exorcising demons[19] or predicting the sex of an unborn child.[20]

II. Lexical

A second aspect of Byzantine materia medica, prominent in itself but also related to the authoritarianism noted earlier, is a concern with nomenclature and other lexicographical issues.[21] That concern with words was not restricted, however, to glossaries, many of which have been published in the past half century.[22] The same attention to orthography, etymology, and synonymies is also evident in the recipe literature and in the descriptions of simplicia that are scattered throughout the several medical texts.

A concern with words might appear, at first glance, somewhat sterile from a medical point of view. But, despite the inevitable excesses of some authors, such a concern was meant to serve a practical purpose. To what extent it did so, however, is difficult to judge in the absence of evidence concerning the use of those lexica by physicians and apothecaries in the course of their daily professional work.[23] But the time and effort expended in preparing lexica, especially the multilingual ones, was probably justified.

Classical, that is, Attic Greek, the *koine*, and some vernacular dialects came together in Byzantium. Over the centuries that polyglot was infiltrated by many non-Greek loan words, some of which were later accepted and functioned as if they were truly Greek.[24]

As a result of this linguistic admixture, various problems arose concerning the vocabulary of materia medica. As older words lost currency and were

[16] In addition to Delatte, *Anecdota*, II (cf. note 4 above), see H. J. Magoulias, "The Lives of Byzantine Saints as Sources of Data for the History of Magic in the Sixth and Seventh Centuries A.D.," *Byzantion*, 37 (1968), 228–69. Means of averting the potential dangers of traveling are indicated in numerous ways. For example, see G. Dagron et J. Rougé, "Trois horoscopes de voyages en mer," *REB*, 40 (1982), 117–33.

[17] Margaret H. Thomson, ed., *Textes grecs inédits relatifs aux plantes* (Paris, 1955), p. 111.9–19.

[18] Kousis, "Quelques considérations" (note 12 above), 213.1–. Three recipes for darkening the hair are given by Alexander, , 453 Puschm

[19] Sp. Kabasilas, Λαογραφικὰ σύλλεκτα, Λαογραφία, 2 (1911), 45–54; Delatte, *Anecdota*, I, 111.1–18. For a modern parallel, cf. L. Arnaud "La baskania ou le mauvais oeil chez les grecs modernes. II. Exorcismes quérisseurs," *EO*, 15 (1912), 518.

[20] E. Legrand, *Bibliothèque* (note 12 above), II (1881), 16.452–5; A. Delatte, *Anecdota*, I, 451.19–22.

[21] Cf. J. Stannard, "Byzantine Botantical Lexicography," *Episteme*, 5 (1971), 168–87.

[22] A. Delatte, "Le Lexique de botanique du Parisinus graecus 2419," *BiblFacPhilosLettLiège*, 44 (1930), 59–101; *id.*, *Anecdota*, II, 273–454 [15 lexica]; Margaret H. Thomson, *Textes grecs* (note 17 above), 125–77 [3 lexica]. Much earlier, Vilh. Lundström published two other lexica: "Botaniska lexika från den grekiska medeltiden," *Göteborgs Högskolas Årsskrift*, 15 (1909), 42–52; *id.*, "Ett persiskt-grekiskt medico-botaniskt lexikonfragment," *Eranos*, 12 (1912), 170–74.

[23] It is tempting to think that the ἐπιστήκων or the πημεντάριοι attached to the Pantokrator Hospital had access to a receptarium in order to prepare the many different compounds for patients of both sexes who suffered from a variety of illnesses and wounds. Cf. P. Gautier, ed., "Le typikon du Christ sauveur Pantocrator," *REB*, 32 (1974), 1–145.

[24] For example, ζουλάπιον (Arab. *julab*), used by Planudes, *De morborum materie*, 30 (Ideler, II, p. 321.9). For similar reasons, μαατζοῦν (Arab. *matzun*) i.e., an electuary, "encore usitée aujourd'hui chez le peuple grec," A. Kousis, "Quelques considérations" (note 12 above), p. 217. Cf. also πημεντάριος (Lat. *pigmentarius*, note 23 above. For further examples, see G. Meyer, "Die lateinischen Lehnworte im Neugriechischen," *SBWien*, Phil.-hist. K., 132/3 (1895), 1–84.

supplemented, then supplanted, by newer words, lexica were required in order to stay abreast of medical terminology, and, at the same time, to read with understanding the older authorities, especially Dioscorides and Galen.

The perplexity created by the fluid vocabulary of materia medica can be illustrated in many ways—for example, the profusion of synonyms, Greek and non-Greek alike,[25] for a single substance, scribal errors based on similarities of sound and/or spelling of words denoting medical substances, and the fact that, even today, the denotata of some of those terms is an educated guess.[26]

The ambiguity that resulted was not a mere academic challenge, for it may well have made a difference whether, for example, *ammoniakon* was thought to mean a mineral or a gum resin,[27] or whether *peristereon* meant a bird or a plant,[28] or *chelidonion* a bird, a mineral *or* a plant.[29]

Under these circumstances and in the absence of a standardized nomenclature, many physicians prepared lexica for their private use. These must be distinguished from the lexica prepared by literary scholars and professional glossators such as Photios and Hesychios. For while the former are grossly inferior in terms of style, and all of them much smaller, they have the merit of being restricted, for the most part, to the vocabulary of medicine and allied disciplines. This should not be taken to mean, however, any significant advance in medical lexicography. For, despite the praisewo thy purpose for which these smaller lexica we compiled, they are studded with errors of one kir or another. The most important, for our pu pose—and I think I also speak for the namele patients of those physicians—is the high numb of incorrect synonyms for plants, portions of whic served as ingredients in literally hundreds of con posita. It would be uncharitable to catalogue tho incorrect synonyms here, but it is worth emphasi ing if only to call attention to the possibility of s rious consequences when the names of medicin substances were misunderstood.

III. Medicinal Substances

It is but a short step from the words used to d note medicinal substances to an examination of th substances themselves. The vast majority d scribed in Byzantine texts on materia medica a having therapeutic properties can be identified wit some assurance.[30] Of that number many are als known today, though not always as medicinal sub stances. Some of them, garlic and parsley for ex ample, are used as seasoning agents. Others, fo example thyme and rue, are grown in modern her gardens, while others, such as the lily or daffodi are known as ornamentals. Still others, such a frankincense and myrrh, are used for liturgica purposes much as they were in Byzantium. Finally of course, many of the substances listed as ingre dients in Byzantine recipes are still employed a primary foodstuffs: tuna, goose, wheat and barle cheese, bread, and wine. Moreover, nearly ever one of the substances mentioned in our texts is als used in modern folk medicine, and often for sim ilar purposes. Most of the substances found in Byz antine materia medica, however, have long sinc been abandoned in modern *Schulmedizin*.

For a variety of reasons—uncertainty concern ing the precise identification of some substances the ambiguities of the lexical synonymies and, above all, the absence in antiquity of any chemical mean of describing mineral substances—it is idle to spec ulate on the number of therapeutic substances re ferred to in our texts. But even if certainty were reached concerning their identification, it is un

[25] Three of the lexica edited by Delatte (*Anecdota*, II, pp. 393–454) contain synonyms in Latin, Italian, Turkish and Arabic.

[26] A selection of neologisms and rare words pertaining to Byzantine materia medica will be found in Jeanselme, "Un aide-memoire" (note 12 above), *passim*.

[27] Ἀμμωνιακόν was a gum resin derived from *Dorema ammoniacum* D. Don. Used medicinally for its pungent odor, the fumes were inhaled; hence the frequent gloss θυμίαμα. Ἅλς ἀμμωνιακός, known to Byzantine glossators as ἅλας ἀμμωνιακόν, was a natural, crystalline, mineral substance whose composition probably varied depending upon the water-soluble impurities. Because the two substances were not always distinguished, a post-classical name, λεοντόγαλα, was coined (Delatte, *Anecdota*, II, 306.12).

[28] Περιστερεών, sometimes περιστερά in the lexica, usually meant a plant, probably *Verbena* sp., hence the synonymy ἱερὰ βοτάνη. The form περιστερεὼν ὄρθιος or ὀρθός was presumably an attempt to distinguish the plant from the common dove. The latter occurs rarely as a medicament and not at all in the medical lexica. Dioscorides' effort to explain the name (IV, 59 Wellmann) was abbreviated by Galen, XII, 98 K. For further details, cf. J. Stannard, "Magiferous Plants and Magic in Medieval Materia Medica," *Maryland Historian*, 8 (1977), 33–46.

[29] Due to the similarity between χελιδών, χελιδόνιον and χελιδονία these terms became confused, as they had in Latin. In fact, the confusion in late Greek reflects the Latin, cf. χελιδόνιον τὸ μικρόν· κελιδώνια μηνόρε and χ. τὸ μέγα· κ. μαϊόρε, Delatte, *Anecdota*, II, 416.16–17.

[30] Cf. B. Langkavel, *Botanik der späteren Griechen vom dritten b driezehnten Jahrhunterte* (Berlin, 1866). (Langkavel's study mus be used with caution). For supplementary data and moder scientific nomenclature, cf. J. Stannard, "Identification of th Plants Described by Albertus Magnus, *De vegetabilibus*, Lib. VI, *Res Publica Litterarum*, 2 (1979), 281–318.

ikely that all of them would have been available at ny one time in a given place.

It is more compendious, therefore, to divide the nedicinal substances into five large groups and rovide a few illustrations of each.

i) Substances of Plant Origin

Since several examples of plant substances that vere used therapeutically have been noted above, t will suffice to make but a few general observaions. Approximately 450 different species of plants an be identified in the materia medica. This inludes all the major taxa, including mosses, ferns, lgae and even fungi. The greatest emphasis, hower, was on the seed-bearing plants whose seeds, oliage and roots accounted for most of the botancals whether used as medicaments, as seasoning gents, or as primary foods. In many instances it ppears that a pronounced aroma or taste was the nost important rationale for its use, unless, of ourse, the older authorities could be cited as havng used that species for a specific purpose. But it as the exotica, transshipped via the Near East, that ominated the recipes—cinnamon, pepper, giner, cloves and sugar. These, the so-called spices, vere usually available only in the dried commercial orm. Various passages suggest that their quality aried considerably. This may have been due to olitical and economic events which periodically isrupted trade. Despite the efforts of the comnercial guilds, uniform quality was probably not lways achieved.

i) Substances of Animal Origin

As in the case of plants, some portion of most of he common animals—sheep, goat, ass, for exmple—was used for some medicinal purposes. This ncluded not only mammals but birds, fish, reptiles nd amphibia. But the use of animal products exended well beyond the vertebrata. A wide range f insects, arachnids, crustacea, molluscs, and other hyla were also employed.[31] By and large, most of he animal substances long ago disappeared from nedical practice, and one suspects that part of the eason for this was that while it may have been ifficult to authenticate, it was all too easy to falsify rocodile dung, hyena gall, rabbit brain, and the ke.

(iii) Substances of Mineral Origin

It is even more difficult to summarize the therapeutic uses of mineral substances for, in many cases, their descriptions permit only a tentative identification. Depending upon the impurities, inorganic as well as organic, plus physical processes such as weathering, what passed for *nitron* or *adarke* in one region may have been quite different with respec to color, odor, texture and weight in another region.[32] It was their physical properties and physiological action that determined the medicinal uses, particularly the water-insoluable minerals and metallic ores. Nonetheless, one can specify a few of the more frequently used mineral substances. Leaving aside coral and amber, whose animal and plant origin respectively were apparently unknown, this includes sulfur, various iron, lead, copper and arsenic compounds, argillaceous clays, and a host of precious and semi-precious gem stones whose virtues were enumerated in lapidaries.[33]

(iv) Praeparata

The fourth class, praeparata, can be dismissed more quickly, but not because those substances were unimportant. It is rather the case that the items falling within this class—wine, vinegar, olive oil, butter, barley meal, cheese, bread and the like—were, and are, household items known to all.[34] Each of the aforementioned was used for therapeutic purposes, either singly or as an ingredient in composita. To some of them specific properties were attributed. Butter, for example, possessed, according to Paul, a digestive and a dispersive or discutient property. For that reason it was useful for buboes and other swellings.[35]

(v) Composita

The last class, composita, can be dealt with even more quickly, for there is no simple way of discussing several hundred compound drugs in one paragraph. Suffice it to say our texts provide accounts

[31] J. Théodoridès, "Intérêt scientifique des miniatures zooloiques d'un manuscrit byzantin," *Acta Biologica Debrecina*, 7/8 969–70), 265–72, 8 figs.

[32] Cf. M. Japhet, "Etude sur les principales eaux minérales de l'Asie mineur," *AnnSocHydrolClimatolMéd de Paris*, 23 (1877–78), 316–80.

[33] Cf. M. Psellos, *De lapidibus*, in Ideler, I, 244–47. (The recent edition by P. Galiani, *Michele Psello, De lapidum virtutibus* [Firenze, 1979] was not available to me.) Extracts from the Ps.-Hippocratic Ἑρμηνεία περὶ ἐνεργῶν λιθῶν are printed by Legrand, *Bibliothèque*, II, (1881), p. xxiii.

[34] E. Jeanselme et L. Oeconomos, "Aliments et recettes culinaires des Byzantins," *ProcInternCongrHistMed, London, 1922*, (Anvers, 1923), 155–68.

[35] Paul (ed. Heiberg [*CMG* IX, 1] I, p. 201).

of how to prepare and administer them, possible modifications, admissible substitutions, and the purpose or purposes served by each. Not infrequently this is accompanied by remarks concerning dosage and potential side effects. Many of those composita, incidentally, bore specific names, for example the Collyrium of Constantine or the Salt of St. Gregory the Theologian.[36] This association of a drug with the name of the person who allegedly introduced it or used it with conspicuous success was a convenient tag for physician and patient alike and acted occasionally, no doubt, as a testimonial.

IV. Dietetics

The fourth aspect concerns dietetics. Throughout antiquity and the Middle Ages, there was an intimate relation between materia medica and dietetics. This was due partially to the dangers inherent in one of the most obvious of alternatives to drug therapy, surgery. Emphasis was thus often placed on the adoption of a regimen, of which diet was only one, albeit a conspicuous, part. But the close relations between materia medica and dietetics were further strengthened by the fact that the same substance—depending, of course, on how it was prepared, administered, and the amount—might function sometimes as a foodstuff or as a seasoning agent, at other times as a medicament. Herbs are good examples, but so too are some of the vegetables, for example onion, beet, or cabbage. Even to fruits and nuts there were attributed specific therapeutic properties. But pride of place went to barley. In the form of a thin soup or a thick gruel flavored with various ingredients, a barley tisane was, one is tempted to say, standard bill of fare.[37]

Because of the emphasis placed on dietetic advice as a means of curing specific diseases as well as averting future complaints, it is not surprising that a sizable literature developed.[38] The writings falling within this class vary considerably, from Anthimus' almost illiterate letter to the Emperor Theodoric[39] to Simeon Seth's painfully scholastic analysis of nearly 200 items, many of which we part of everyday living.[40]

There is little need at this point to summari the substances discussed by the medico-culina writers, for there is a high correspondence b tween substances eaten for nourishment and tho eaten to correct an imbalance or cure a disease. the process of describing these substances ar enumerating their virtues, attention was som times directed to the proper season to eat nuts, f instance, or certain species of fish or fowl. Wh reduced to rules, that information could be co nected with calendaric and astrological data.[41] I so doing, a mystique *and* a rationale was provide not only for dietetic advice but for a wide class magical recipes and rituals designed to allevia suffering.

V. Paramedical Data

The last aspect, which for want of a better ter I shall call "paramedical data," is also the most di ficult to keep within bounds. Those paramedic data that I shall concentrate on here include am lets, talismans, incantations, and other technique all of which had, in addition to these medicinal use a further range of religious and magical uses. Admittedly, only a thin line separates magic an religion, and an even more tenuous line separat folk beliefs from the grosser forms of superstitio in which supernatural agents are either involve or implicated. My purpose here is not to settle th long-standing controversy of where to draw the lin but to call attention to the medical aspects thereo A simple example will illustrate the problem. Fo both epilepsy and gout there was a large assor ment of simplicia and composita from which t choose;[43] moreover, various diets and regiminar

[36] For τὸ Κωνσταντίνου κολλούριον cf. Ermerins, 135–37. For St. Gregory's salt, cf. E. Jeanselme, "Sels médicamenteux et aromates pris par les Byzantins au cours des repas," *BullSocFrançHistMédical*, 16 (1922), 327. Gregory's role as a healer even extended to praying to him for health and prosperity, cf. Legrand, *Bibliothèque*, II (1881), p. xx.

[37] Cf. E. Darmstädter, "Ptisana," *Archeion*, 15 (1933), 181–201.

[38] Cf. Delatte, *Anecdota*, II, pp. 455–99.

[39] *Anthimi De observatione ciborum ad Theodoricum regem francorum epistula*, ed. E. Liechtenhan (Berlin, 1963) [*CMG* VIII, 1].

[40] *Simeonis Sethii Syntagma de alimentorum facultatibus*, ed. Langkavel (Leipzig, 1868).

[41] Cf. Hierophilus, πῶς ὀφείλει διαιτᾶσθαι ἄνθρωπος ἐ ἑκάστῳ μηνί, in Delatte, *Anecdota*, II, 456–66. (A different ve sion is printed by Ideler, I, 409–17.) The same calendaric o entation underlies Theodore Prodromus, *Mensium adorna* (Ideler, I, 418–20).

[42] Cf. G. Schlumberger, "Amulettes byzantins anciens de tinés à combattre les malefices et maladies," *REG*, 5 (1892), 7 93; V. Laurent, "Amulettes byzantines et formulaires m giques," *BZ*, 36 (1936), 300–15; and F. Pradel, "Griechische u süditalienische Gebete, Beschwörungen und Rezepte des M telalters," RVV, III:3 (1907), 253–403.

[43] For epilepsy, the peony (*Paeonia officinalis* Retz.) was the m popular simplex, partly due to Galen's endorsement (XI, 8 K.), partly because of its prominent role in late Greco-Roma astrology. (cf. A. Olivieri, "La peonia nell'astrologia greca," *RIG* 21 [1937], 139–56.) Among the many recipes for composita, c Alexander (ed. Puschmann), I, 545–49; Kousis, "Quelques co

rocedures such as baths were also advised.[44] But n addition, amulets, periapts, and semi-precious tones carried on one's person were also thought to e of avail.[45] Generalizing from this example, it is lear that some people regarded the employment f amulets and the like as not essentially different rom the use of poultices, decoctions, and clysers.[46] This was especially the case when the amulet vas fashioned of a natural product that, in other nstances, was regarded as a common medicament r foodstuff. This explains in part, I believe, the requency with which laurel and olive leaves were employed in medico-magical rituals, some of which had a healing function.

To conclude, it has been my intention to emphasize the most notable aspects of Byzantine materia medica and to provide a few illustrations of each. If space permitted, several other less obvious features would have been included, for example the several different, but ultimately interrelated philosophic theories that provided the rationale of drug action and hence explained why a particular substance was thought to be efficacious for a specific purpose. But since a discussion of such theoretical issues is restricted, on the whole, to a limited range of texts, it has been excluded here. It would also have been desirable to enlarge upon the physical operations and technological processes by means of which the raw or crude substances were prepared for administration and/or capable of being stored away for an indefinite period of time. Finally, I have not thought it necessary to enumerate the diseases and complaints for which drug therapy was recommended, the wide range of therapeutic forms available at the time, or the many alternatives to drug therapy. It is sufficient to say that the Byzantine physician had at his disposal one or more means of treating nearly every complaint with which he was faced. For that purpose texts were required, and it is presumably to such texts that Alexander of Tralles was referring when he stated that by following his recipes it was unnecessary to search further.[47]

---

dérations," 212.20; 213.2; 214.23 etc. For gout, many simplicia ere employed, including *Colchicum autumnale* L., the source of olchicine, clinically recognized as specifically effective for gout. here are also many recipes for the preparation of composita, g., Alexander (ed. Puschmann), II, 517–47; Kousis, 214.7; . H. Thomson (note 17 above), p. 73, etc.

[44] Alexander (ed. Puschmann), I, 541–45 recommended a wide ange of regiminary measures for epilepsy. A few dietary counels are recorded by S. Seth (ed. Langkavel) pp. 29.20; 47.15; 09.3; 123.13. For gout, Alexander recommended baths (II, 511–3 Puschm.) and some dietary restrictions (II, 508–10 Puschm.). lexander, II, 531 Puschm. and Seth, 38.14–19 favor the use f chickpeas (*Cicer arietinum* L.) in the diet. For further details, . G. Schmalzbauer, "Medizinisch-Diätetisches über die Podara aus spätbyzantinischer Zeit," *JÖBG*, 23 (1974), 229–43.

[45] Because the neuro-physiological cause of epilepsy was unnown, a supernatural origin was accepted, on the basis of which pileptics were shunned (cf. F. J. Dölger, "Der Ausschluss der esessenen von Oblation und Kommunion," *Antike und Christenm*, 4 (1974), 110–29). As a consequence, magical devices were ommonly recommended (cf. Alexander I, 557, 561, 567 Puschm.]; Delatte, *Anecdota*, I, 486.11; 618.17), though rational erapy was still attempted (A. Philipsborn, "IEPA ΝΟΣΟΣ und ie Spezial-Anstalt des Pantokrator-Krankenhauses," *Byzantion*, 3 (1963), 223–30). In general, see E. Jeanselme, "L'epilepsie ur le trône de Byzance," *BullSocFrançHistMédical*, 18 (1924), 225–4.

[46] Anna Chatzenikolaou, Μετάλλινα μαγικὰ εἰκονίδια ωνσταντίνου καὶ Ἑλένης, Ἐπ.Ἑτ.Βυζ.Σπ., 23 (1953), 508–8.

[47] Alexander, II, 109 Puschm.

IE *Bulletin de l'Institut d'Egypte* (before 1919: *Egyptien*)
Z *Byzantinische Zeitschrift*
MG *Corpus Medicorum Graecorum*
HB *Journal of the History of Biology*
EG *Revue des Etudes Grecques*
hM *Rheinisches Museum für Philologie*
IGI *Rivista Indo-Greco-Italica di Filologia, Lingua, Antichità*
B **Sitzungsberichte (followed by name of Academy, abbreviated, e.g., SB Berl)**

X

Benedictus Crispus, an Eighth Century Medical Poet*

OUR knowledge of early Western Latin medicine is slight when compared with what we know of the medicine of the Graeco-Roman period, contemporary Byzantine medicine, and the later period of Western medicine as well as Arabic medicine. The reason for this is the paucity of medical texts from the 7th to the 11th century. Singer termed this period the Dark Age of Medicine and dated its inception from the death of Beda in 735.[1] Apart from the numerous MS fragments and glosses and allusions to medicine in non-medical writers, little has survived from this period which could be called medical.[2] By Singer's reckoning, such writings as Isidore of Seville's *De medicina* (*Etymologiae* Bk. IV), Anthimus' *De observatio ciborum,* Beda's *De minutione sanguinis,* the *Lorica* of Gildas, etc., are excluded as too early. This leaves little more than some translations from the Greek,[3] later redactions of Pseudo-Dioscorides[4] and Pseudo-Apuleius,[5] some antidotaria,[6]

* A grant from the Medical School, University of California at Los Angeles, during the summer of 1962, facilitated the initial research for this study, To Dean H. W. Magoun, to Prof. John Thibault for his assistance in polishing the translation of the verses, and especially to Prof. C. D. O'Malley for his advice *in rebus medicis arcanis,* I am most grateful. The remainder of the research was supported by USPHS Research Grant 9361 and carried out at Yale University.

1 Charles Singer, A Review of the Medical Literature of the Dark Ages, with a New Text of about 1100. *Proc Roy. Soc. Med., Hist. Sec.,* 1916/17, *10,* 107-160.

2 A good example of a non-medical writer whose testimony is important in the history of medicine is Richer of Rheims. Cf. L. C. MacKinney, Tenth Century Medicine as seen in the Historia of Richer of Rheims. *Bull. Inst. Hist. Med.,* 1934, *2,* 347-375.

3 An excellent summary of translations, paraphrases, and commentaries of Greek medical writings available in early Western Latin versions is contained in Walter Puhlmann, Die lateinische medizinische Literatur des frühen Mittelalters. *Kyklos,* 1930, *3,* 347-375.

4 Cf. H. F. Kästner, Pseudo-Dioscoridis De herbis femininis. *Hermes,* 1896, *31,* 578-636 and Henry E. Sigerist, Materia Medica in the Middle Ages. *Bull. Hist. Med.,* 1939, *7,* 417-423.

5 Cf. E. Landgraf, Ein frühmittelalterlicher Botanicus. *Kyklos,* 1928, *1,* 114-146. For the text tradition, cf. the Preface to the edition of Pseudo-Apuleius by E. Howald and Henry E. Howald and Henry E. Sigerist, *Antonii Musae de herba vetonica liber, Pseudo apulei herbarius. . . .* Leipzig, Teubner 1927 (Corp. Med. Lat. IV).

6 Cf. Henry S. Sigerist, Studien und Texte zur frühmittelalterlichen Rezeptliteratur. *Stud. Gesch. Med.,* 1923, *13,* vii + 220 pp. Several of the antidotaria fall within our period: the London Antidotarium (sc. VII/VIII) pp. 17-21; the Bamberg Antidot. (s. IX/X) pp. 21-39; the Berlin Antidot. (s. IX/X) pp. 65-77 etc. Several early Latin receipts are contained in K. Sudhoff, Die gedruckten mittelalterlichen medizinischen Texte in germanischen Sprachen. *Arch. Gesch. Med.,* 1910, *3,* 273-303. Cf. also the 10th century MS containing both Latin and ME receipts edited by Chr. Ferckel, Medizinische Marginalien aus dem Cod. Trevirens Nr. 40. *Arch. Gesch. Med.,* 1913/1914, *7,* 129-143.

Reprinted from the *Journal of the History of Medicine and Allied Sciences,* 1966, Volume XXI, No 1, by kind permission of Oxford University Press.

Walafrid Strabus' *Hortulus,*[7] the writings of Hildegard of Bingen,[8] and several anonymous gynaecological writings.[9] Even if one included Hrabanus Maurus' *De homine et partibus eius,* Charlegagne's *Capitulare de Villis,* and the Anglo-Saxon *Book of Bald,* our resources are still meagre.

Yet to this list one ought to add the *Commentarium medicinale* or, as it is sometimes called, the *Poematium medicum* of Benedictus Crispus, Archbishop of Milan. His death in 725 or 735 does not support Singer's exclusion of this poem from the Dark Age of Medicine on the grounds that the date is one "hardly falling within our period."[10] Because of the date of Benedictus' death, if for no other reason, his poem deserves a recognition that it has not yet received from students of early Western Latin medicine. It was to supplement our knowledge of this period that the present study was undertaken.

If Benedictus Crispus is a name infrequently encountered in the history of medicine, it is otherwise with the history of early Western Latin literature.[11] For here his position is secure, and his reputation as one of the more noteworthy teachers and savants in the century before the Carolingian revival is duly recorded by literary historians.[12]

7 We have consulted the edition printed in J.-P. Migne, *Patrologia Latina,* vol. 114 (Paris 1879), col. 1119-1130.

8 From a medical standpoint, the most important of her works is the *Physica,* printed in J.-P. Migne, *Patrologia Latina,* vol. 197 (Paris 1882), cols. 1125-1352, A German translation, with notes and bibliography, was prepared by Heinrich Schipperges, *Hildegard von Bingen, Heilkunde.* Salzburg, 1957.

9 Cf. F. P. Egert, Gynäkologische Fragmente aus dem frühen Mittelalter. Nach einer Petersburger Handschrift aus dem VIII.-IX. Jahrhundert zum ersten Mal gedruckt. *Abh. Gesch. Med. Naturw.,* 1936, *11,* 1-59; A Dold, Donaueschinger Fragmente eines mehrere Bücher umfassenden medizinischen Rezeptars. (in Unzialschrift des 7./8. Jahrhundert). *Arch. Gesch. Med.,* 1931, *24,* 205-219 and Paul Diepgen, Reste antiker Gynäkologie im frühen Mittelalter. *Quell. Gesch. Naturw.,* 1933, *3,* 18-34.

10 Singer (n. 1), p. 112.

11 Generally, a paragraph is devoted to Benedictus, and all of the histories of medicine mentioned below, with the exception of Diepgen, follow with considerable uniformity the account given by Ludwig Choulant, *Handbuch der Bücherkunde für die ältere Medizin* (1841), reprint edition, Graz 1956, pp. 226-228. Cf. H. Haeser, *Lehrbuch d. Gesch. der Medicin. . . .* 3. Aufl. I. (Jena 1815), pp. 634-635; Th. Puschmann, *Handbuch der Gesch. der Medizin,* ed. Neuburger u. Pagel. I (Jena 1902), pp. 629-630; Max Neuburger, *Gesch. der Medizin* II. (Stuttgart 1911), pp. 253-254; F. H. Garrison, *An Introduction to the History of Medicine,* 4th ed. (Philadelphia 1929) p. 146; Arturo Castiglioni, *A History of Medicine.* Trans. E. B. Krumbhaar. (New York 1941), p. 297; Paul Diepgen, *Gesch. der Medizin.* I. (Berlin 1949) pp. 193, 197. The biographical details in Meyer's account likewise go back to Choulant, E. H. F. Meyer, *Gesch. der Botanik.* II. (Königsberg 1855), pp. 421-423.

12 Max Manitius, *Geschichte der Christlich-lateinischen Poesie bis zur Mitte des 8. Jahrhunderts.* (Stuttgart, Cotta, 1891), pp. 396-397; F. J. E. Raby, *A History of Secular Latin Poetry in the Middle Ages* I2. (Oxford 1957), p. 159; Antonio Viscardi, *Storia Letteraria d'Italia.* vol. I, *Le Origini* (Milano 1939), pp. 18-19.

So few details of Benedictus' life are available that Angelo Cardinal Mai was able to summarize his life in one paragraph.[13] Pertinent to the present study is the fact that Benedictus was ordained by Pope Sergius I in 681 as Archbishop and, as such, was involved in some litigation in 689 with the representatives from Pavia. Regarding this matter, Paul the Deacon stated that despite his probity of life and gift of speech, he lost his case against Pavia.[14] There is, in fact, some doubt as to the authenticity of the speech he is reported to have delivered on that occasion, and Muratori has suggested that it may have been written by Landulf, the historian.[15] There remain two other writings of Benedictus, neither of which, unfortunately, can be dated with certainty. Mai reported that there was extant in the 15th century an inscription in St. Peter's to the effect that, while in Rome, Benedictus wrote the epitaph for Caedualla, King of the West Saxons.[16] Although the inscription has disappeared, the epitaph has been preserved by both Beda and Paul the Deacon.[17] There is also the tradition, recorded by Joannes de Deis, that Benedictus wrote "some commentaries worthy of note" (*nonnula commentaria laude digna*), but of these, nothing further is known.[18] There remains, then, his medical poem, in which we are chiefly interested.

This poem appears to have been mentioned for the first time in the latter half of the 15th century by Peter Cenninus, although it remained unedited until 1833 when Mai, using a Vatican MS, the only one then known, published the *editio princeps*. As a specimen of Latin poetry, little need be added

13 Mai, Observationes Praeviae, reprinted in J.-P. Migne, *Patrologia Latina*, vol. 89 (Paris 1850), cols. 367-368. The earlier literature, inaccessible to us, is cited by Choulant (n. 11) and J. V. Ullrich, *S. Benedicti Crispi commentarium medicinale ad fidem Codicis Vindobonensis. . . .* (Kizingae 1835). I am grateful to the National Library of Medicine for supplying me with a copy of the latter.

14 *Historia Langobardorum* VI. 29, ed. G. Waitz (Hannoverae 1878), p. 225.

15 Cited by Mai (n. 13), 89, 368.

16 *Ibid.*

17 Bede, *Historia ecclesiastica gentis anglorum* V. 7, ed. Car, Plummer. I (Oxon. 1896), pp. 292-294. Paul Diacon., *Hist. Langob.* VI. 15, pp. 217-218, ed. Waitz. The *Epitaphium ceadual*, printed in Migne (n. 13), 89, cols. 375-376, follows the text of the editio princeps, Angelo Mai, ed. *Classicorum auctorum e vaticanis codicibus. . . .* t. V (Romae 1833), pp. 391-404, Cf. also Jacob Hammer, An Unrecorded "Epitaphium Ceadwallae," *Speculum*, 1931, *6*, 607-608.

18 Mai (n. 13) 89, 369. According to a certain Ciacconio, Benedictus also wrote a *De ineffabili trinitate* and a *De rebellione Angelica* but Mazzuchelli, the source of this information, thinks that these are nothing more than the aforementioned commentaries. Giammaria Mazzuchelli, *Gli Scrittori d'Italia cioe notizie storiche, e critiche intorno alle vite, e agli dei letterati italiana.* Vol. II, parte ii (Brescia, Giambatista Bossini, 1760), p. 827. I am grateful to Professor O'Malley for permitting me to use his personal copy of this rare book.

to the remarks by Manitius and other historians of literature beginning with the aforementioned Cenninus, who formed a low opinion of Benedictus' poetical abilities.[19] It is composed of 241 hexameters, some of which fail to scan, and is replete with late Latin constructions, vocabulary, and orthography. His choice of genre, didactic poetry, is probably best explained in his prefatory letter to Maurus, though the fact that his principal source was also a didactic poem must not be discounted. But since our interest in the poem is medical, we leave further literary criticism to others to concentrate on its content.

The most important fact regarding the medical contents of the poem is the absence of any direct Greek influence. Although its date of composition is uncertain, it is probable that it was written early in his career when, as a deacon, he was still active as a teacher.[20] This means that it was composed almost certainly before 689, when he became involved in the Pavian controversy, and probably before 681, when he was ordained as Archbishop. At this date, first-hand knowledge of Greek texts, especially secular and medical texts, is not to be expected.[21] This is confirmed in two different ways. The occasional words of Greek origin, probably derived from his Latin sources, are badly garbled and indicate no knowledge of the Greek original.[22] A second reason for minimizing Greek influence stems from the fact that his materia medica follows that of the Latin authors on whom he depended. It is fortunate that he explicitly names his two primary sources. As far as the textual evidence permits a decision, Quintus Serenus and Pliny are the only written sources utilized by him.[23]

In the course of describing a remedy for anthrax, he states (v. 209) that it is repelled by turnip seeds "as Quintus Serenus asserted with good reason." Q. Serenus or, as he is sometimes

19 Cf. the colophon, translated below. Benedictus is also censured by Manitius, *Geschichte der lateinischen Literatur der Mittelalters,* I (München, Beck, 1911), p. 198, n. 2 who calls attention to the lack of scansion in v. 238. Likewise, Adolf Ebert, *Allgemeine Geschichte der Literatur des Mittelalters in Abendlande.* I (Leipzig, Vogel, 1889), p. 619, n. 1 who, besides noting the "fehlerhaften Hexametern" states that the *Epitaphium ceadual* "etwas besser als diese trockenen versificirten Recepte ist. . . ."

20 Mai (n. 13) 89, 369 and repeated thereafter in the standard histories of medicine, cf. Castiglioni (n. 11), p. 297.

21 Despite Cassiodorus' encouragement to the study of Greek medical texts, there is little evidence of this practice in the early medical writings cited in notes 9, 33, and 38. Cf. Cassiodorus, De institutione divinarum litterarum, I 31 in J.-P. Migne, *Patrologia Latina*, vol. 70 (Paris 1847), col. 1146.

22 Cf. the notes to verses 51-52 below.

23 On several occasions Benedictus refers to unspecified written sources: v. 158 "fertur ut in libris"; v. 198 "referunt quod (sc. stercus) carmine plures." Cf. also v. 228. But these passages, like the reference to "cura vetusta" in v. 23 do not permit an identification.

called, Serenus Sammonicus, was the author of a medical poem, *Liber medicinalis,* in 1107 verses.[24] The date of Q. Serenus and whether the poem was composed by father or son, who bore the same name and who are both credited with medical writings, are equally uncertain.[25] Pépin, the most recent editor, believes that the poem was written by the son early in the 3d century.[26] But whatever the uncertainty is on these matters, we possess the poem and there is little doubt that it was Benedictus' primary source. This can be shown by a comparison of the remedies used by the two writers for a given complaint.[27] Benedictus' dependence upon Q. Serenus becomes even more certain when one compares the textual parallels, far too numerous to be ascribed to accident or to literary convention.[28]

His other source is not so readily determined. Although he mentions (v. 107), when describing remedies for intestinal disorders, that "Pliny records it, we know that he spoke the truth," there is some uncertainty about the identity of this Pliny. It may refer to Caius Plinius Secundus, better known as the author of the influential and encyclopedic *Historia naturalis.* It is rather more probable, however, that it refers to the anonymous *Medicina Plinii,* an early 4th century eclectic compilation of passages from Pliny and other later writers.[29]

From these two sources, Benedictus derived the bulk of his material. The assertion that he also utilized Dioscorides is without foundation.[30] However, since both Pliny and Q. Serenus borrowed from Dioscorides, parallels between Benedictus and Dioscorides are to be expected.[30a] The fact that his sources were purely Latin suggests a point which may be overlooked. Except for the translations and paraphrases of Greek writings, his poem is one of the latest Latin medical writings in which the materia

24 Quintus Serenus (Serenus Sammonicus), *Liber medicinalis,* ed R. Pépin. (Paris, Presses Universitaires de France, 1950).

25 *Ibid.,* Introduction pp. vi-viii.

26 *Ibid.,* p. ix.

27 Cf. the notes below to verses 3, 28, 37, 54, 122, 132, 134, 143, 154, 156, 193, 194, and 208.

28 Cf. the notes to verses 19, 53, 92, 109, 139, 150, 155, 156, 161, 172, 223, 232 and 241.

29 Cf. V. Rose, Über die Medicina Plinii. *Hermes,* 1874, *8,* 18-66; A. Köhler, Handschriften römischer Mediciner. *Hermes,* 1883, *18,* 382-392 and Pépin (n. 24), pp. xvi-xix. Neuburger (n. 11), II, 253 explicitly asserts that Benedictus borrowed from Pseudo-Pliny but presents no evidence.

30 Choulant (n. 11), p. 226 and Castiglioni (n. 11) both assert, but with no documentation, that Benedictus borrowed from Dioscorides. A specific case which makes borrowing unlikely is noted below in reference to v. 21.

30a For the relation between Pliny and Dioscorides, cf. Jerry Stannard, Pliny and Roman Botany. *Isis,* 1966 (in press).

medica shows no traces of Greek influence other than that derived from his intermediary sources. This is of considerable importance for determining the date at which Greek, and later Arabic, drugs and medical ideas entered the West, by what channels, and under what names.

Benedictus' medical knowledge, like his materia medica, is bookish and was probably derived from the same sources as those noted above. It has been suggested that there are elements of folk medicine in his poem, but while this may be true, they are of secondary importance.[31] It will be noted that he makes no mention of the amulets, charms, and other superstitious practices which occur in Q. Serenus and Pliny.[32] There are, naturally, traces of monastic medicine, but beliefs in supernatural cures and a supernaturalistic etiology are less pronounced than in other writings of a similar type.[33] There is no indication of knowledge on his part of urinoscopy or the doctrine of pulses. If these are lacking, so too are such well-known medical doctrines as humoral physiology and pathology and the theory of the four temperaments.[34] It may be said, then, that his medical knowledge is confined to naming a disease, a crude delineation of symptoms plus some stray bits of erudition, all of which are versified in a rather dramatic form by a mixture of poetical phrases and quasi-technical vocabulary.[35] Since his aim was to supply the younger Maurus with a select set of remedies, it is not surprising that the bulk of the poem is concerned with a description of the drugs for a given disease, their preparation, and directions for their administration. The absence of dietetic advice, except for one verse (v. 231) is noteworthy. This may be ascribed to his primary source, Q. Serenus, who was more concerned with therapeutic recommendations than with preventive measures.

31 Singer (n. 1), p. 112. Elements of folk-medicine may underline verses 123 and 183.

32 Quintus Serenus (v. 935) employs the well-known *abracadabra* charm for fever. For other similar occurrences cf. Pépin's note ad. loc.

33 Cf. v. 142 and the phrase "prudens medicina Dei." Other passages indicating Christian influence are vv. 89-89, 160, 170, and 212. For the extent to which a naturalistic etiology and therapy are subordinated to religious dogma, cf. the 8th Century MS edited by Sudhoff, Eine Verteidigung der Heilkunde aus den Zeiten der "Mönchsmedizin." *Arch. Gesch. Med.*, 1913/14, 7, 223-237. The combination of religious dogma and formalized medical practices underlie such usages as that reported by Beda (n. 17), III 17, p. 161 where a cure is effected by drinking a potion in which chips of a holy wooden relic have been steeped.

34 The nearest approximation to physiological theory is the statement in v. 101 that a dung-plaster is effective because "it closes the narrow passages (*claudit angustos . . . cursus*).

35 An etymology is given in v. 20; prognoses are attempted in vv. 47-148 and 205-206. There are several anatomical observations, vv. 18, 41, 84, and there appears to be a genuine botanical observation in v. 76.

The later history of Benedictus' poem is unknown at present. Earlier Latin writers, like Beda, who knew his non-medical writings, do not mention his poem. Nor is there any evidence of his poem being known at Salerno. This is not surprising in view of the Salernitan practice of favoring Greek and Arabic drugs over native Western drugs. Accordingly, the greatly increased resources of Salernitan materia medica would make superfluous any return to the drug lore of a Christian poem then several centuries old. For reasons such as these, Singer rightly placed Benedictus' poem in a class termed by him, "Latin extra-Salernitan Literature."[36] De Renzi's inclusion of the poem in his edition of Salernitan texts should not be taken to indicate any historical connection with Salernitan medicine.[37] Rather, as he stated in his introduction to the poem, his edition marks an improvement over Mai's edition. De Renzi's choice to include Benedictus' poem is fully justified when one seeks for pre-Salernitan Western medical texts. Among the little published material in this area, Benedictus' poem may be taken as typical of the Latin-derived, local Italian medicine prior to the Salernitan renaissance.[38]

The present translation is based on De Renzi's edition primarily. Most of the significant departures from this edition have been recorded in the notes following the translation. None of the MSS has been examined.[39] A critical text is required since there is a considerable difference of opinion as to the variant readings as reported in the editions of Ullrich, De Renzi, and Brunhölzl. In order to keep the present study within manageable size, matters of philological, metrical, and orthographical interest have been omitted save where they affect the meaning. The only

36 Singer (n. 1), pp. 111-115.

37 Salvatore De Renzi, Ed., *Collectio Salernitana,* 5 vols. (Napoli, Filiatre-Sebezio, 1852-1859), I, pp. 73-87. Elsewhere, De Renzi explicitly notes the importance of Benedictus' poem "come uno de'pochi documenti che il tempo ci ha trasmesso, della continuazione della medicina latina in un'epoca in cui non ancora era cominciata la medicina araba." *Storia Documentata della Scuola Medica di Salerno.* 2nd ed. (Napoli, Gaetano Nobile, 1857), p. 74.

38 The available *edited* material falling within this class is scarce. Cf. G. Carbonelli, Frammento medico del sec. VII. *Rass. Clin. ter. sci. aff.,* 1921, *20,* 5-21 and R. Simonini, Herbolarium et materia medica Codice Ms n. 296 de la Biblioteca Governativa di Lucca. *Rass. Clin. ter. sci aff.,* 1936, *35,* 268-286, 321-331. Much has been written on Salernitan chronology; a conservative and well-documented study is P. O. Kristeller, The School of Salerno. *Bull. Hist. Med.,* 1945, *17,* 138-194.

39 Cf. Franz Brunhölzl, Benedetto di Milano ed il "Carmen Medicinale" di Crispo. *Aevum,* 1959, *33,* 25-67. I am indebted to Professor Luigi Belloni for calling my attention to this study. In addition to the three previously known MSS:

Vat. Pal. Lat. 1587 (ed. Mai, cf. n. 17)
Paris. B. N. Lat. 6864 (ed. De Renzi, cf. n. 38)
Vindobon. Bibl. Nat. 4772 (ed. Ullrich, cf. n. 13)

Brunhölzl has discovered another MS: Vat. Urb. Lat. 668. Its readings indicate affiliation with the Vat. Pal. and Paris MSS.

exception to this practice has been to note the more striking parallels with Q. Serenus. On the other hand, we have attempted to identify all of the drugs mentioned by Benedictus. Since one of the purposes of this study is to determine the nature and extent of his materia medica, references have been made to other contemporary and earlier medical writers.

A versified translation appeared warranted in order to facilitate line reference and to recapture something of the artless didacticism of the original. Needless to say, sense rather than rhythm or a duplication of the imagery has guided our efforts. The chapter numbers, introduced by Mai, have been retained for the sake of convenience, while the headings appear in the Vienna Ms.[40]

A Book on Medicine by Crispus, Dean of Milan, to Maurus, Vicar of Mantua

Preface

Since I have instructed you, my dear Maurus, almost from your earliest childhood and have endowed you with the gift of sevenfold eloquence, only one thing is lacking to you which you always wanted to hold displeasing while you were still established in your growing years; that is, you claimed that the knowledge of good medicine—which you thought of little value for anything—held no place or share in a liberal education. Now, however, since the irregularity of various illnesses more frequently overpowers you, you compel that very irregularity to broaden your knowledge of the arts, which formerly you did not blush to call wicked and vile. You require of me that I make you a rustic and teach you the powers of medicinal herbs, displaying them to you one by one in their established season, and that I unerringly show you the different sorts of cures according to the manifold nature of the diseases. This is a task I gladly undertake in compliance with your wish. For the sake of brevity in the present little work, I want to feed you on heroic honey, so that I may gradually lead you to a love of the art. Else, by opening my mouth I will reveal too many matters to begin with, the number of which may arouse repugnance, repugnance then taking away desire. Therefore, I have limited to a select number what I hold to be most excellent; in order that, while you approve them as true by the evidence of this work, you

40 I am indebted to Dr. Franz Unterkircher, Direktor der Handschriftensammlung, Österreichische Nationalbibliothek for replying to our queries concerning the Vienna MS 4772 and for supplying us with a photostat. The MS, we are informed, was written partially by the Viennese physician and humanist Johannes Cuspinianus (1473-1529) shortly after 1500.

may with greater eagerness and earnestness pursue the many things that remain.

I. THE HEAD

If an ailing head is disturbed by countless throbbings,
And is troubled by a kind of heaviness all about
Take care at once to wrap it in a chaplet of ivy.
Frankincense is mingled with pungent vinegar,
Myrrh, too, is similarly mixed with olive oil
And famous *cephalote* with savine salt;
With these you will be able to assuage more quickly the sharp pain.
Nevertheless, if, when roused [only] part of that illness is shaken off,
Take a healthy mole, cooked whole in oil,
And apply to the head and face an ointment made of it.

II. TOOTHACHE

If the omnivorous teeth are upset by a storm of pain,
Potent henbane is pressed upon the tooth with the thumb.
At once all maddening pain will leave.
But I warn you not to swallow it, for a deadly disease
Arises and takes away consciousness as if seized by the image of death.
If a heedless throat takes it and agony strikes
Spignel will be the drink, a health-giving medicine, as well as an ointment.

III. DELIRIUM

If a terrible pain should pierce the dura mater,
And cursed madness, rampaging, infect the brain,
The lung taken from the viscera of a stag will be pleasing,
Mixed with *cephalote;* or, if you prefer, drink coriander
Or the rind of a radish softened by harsh vinegar:
It is extremely pungent when taken, but really an ancient remedy.

IV. THE EYES

There is no doubt that a weakness aroused affects the lively eyes,
And Fortune disturbed in various ways distresses them;
For it afflicts them, but a friend attends to them at a fixed time.
Come, then, cook together at once horehound,
Betony, folded in, on moistened wool.
It removes every evil more quickly and restores good eyes.
If there is pain in the eyes, caused by a red humor,
Then powerful verbena is drunk in a mild liquid.

V. A SPOT ON THE EYE

If you have a spot on the eye, look at it while the eye turns inward:
Take the red snail that you find when you walk in the fields
And place it unerringly on the middle of the eye.

VI. SORE THROAT

If pain spreads throughout the throat,
And shutting it off, fills it with swollen folds,
Arsenic is ground up and mixed with pepper, and a bit of it,
Placed on a feather, is breathed in by the hoarse throat.

VII. SORE PALATE

When the moist palate loses its shape,
Heavy putrid drops drip into the body,
And uncleanly and evilly bore through the light wings of the lung;
Despite their extension, they are quickly dried by a similar powder.

VIII. THE CURE OF THE LUNG

Often the remarkable vigor of a man's chest is shaken,
And it threatens horrid death to many.
Therefore if you have an ailing chest to cleanse by remedies,
Let it not shame you to understand the different remedies of herbs.
Wool moistened with oil, also the juice of fenugreek,
These refresh the stomach, and a draught of tamarix given too.
A large cabbage is mixed with the renowned betony;
Radish-flavored hydromel, as well as the juice of horehound ought to be administered.
That worthy herb, the famous sage, is also taken.
Finally, the large aloes will be given.

IX. PAIN OF THE HEART

When furious pain strikes the diaphragm in the side,
And bitter anguish greatly torments the fibers of the heart
Strong fluids are a help, joined by the power of coral:
Next *pipinella* is mixed with a broth of fennel;
Zedoary then is taken, also potent tansy
Which calms fevers and repels poisons of the heart.

X. CURE FOR PLEURISY

But if a great sickness should strike the tender side,
And life is endangered by a double peril,
Take care to hasten with a wolf-skin and apply it warm.
Let potent stag-marrow with a goat's hoof
Straightway be put in a cooking pot over a slow fire.
Pitch is mingled with *britia,* and meal is cooked thoroughly,
With lupine as well as with saffron,
And thus a good poultice is applied to the bitter disease
By which relief may be given to one suffering excessively.
The renowned *medicina Tonantis* drives off the illness.

XI. CURE FOR THE SPLEEN

When a swollen spleen bothers a wagon-driver,
Cook betony and add the droppings of a dove;
These two ingredients united require at the same time to be joined to *cephalote*
And to rose-oil; let the ingredients be mixed when moist.
Gather a little savory and mix it with lupine;
This is administered rather often; ceterach is also drunk
Whose virtue destroys the disease but leaves a pleasant face wrinkled.
Humble, it grows among the rocks, but is excellent for the spleen
And friendly to man, but scarcely suited to the eyes.
There are as many diseases in the body as there are parts to man;
But a strong medicine curbs many of them by its power.
When the liver is diseased and the hardness spreads, as it usually does,
Pungent costmary is drunk, also that remarkable plant, rhubarb.
These, ground together with salt, check the swelling.
Lupine fibers are strong and when drunk down help the sufferer.
Although the liver causes pain on the right side and the spleen on the left,
Their cures are alike.

XII. STOMACH AND INTESTINAL PAIN

The inmost vital parts are racked by many sorts of pains
And their painful twisting is tormented by threatening feebleness.
Nor does any hope of life remain, when the soul itself
Seeks the heavens, departs as a traveller, and abandons every good thing.
This is called *tortura,* because it seriously torments the bowels
And twists them over and over; at the same time the very nature of the whole body
Is also afflicted, for which a hot bath is a restorative.
Sluggish limbs are thus warmed by the liquid,
And all that is cold is driven out by the hot fluids.
Afterwards goose fat, broken up, is melted,
Taking care to mingle with it the virtue of fenugreek,
The power of these remedies being augmented by cooking in the fluids.
Forthwith a clyster in anum is given to the anxious sufferer
By virtue of which he will be healed of his pain.
The dung plasters which they make, though not a healthy cure,
Close the narrow passages of the patient to jaundice.
White ivory, too, is scraped and drunk at once;
The potion unwinds the twisted organs more quickly.
The tender entrails of a blind whelp are, it is reported, eaten;

Thus can the soft little members be used for the tender fibers;
The pain is gone, and passes into the inward parts of the fissure.
Pliny records it, we know that he spoke the truth.

XIII. CURE FOR DROPSY

From the hardening of the spleen arises a fatal fluid,
And separates the wretched skin from the miserable viscera.
So great is the power of the disease, that the sufferer cannot slake his thirst.
Take cinders, into which the tall ash tree burns,
From these make lye, and crush hard barleycorns;
When these have been cooked together, apply them faithfully to the sufferer.
Let a radish plant be mixed with the finest honey;
From its leaves let a poultice be applied.
Warm sulphur is drunk with an old wine.
From these apply a poultice to the swelling.
They dissolve the hardness of the spleen, and also dry up the liquid.

XIV. WORMS

If putrid wounds spread with bloody diseases,
It will be suitable to drink hot goat's milk.
They cure the awful abscess with asses' milk.
Ashes of the Aminaean vine are moistened with olive oil.
For centuries this was the appropriate potion for many foul illnesses.
But celery and dried figs together, also laurel berries,
The Illyrian iris, the large centaury and rue,
These can rather quickly clean off the purulent sores,
And restore a renewed man to happy days.

XV. THE EAR WORM

It is characteristic of imprudent persons to allege sure cures.
In the same way bodies may be disturbed by a rude sense.
When a tremendous lassitude possesses a man's limbs,
Then is an animal accustomed to enter the afflicted ears.
Quickly take mouse-gall with harsh vinegar.
Make no delay, and you will be able to drive off the worm.

XVI. DEAFNESS

But if you suffer deafness, gather the red worms
Of an ancient tree, mingle them with pure olive oil:
Pour it into the ears, and restore your desired health.
Goose fat is helpful, and the fat of the black vulture.
Onion juice helps the ears and offers good results,
When water has a way of getting into incautious ears.

XVII. DYSENTERY

Often the stomach's course hastens excessively to its end,
Gradually to destroy the tender flesh of many men,
Which a provident remedy of God knows how to check.
Asian gall-nut is an aid, also the long fruit of the cherry.
Cheese will be suitable with the sweet gift of honey;
Helpful, too, will be the warm suet taken from a goat's entrails
And applied to the stomach, this is a recognized cure.
But if he loses his nails,
It means that the life desired is now perishing quickly.

XVIII. INDURATION OF THE STOMACH

If the belly is disturbed by an immense weight,
Look for the olive where it grows from its venerable shoot.
When pure, this is said to be a safe friend to all.
Straightway take it, a good influence grows in the stomach.
Hellebore is taken, ground up, and hydromel is drunk.
Always good for this illness too is danewort
Diluted by old wine which is to be drunk at once.
Advantageous too, is a draught of mare's milk without delay.
Agrimony, horehound and meadow saffron,
As reported in books, are always friends of the stomach.

XIX. THE BLADDER

Many remedies are helpful in many illnesses because they are their opposites.
Disease is conquered by the antidote of the high god.
There is a kind of ailment and a disease hostile to many,
When a burning, maddening bladder stone rages,
Begetting stones by which the urine runs slowly;
Nor is the suffering man able to unloose the benign liquids.
Tall ivy will provide berries from its lofty tip;
Drink the juice of this with warm wine.
Without delay, let a plantain be broken into kernels, and
Eaten, while kindly alexanders is drunk.
The stones submit to *saxifraga* and celery, too.
You will be sustained by these opinions, and you will seek the Lord.

XX. THE FLOW OF URINE

But if flowing contrariwise, an excessive amount pours forth
A rabbit's brain will check the flow at once.
Take it with old wines;
A vile thing to drink indeed, but not vile as a remedy.

XXI. NOSEBLEED

The sure hope for health must be reported to all,
When a polyp of the nostril pours out in a wave of blood.
Rue provides a drink together with a vapor when crushed,
Horse-tail from marshes is helpful and it stops the blood
When it is applied to the nostril crushed with cumin.
The powder of the mushroom, which they say is lupine,
It also applied to the nose with sprigs of the lowly mint.

XXII. HEMORRHOIDS

If a repulsive disease makes trial of the rectum,
Let potent nightshade be sought, unfriendly to the mole;
Also henbane: of these a healthful remedy is made.
These things drunk together halt the blood at once.
Burnt lead, in conjunction with moist olive oil, is a check.
The surge of blood halts from both these used jointly.

XXIII. ERYSIPELAS

A piercing flame rages, continually burning the bowels.
It spreads slowly and wanders, ferociously sending flame beneath the skin,
And again leaps up, spreading in the likeness of a fountain.
Let the mind of man always meet its internal trickery,
To wit, the herb celandine is helpful for anointing;
And the interior white of a snowy egg is added.
Rue helps and puts the disease to flight and also checks the flame
If it is joined at the same time with warm olive oil.
Do not imagine that what I say is false; since to tell the truth
Is what truth itself commands; apply to the disease the dung
Of a sheep, when warm, which many mention in verse.

XXIV. ANTHRAX

A hostile force flourishes when an awful disease rages,
When a frightful tumor is destroying the weakened viscera.
For this the juice of *barba iovis* is beneficial,
And at the same time hostile; powdered savine, too
With the droppings of a white dove, is drunk as a broth.
When consumed, if they remain in the body's trembling belly, after being taken,
Life's salvation is then promised to the frightened patient.
But if he seems to reject the potion after it has been drunk,
Believe me, it appears to disclose a swift death.
That this terrible disease is repelled by turnip seeds

Quintus Serenus asserted with good reason.
This serious illness is also helped by cautery
When quicklime is sprinkled immediately
According to God's will, upon the reddened scar.
Then fearlessly apply the fearful remedy
By which it may suffice to draw off the virulent poisons.
But let this be stated too, know how to avoid sleep,
Lest the savage disease, reviving, may restore its strength.

XXV. ARTHRITIS

If the hip has betrayed anyone's step
And a maddening fluid that wanders through the bones has increased,
Take polypody, which they call *filicla*
With which crushed black cloves are well mixed
And dates, pepper and cinnamon—pleasing to all.
Thus make the potion; administer it, with the addition of honey.
Learn to search diligently for the inner bark of the elm tree,
When madder is mixed with it, they make a pleasant drink.
Take the larger nettle of the countryside,
Which, when drunk for a time, knows how to burn out the disease.
Hold this certain, firm, and true, my son.

XXVI. CURE FOR GOUT

Despite many valuable writings of many men,
Inactive gout is said to have no cure.
If you want my opinion and if you wish help against the illness,
Begin as I advise; give up the pleasure of drinking milk,
Enormous pain counting for more than deprivation.
An application of the ointment of polypody provides rest for the step.
Potent tamarix, and the living *barbula iovis,*
When crushed together, and bound to the feet,
Instantly provide a portion for the lame one.
Do you wish now to command a marvelous deliverance for yourself?
Take the spines of the little fish called *ruccas,*
Arrange them on a tile and place in boiling plums,
Then with sureness apply, when hot, to the sufferer for his walking.
They check all evil and the dreadful disease takes flight.

The Book of Crispus of Milan to Maurus, Vicar of Mantua, is ended. But that Crispus is surely neither a poet nor a good versifier, inasmuch as he constructed lines not according to the law of meter, but to the sound of his ears, that is, according to rhythm alone.

The End.

Peter Cenninus wrote this at Florence, November 19, 1468.

COMMENTARY

Line 1. The enumeration of diseases or parts of the body affected, and their corresponding remedies *a capite ad calcem* is the common order in Latin medical writers. Besides Quintus Serenus (hereafter QS) whom Benedictus followed cf. Pliny, *Naturalis historia*, ed. Janus (Leipzig 1860-80) XXV. 83, 132-110, 175; Celsus, *De medicina*, ed. W. G. Spencer (London 1935-38) IV. 2, 1; 6, 1; Theodorus Priscianus, *Euporiston libri III* (ed. Val Rose, Leipzig 1894) lib. I passim; Cassius Felix, *De medicina* (ed. Val Rose, Leipzig 1879) lib. I passim and Scribonius Largus, *Conpositiones* (ed. G. Helmreich, Leipzig 1887) who states in the prefatory letter (p. 6): Quarum initium (scil. partes corporum) a capite fasciemus. . . .

3. *Hedera*, or ivy (*Hedera helix* L.). The use of a chaplet of ivy was considered especially beneficial for the brain, Pliny XXIV. 47, 75 while QS 20 recommends ivy berries, macerated in oil and applied to the head wth a sponge. Benedictus 165 also uses ivy berries as a diuretic.

4. *Thus*, or frankincense, from an unidentifiable species of *Boswellia*, probably including *B. carterii* Birdw. Along with myrrh, cinnamon, cloves and costus, it was one of the items imported from the Near East. Cf. Isidore, *Etymologiarum sive originum* libri XX, ed W. Lindsay (Oxford 1911) XVII.8, 2 who mentions a *tus arbor Arabiae* and Pliny XII.30, 51. Among the uses mentioned by Dioscorides, *De materia medica libri quinque*, ed. Max Wellmann (Berlin 1907-1914) I.68 are several connected with the head. QS 1014 states that taken with wine it is effective for epilepsy.

5. *Myrrha*, or myrrh, from *Commiphora* spp. Several species have been advanced as the source of the ancient myrrh. J. André, *Lexique des termes de botanique en Latin* (Paris 1956) s.v. myrrha suggests *C. Schimperi* Engl. and *C. simplicifolia* Engl., while *C. myrrha* (Nees) Engl. (= *Balsamodendron myrrha* Nees) is suggested by others. Myrrh and frankincense, probably because of their fragrant odor, were often associated, cf. Pliny XXIV.118, 179; XXV.18, 41. A combination of linseed oil, myrrh, and water is held to be useful for complaints of the eyes and ears, Pliny XX.92, 249.

6. *Cephalote*, or savory is *Thymus capitatus* Hoffm. (= *Satureia capitata* L.). Also mentioned below in vv. 21, 71, and 73 as *thymbra*. For the identification cf. QS 423: thymbrae specimen, quem commemorant cephaloten. As with most of the Labiatae, it is difficulty to determine whether the *thymbra* of Diosc. III. 37, 2. Virgil, *Georgics*, IV 31, etc. are the same plants; at any rate it was infrequently referred to in ancient medical literature.

9. *Talpa*, the common European mole (*Talpa europaea* L.) is infrequently used in ancient medicine. Pliny XXV.3, 20 mentions it in connection with toothache and in a 9th century MS from St. Gall it occurs as a vermifuge, E. Landgraf (n. 5), p. 122. For animal-derived drugs in early Latin medicine, cf. W. B. McDaniel, The Medical and Magical Significance in Ancient Medicine of Things Connected with Reproduction and its Organs. *J. Hist. Med.*, 1948, *3*, 525-546 and for the period contemporaneous with Benedictus, cf. R. Simonini (n. 38), esp. pp. 326 sq.

12. *Symphoniaca*, or henbane, is a species of *Hyoscyamus*, perhaps *H. niger* L. or *H. reticulatus* L. This identification, maintained by André, Lexique s.v. symphoniaca, is confirmed by *Alphita*, ed. J. L. G. Mowat (Oxford 1887) p. 173; symphoniaca, insquiamus. . . . idem. g[allice] simphonie, an[glice] hennebane. The name is a transcription of συμφωνιακή which occurs very rarely, e.g. Pseudo-Apuleius IV Tit. p. 32 (n. 5) and, as useful for tumors, in a 10th century MS, cf. Chr. Ferckel (n. 6), p. 130. This is one of the few cases where Benedictus' remedies may have been of avail. The analgesic (v. 13) and narcotic (vv. 14-15) properties of henbane were well known to the ancients, cf. Celsus V.25, 2 and Benedictus 184 where its analgesic properties are probably in question.

17. *Sister*, or spignel (*Meum athamanticum* Jacq.). The identification is made likely by Simon Januensis: Meu, herbam. . . quod vulgus sistram vocant (ap. Meyer (n. 11), p. 422). Pliny XX.94, 253 states that it was grown especially for medicinal purposes and includes it as an ingredient in an antidote for the bites of venomous animals, XX.100, 264.

18. "Delirium," "madness," "frenzy" all fail to capture the many nuances of *phrenesis*. Literally, it means a disease of the φρην. Since the *phrēn* or diaphragm was thought to be the seat of consciousness or mind, a disease so localized was felt to result in some form of mental derangement. Cf. Caelius Aurelianus, *Morb. acut.*, I praef. 4, ed. I. E. Drabkin, *On Acute Diseases and On Chronic Diseases* (Chicago 1950). p.2: phrenas. . . graece mentes vocaverunt, quarum. . . impedimentum phrenitica ingerit passio. Homeric psychopathology was based on this belief, cf. R. B. Onians, *The Origins of European Thought* (Cambridge 1951) pp. 23-32 which became amalgamated with

later explanations covering a wide variety of mental aberrations. Cf. I. E. Drabkin, Remarks on Ancient Psychopathology. *Isis,* 1955, *46,* 223-234.

19. This verse appears to have been based on QS 87.

21. *Coriandrum,* or coriander (*Coriandrum sativum* L.). It is difficult to understand why Benedictus advised coriander as effective against delirium when Diosc. III.63 expressly states that an overdose may cause mental derangement. QS uses it for worms (570) and for eruptions (742), the latter based on Diosc. *loc. cit.*

22. *Raphanus,* or radish (*Raphanus sativus* L. et varr.). The widespread use of radishes (cf. Benedictus 50, 114; QS 117, 294, 309 etc.) may explain the reference in v. 23. As stated in the Preface, Benedictus selected what he believed to be the most useful remedies. For this reason it is not always possible to identify the source of a particular remedy.

27. *Marrubium,* or horehound (*Marrubium vulgare* L. and perhaps *M. creticum* Mill.). Frequently used for cough, but Pliny devotes an entire section to its uses (XX.89, 241-44) and mentions that its juice improves the eyesight (244) as does Diosc. III.105. QS mentions it seven times but not in this connection. The plant called *prasion* in Benedictus 50 and 157 is a synonym for marrubium, cf. Pliny XX.89, 241: marrubium. . . quod Graeci prasion vocant..

28. *Bet (t) onica, vet (t) onica* or betony (*Betonica officinalis* L.). QS 201 also advises its use to remove eye pain. Benedictus recommends betony for the lungs (49) and for the spleen (70), cf. Pliny XXVI.48, 75. The *Liber de herba vettonica,* attributed to Anontius Musa and usually printed following Pseudo-Apuleius (n. 5) mentions among the *virtutes vettonicae* that of removing eye pain.

31. *Verbena, berbena,* or vervain (*Verbena officinalis* L.). Also known as verbenaca or peristereon, cf. André, *Lexique. . .* s.v. verbenaca. Pliny XXV.91, 141 mentions it as effective for fluxes of the eye and Theodorus Priscianus (n. 1), II.14, 44, p. 144 for migraine.

33. *Limax* may refer to a number of different land snails or slugs. Cf. Diosc. II.9 and J. Berendes, *Des Pedanios Dioskurides Arzneimittellehre* (Stuttgart 1902), p. 156. Dioscorides (n. 4) recommends burnt whole snails applied to the eyes with honey and QS suggests that a raw snail applied to the head is useful. The flesh of a snail is also stated by Diosc. I.64 as useful when applied to contusions or fractures.

37. *Arsenicon* or yellow trisulphide of arsenic (AS_2S_3), also called auripigmentum, cf. Celsus V.5; auripigmentum, quod arsenicon a Graecis nominatur. Celsus VI.15, 1 used it as a caustic for ulcers in the mouth. The disulphide of arsenic (AS_2S_2), known also as sandaraca, is likewise used by Diosc. V. 105 for cough and other throat conditions.

Piper, or pepper (*Piper nigrum* L.). Much used both as a condiment and as a drug. Among its many uses, QS 228 recommends it with nitre and wine for toothache. Benedictus 221 also advises pepper in a potion for arthritic pain, a usage based perhaps on QS 462 where pepper corns are taken in a potion for lumbar pains.

46. Cf. Benedictus' Preface.

47. *Fenumgraecum,* or fenugreek (*Trigonella fenum graecum* L.). Also used externally (v. 96) and as a clyster (v. 98). Celsus II.12 E includes fenugreek in a mild enema. In a potion it is reported to be effective for colic (Diosc. II102).

Wool, wetted with oil, is used in a variety of ways, QS 194 using it to daub on a drug, and Benedictus 28 mentions it in connection with a topical application. For other uses cf. Diosc. II.73. Here it is apparently to be held near the nose and the pungent fenugreek juice inhaled. Cf. Cassius Felix (n. 1) LXII, p. 155 for a similar usage.

48. *Myrica,* or tamarix (*Tamarix gallica* L. or a related species). Tamarix was generally given in a drink for splenic conditions (QS 408; Pliny XXIV.41, 67; Diosc. I.87). Also mentioned by Diosc. I.87 as useful in a cataplasm for swellings, a use which may underlie Benedictus' reference in v. 234 for gout.

49. *Brassica,* or cabbage (*Brassica oleracea* L.) enjoyed much prominence in Roman medicine owing largely to the many therapeutic powers attributed to it by Cato, *De agricultura,* Chapters 156-157 ed. W. D. Hooper (London 1934) pp. 140-150. Pliny XX.33, 78 sq. states that Chrysippus wrote an entire book on its medicinal uses. For Chrysippus, cf. Jerry Stannard, Lost Botanical Writings of Antiquity (in press).

51. The word translated as "sage" appears in various forms in the MSS: *lelisfagus* (Cod. Vat., Pal. Lat., and Vat. Urb.), *liliffagus* (Cod. Vind.), *lelis fagus* (Cod. Paris). The former spelling, as the more conventional, is preferable but all are debased forms of ἐλελίσφακος or *Salvia* spp. QS 381 evidently knew the Greekk form for he states elelisphacus herba est. The Greek continued to be garbled throughout later botanical writings, cf. J. Schmidt, Das medicinisch–botanische Glossar von Siena. *Hermes,* 1883, *18,* 521-543 and especially p. 537 s.v. elilisfago.

52. The emendation of *picrae,* the latinized form πίκραι (bitter), for the MSS *pigrae,* is due to Ullrich. De Renzi follows Ullrich and defends the reading *picrae* on the

grounds that "bitter" refers to aloes. The bitter juice of *Aloe vera* L., for example, was well known, cf. Pliny XXVII.5, 14 and became the chief ingredient in the famous medieval compound known as *Picra Galieni* or *yera pigra*, cf. *Secreti medicinale di magistro Guasparino da Vienexia C. Castellani* ed. (Milan 1959), Sections 203, 321.

53. An echo of QS 399; cf. also QS 473.

54. *Fibras cordis* refers to the supposed lobes of the heart, a belief which goes back ultimately to Rufus of Ephesus and his theory of the lobulated liver. Cf. *Oeuvres de Rufus d' Éphèse*, edd. C. Daremberg et C. E. Ruelle (Paris 1879), pp. 158, 175. QS 380 writes *jecoris fibras* and this passage was probably the direct source of Benedictus' phrase.

55. A coral and emerald necklace is recommended by QS 942 for an attack of the *hemitritaion* fever. For the medicinal properties attributed to coral cf. Diosc. V.121 and Celsus V.18, 26.

56. *Pipinella*, a late form of *pimpinella* (*Sanguisorba minor* Scop. = *Poterium Sanguisorba* L.) is rather uncommon in early Latin medical literature unless it be identified with *sideritis* or *achilleos*, as André (n. 5) s.v. achilleos believes. If the identity can be maintained, its use is confined to topical applications for wounds according to Diosc. IV.34. As an ingredient in an unguent it is said to be beneficial ad dolorem. . . pectoris et stomachi, W. S. Van Den Berg, *Eene Middelnederlandsche Vertaling van het Antidotarium Nicolaï* (Leiden 1917) p. 171 and in a powder it is used *contra omnes angustias cordis*, K. Sudhoff (n. 6), p. 275.

Marathrum, or fennel (*Foeniculum vulgare* Gaertn.). Fennel seeds with goat's milk are reported by QS 312 as a carminative and Diosc. III.70 reports further therapeutic uses.

57. *Zeduar*, or zedoary (*Curcuma Zedoaria* Rosc.). The name derives from arab. *zadvar* from pers. *zeduar*, Eguilaz y Yanguas, *Glosario Etimológico de las Palabras Españolas. . . de origen oriental* (Granada 1886) p. 529. It was not known to early Greek writers despite Meyer's contention (n. 11), II 423 that it is found in Aetius and Paul of Aegina. Meyer was perpetuating an error which began with Serapion (cf. F. Adams, *The Seven Books of Paulus Aegineta* (London 1844-1857) vol. III, 435) who thought zedoary occurred in Galen and other Greek writers. It occurs in Near Eastern materia medica, e.g. Abu Mansur Muwaffak, Die pharmacologischen Grundsätze. . . ubers. von Abdul-Chalig Achundow, *Historische Studien a.d. pharmakol. Inst. d. k. univ. Dorpat*, 1893, 3, 215 s.v. *zarwâr* and entered the main stream of European medicine before the 12th century as it is found in Macer Floridus vv. 2131-40 ed. L. Choulant, Macer Floridus, *De viribus herbarum*. . . (Leipzig, L. Voss, 1832) p. 117. Zedoary was often confused with *C. Zerumbet* L., e.g. John Gerarde, *The Herball* edited by Thomas Johnson (London, Norton, 1633), p. 34 and continued to be even in the Byzantine world cf. M. H. Thomson, *Textes Grecs inédits relatifs aux plantes* (Paris 1955), p. 151 where **Ζουδάραν** is glossed by ΖουΖουμπέδ. This passage in Benedictus is noteworthy as being the earliest Latin reference to zedoary. H. Fischer, *Mittelalterliche Pflanzenkunde* (München 1929), p. 266 does not mention this passage and cites nothing earlier than Hildegard of Bingen. In point of fact, it occurs twice again in a 9th century MS edited by Sudhoff (n. 6), p. 275 (zaduar) and p. 277 (zitwar).

Tanaceta, or tansy (*Tanacetum vulgare* L.) does not occur in Dioscorides, Pliny, or QS. Ps-Apuleius XII (n 5), p. 45 mentions it as a febrifuge. In the MS edited by Sudhoff (n. 6) p. 275 it occurs as an ingredient in a powder "contra omnia venena. . . et contra omnes angustias cordis," a usage close to the present one. It becomes more common later, appearing in the list of plants which Charlemagne ordered to be planted for medicinal purposes, *Capitulare de villis* ed. Alfred Boretius, *Capitularia Regum Francorum* vol. I (Hannover 1883) p. 90 s.v. tanazita and finally ending up in the semi- official Alphita Salernitanae, ed. S. De Renzi (n. 37), vol. III (Napoli 1854), p. 317.

61. No exact parallel for this use has been found. It may be local folklore but a medicinally-inspired prayer from Southern Italy designates portions of a wolf to be used along with incantations, cf. Fritz Pradel, Griechische und süditalienische Gebete, Beschwörungen und Rezepte des Mittelalters. *Religionsgeschichtliche Versuche und Vorarbeiten* III/3 (Giessen 1907), pp. 286, 372. In Roman medicine, a wolf-liver is frequently mentioned in connection with liver complaints. Cf. QS 395 and Theod. Priscianus (n. 1), II.26, 75 pp. 177 and the criticism of this usage by Caelius Aurelianus (n. 18), III 16, 137, pp. 386, 388.

62. QS thrice (116, 285, 467) mentions *medulla cervina* as an ingredient in various preparations but not for pleurisy. Cf. Diosc. II.77 for the uses of different sorts of marrow. Goat's hoof is mentioned in Disc. II.44 but only for alopecia.

64. *Pix*, or pitch, the resin obtained from various conifers, especially the genus *Pinus*: used as a base for poultices and plasters (cf. v. 66). For the various types of pitch and their uses cf. Diosc. I 72.

Britia (v. 1 *bricia*) is a synonym for *pix*, cf. Pliny XVI.22, 53; XXIV.23, 37 and J. André, Contribution à l'explication des gloses de médecine. *Archivum Latinitatis Medii Aevi*, 1956, *26*, 194 s.v. pice.

Far usually refers to emmer or two-grained wheat (*Triticum dicoccum* Schrk.), cf. Pliny XXVI.22, 39 and N. Jasny, *The Wheats of Classical Antiquity* (Baltimore 1944), pp. 13, 112. Here it refers to some form of ground meal added to give bulk to the poultice mentioned in v. 66.

65. *Crocus*, or saffron (*Crocus sativus* L.). One of the most common medicinal plants of antiquity. Cf. Maria Tscholakowa, Zur Geschichte der medizinischen Verwendung des Safran. *Kyklos*, 1929, *2*, 179-190.

68. This verse does not appear in the Paris MS. Whatever be the reason for its absence, the meaning of *medicina tonantis* is not immediately clear, De Renzi ad. loc. understood it as a synonym for *barba iovis*, i.e., *Sempervivum tectorum* L., the common house leek. The more common name *barba iovis* occurs at 201 while at 234 the variant *barbula iovis* occurs. *Medicina tonantis* is apparently unrecorded; the closest parallel seems to be *tonerburz*, a gloss recorded by Fischer (n. 57), p. 284, as a synonym for *S. tectorum*.

70. Excrement of various kinds was used throughout ancient medicine beginning with Hippocrates. Benedictus also lists pigeon dung for internal use (203), dung-plasters for jaundice (100) and sheep-dung plasters for erysipelas (197-198). The latter is used by QS 964 as a plaster for a fracture which has become ulcerated. *Stercus columbae* is classed among the epispastics by Celsus V.12.

72. Directions for the preparation of rose oil and its uses are given by Theophrastus, *De odoribus*, 25, 47 ed. A. Hort (London 1916 in vol. II of Theophrastus, *Enquiry into Plants*), pp. 348, 368 and Diosc. I.43.

74. There is some question as to the correct reading and consequently on the identity of the plant in question. De Renzi, following the Vat. Pal. and Paris MSS prints *linguula cervi*, Vat. Urb. and Cod. Vind. have *lingula cervi* while Meyer (n. 11), II422 conjectures *lingula cervi*. Presumably these are all variants of an attested plant name *lingua cervina*. Therefore we follow the Cod. Vind. and understand *lingula* = *lingua*, a usage paralleled by Pliny XXV. 84, 133 where *lingulaca* = *lingua*. *Lingua cervi* is recorded however by André, Lexique s.vv. lingua ceruina, asplenos as occurring in the *Corp. Gloss. Lat.* and is identified as *Ceterach officinarum* Willd. If *lingua cervi* is identical with *lingua cervina*, it is hart's tongue fern, *Scolopendrium vulgare* Sm., so Fischer (n. 57) p. 203. Whatever be the correct identification, lingua cervina (Alphita ed. Mowat p. 103), the ἄσπληνον of Diosc. III 134 and the *asplenon* of Pliny XXVII.17, 34 are all said to be beneficial for the spleen.

81. *Costus*, or costmary, is the root of *Saussurea lappa* Clarke. The dried root, imported from the Near East, was valued for imparting fragrance to unguents and perfumes (Pliny XII.24, 41) and as a condiment, cf. Anthimus, De observatione ciborum 3, ed. Val. Rose (Leipzig 1877), p. 9. Pliny *l.c.* states that it has no medicinal use which is surprising since Diosc. I 16 lists several uses but fails to mention it in connection with the spleen. QS 396 mentions it in a drink for liver pains and Scribonius Largus 70, p. 30 Helmr. states that it is useful *ad anginam*. Although Costus played a small part in Western medicine, it was extensively used in the East, cf. Moses ben Maimon, *Sarh asmā' al uqqar (L'explication des noms de drogues)* ed. Max Meyerhof, Le Caire 1940 s.v. qust. This costus must not be confused with *Chrysanthemum balsamita* L. = *Tanacetum balsamita* L., also called costus and frequently employed in later Western medicine, e.g. the *Agnus Castus* ed. G. Brodin (Upsala 1950), p. 144. For an account, cf. H. Marzell, Zur Geschichte des Frauenblattes. *Centaurus*, 1951, *1*, 235-241.

Rheubarbarum is an unidentified species of the genus *Rheum*. Our rhubarb, *R. palmatum* L. and *R. officinale* L., was unknown in the Graeco-Roman world. When it entered the West is a matter of dispute, cf. B. Laufer, *Sino-Iranica* (Chicago 1919), pp. 547-550. For a discussion of Dioscordes' ρα (III.2) and Pliny's rhecoma (XXVII. 105, 128) —both suggested at times as *our* rhubarb—cf. H. Stadler, Rhabarber, *Pauly-Wissowa Realencyclopädie* I A/1 (Stuttgart 1914), col. 726-728, and Jerry Stannard, Dioscorides and Renaissance Materia Medica (in press).

82. Common salt, in the form of ocean water or brine was a common ingredient in potions and in the form of semi-purified crystals it was used in external applications. Cf. Diosc. V.109.

83. *Lupinus*, or lupine (*Lupinus hirsutus* L., *L. albus* L. etc.). Ground lupine meal was employed as a foodstuff (Pliny XVIII. 36, 136) and as a poultice. In this latter form it was used by Benedictus 65 as a poultice, by QS 82 for skin diseases (cf. Pliny XXII.74, 158) and added to a drink, it was said to be effective for disorders of the spleen, Benedictus vv. 73, 83 and Pliny XXII.74, 157.

90. Benedictus' etymology (torta. . . quia viscera torquet) is the usual Roman explanation of colic, called *tormina* (Pliny XXVI.47, 74) or *tormentum* (Cael. Aurelianus (n. 18), III.138 s.q. pp. 388 s.q. Drabkin). Cf. also Benedictus v. 87 where "twisting" translates *tortus*, all of which stem from the belief that the intestines were twisted (from *torqueo*, cf. Celsus II.7, 6). The Roman etymology follows the Greek ἰλεός or εἰλέος from ἕλω "twist" and goes back at least to Diocles Fr. 73, cf. Max Wellmann, *Fragmentsammlung der griechischen Aerzte* Vol. I (Berlin 1901), p. 149.

92. *Fervida lympha*, or warm water, is a poetical phrase borrowed from QS who frequently employs such terminology cf. QS 118, *tepefacta lympha*; 245, *gelida lympha*, etc. A related expression, *dira fluenta* occurs in Benedictus v. 55, and see note on v. 241.

95 Goose grease was a common source of the fats used as a base for ointments. Benedictus employs it here and in v. 98. In v. 137 it is combined with the fat of a black vulture. QS uses the fat of a hog (1048), vulture (971), swan (150) as well as a goose (178). On the vulture cf. note to v. 137.

100. *Squibala*, or dung-plasters, from σκύβαλον, dung or excrement. A recipe for a collirium contains stercora dura et squibalci, Singer (n. 1), p. 146. Celsus III.24, 5 includes poultices for jaundice but does not mention dung-plasters. Cf. note to v. 70.

102. Ivory, like deer antler and horns from various animals, was often powdered or calcined. Celsus V.5 lists powdered ivory and burnt stagshorn as possessing a cleansing effect, possibly because the latter was a source of quick lime (cf. note on v. 211 and, in general E. Trosse, Burnt Substances. *Janus*, 1896, *1*, 143-149.) QS 547 recommends ivory shavings as a remedy for dysentery, following Diosc. II.57. Ivory continued to be an important commodity and was still obtained by Genoese merchants in Constantinople in the early 14th century, cf. P. Halász, Pharmakognostisches aus dem Codex Cumanicus. *Arch. Gesch. Med.*, 1918/19, *11*, p. 280.

104. Viscera of various animals were employed both internally and externally. Cf. Benedictus vv. 20, 145 and QS 950 where the brain of a friendly dog (canis blandis) is applied in case of a fracture. For rabbit brain used internally cf. Benedictus 172 and note.

109. This line borrowed from QS 497.

111. *Fraxinus*, or ash (*Fraxinus excelsior* L.). The seeds, given in wine, are of use against dropsy according to QS 500 and Pliny XXIV.30, 46. Ashes of different vegetable products were often used, cf. QS 1042 *platani cineres* and Benedictus 122, 202 likewise refers to them.

112. *Hordeum*, or barley (*Hordeum vulgare* L.) was, like other cereal plants, used in the form of meal for poultices, cf. QS 710, 1047; Pliny XXIV.38, 63. A pitch-plaster is recommended for dropsy by QS 501. This verse is a translation of Cod. Vind.: ex his lixiviam facito, frange hordea dura. The other MSS present varied readings, cf. Ullrich and De Renzi ad loc. Brunhölzl prints *lisciviam* which may be the original spelling; it is listed as a *v. l.* by A. Forcellini, *Totius Latinitatis Lexicon*, 4 vols. (Patavii, Typis Seminarii, 1827-31) vol. II, 932 b. s.v. *lixivia*.

116. Crude native sulphur is common on the volcanic islands of the Aegean and has been used for medicinal purposes since Hippocrates, cf. Jerry Stannard, Hippocratic Pharmacology. *Bull. Hist. Med.*, 1961, *35*, 507-508. Usually it is applied externally but internal uses are recorded, QS 451, 456 recommends it in a potion for lumbar pains and Celsus IV. 10 advises its use for cough. A poultice containing both nitron and sulphur and used for dropsy occurs in A. Dold (n. 9), p. 213. For other early uses of sulphur, cf. E. von Lippmann, Die chemischen Kenntnisse des Dioskurides. *Ztsch. angew. Chemie*, 1905, *18*, 1210.

120. Goat's milk is used by QS 524 to settle an upset stomach. Benedictus, likewise, recommends asses' milk (121) and mare's milk (156).

122. *Vitis amineae*, i.e., the grapes from Aminaea. They were especially praised by Pliny XIV.4, 21 and Virgil, *Georg.*, II.97. QS 539 recommends sucus Aminei for dysentery, a usage also mentined by Celsus IV.26, 3.

124. A poultice often was employed for insect bites and localized skin infections. Here, presumably, it is composed of dried figs (*Ficus Carica* L.)—often used as the basis of a poultice, QS 974; Pliny XXIV.10, 16; Diosc. I.128—moistened with an aqueous solution containing celery and laurel berries, from *Laurus nobilis* L. The pungent oil of the latter was thought to be a repellent (Pliny XXIV.92, 148) while the macerated leaves sometimes were used alone for relief from insect bites, Diosc. I. 78.

125. *Iris illyrica*, or iris (*Iris Germanica* L.) The rhizome, ground up and applied as a poultice, was employed for sores and localized infections, Diosc. I.1; Celsus V.5; Scribon. Largus 44, p. 21 Helmreich.

Centaurea, or the greater centaury (*Centaurea Centaurium* L.). It was used for purposes similar to iris, Dioc. III.6, Pliny XXV.30, 67.

132. This remedy is copied from QS 177 who uses it for an insect embedded in the ear.

134. This recipe is borrowed from QS 169-170 who, in turn, followed Pliny XXIX.39.

135. The *Vermes* are probably larvae of some insect; Pépin *ad loc.* suggests larva of the order diptera but millepedes may also be included, cf. Adams on Paul of Aegina (n. 57), III, 277 s.v. ὄνοι.

137. For the medicinal uses of the vulture, cf. L. C. MacKinney, The Vulture in Medical Lore. *Ciba Symp.*, 1942, *4*, 1258-1292. Bile of the black vulture is mentioned by QS 204 as good for eye pain. Cf. note to v. 95.

138. *Cepa,* or onion *(Allium cepa* L.). Pliny XX.20, 40, among others, reports that onion juice, mixed with goose grease or honey, is useful for deafness. In general, cf. Kurt Heyser, Die Alliumarten als Arzneimittel im Gebrauch der abendländischen Medizin. *Kyklos,* 1928, *1*, 64-102.

139. Borrowed from QS 178-180.

143. *Galla asiana,* or Asian gall-nut, is the dried gall from *Quercus infectoria* Oliv. produced by several species of wasps of the genus *Cynips.* Its high tannin content, useful for tanning leather (Pliny XXIV.67, 109), determined its use as an astringent, Celsus V.6. It is explicitly mentioned as useful for dysentery by Cassius Felix 48, p. 123 Rose. Cerasus, or cherry (probably *Prunus cerasus* L., the sour cherry but it may also include *P. avium* L., the sweet cherry). Its use here derives from QS 538.

144. Cheese was similarly employed for dysentery by Theod. Priscianus II 31, 101 p. 201 Rose and Marcellus, *De medicamentis,* XXVII 66, p. 284, ed. Georg Helmreich (Leipzig 1889). QS 872 advised it as a cataplasm for spider bite.

150. Literally, "where pallas grows. . ." The text reads "Aspice quo (De Renzi, presumably by a slip, wrote "que") crescit. . ." Pallas or arbor Palladis is a poetical expression for olive *(Olea europaea* L.) reputedly first cultivated by Pallas or Minerva (Ovid *Met.* VIII.265). QS twice uses the expression succus Palladis for olive oil (457, 986). A different explanation is given by Meyer (n. 11), II 422 who, on the basis of a reference to Pallas Athene in the anonymous *Carmen de herbis* v. 15, ed. F. S. Lehrs in: *Poetae Buccolici et Didactici,* ed. Fr. Dübner et U. C. Bussemaker (Paris 1862), p. 173 identified it as *Rhamnus oleoides.* L.

153. *Helleborus,* or hellebore, including the white hellebore, *Veratrum album* L. and the black hellebore, *Helleborus niger* L., *H. viridis* L. etc. was often used as a purgative and emetic, cf. Diosc. IV.148 and IV. 162. This use may be indicated here by Benedictus' phrase *duritia ventris,* QS 506 uses it only for dropsy. Cf. the long discussion of the various types and their uses in Pliny XXV.21, 47-25, 61. For the identity of the various species, cf. J. André, Les noms latins de l'hellébore. *Rev. etudes latines,* 1955, *32*, 174 sq.

154. *Chamaeactis,* or danewort *(Sambucus ebulus* L.), a transliteration of καμαιάκτη, cf. Pliny XXIV.35, 51: Sambucus habet alterum genus magis silvestre, quod Graeci chamaeacten. . . vocant; elsewhere he calls it ebulus or acte, cf. XXVI.73, 120. QS uses it often (97, 333, 498 etc.) and in 523 it is used for a purpose similar to the present.

155. This verse is a close paraphrase of QS 397 who writes *sicci Lyaei* where Benedictus writes *veteri . . . lyaeo.* Lyaeus, by metonomy stands for wine, a usage of which QS is fond (150, 325 etc.). A similar usage is noted above in note to v. 150.

156. Borrowed from QS 525 and used for a similar purpose.

157. *Eupator,* apparently an unrecorded variant for *eupatoria* or *eupatorium,* agrimony *(Agrimonia eupatorium* L.). Frequently used for dysentery, Pliny XXV.29, 65; Diosc. IV. 41 and recommended by Ps-Apuleius (n. 5), XXXI, p. 73 for stomach pains.

Hermodactylus, or meadow saffron *(Colchicum autumnale* L.). A violent purge and, in some cases, toxic, cf. Diosc. IV.83. For later uses, cf. Jerry Stannard, Materia Medica in the Locke-Clarke Correspondence. *Bull. Hist. Med.,* 1963, *37*, 211.

161. This verse follows QS 754 who, however, is speaking about ignis sacer, i.e., erysipelas, the subject of ch. XXIII of the present poem.

167. *Plantago,* or plantain, an unidentified species of *Plantago* perhaps *P. maior* L. Recommended by Pliny XXVI.49, 78 for bladder complaints and as food for those suffering from dropsy ib. XXVI.73, 119. Kernels *(glandes)* refer to the knotty and jointed tap root which is easily fractured into equally-sized pieces.

168. *Olusatrum,* or alexanders *(Smyrnium olusatrum* L.). Also called hipposelinon and used as a diuretic by Pliny XX.46, 117 cf. Diosc. III.67.

169. Many different plants are included under this name, cf. André (n. 5) s.v. saxifraga. To which one Benedictus was referring is impossible to say. As the etymology suggests, the plant was supposed to possess lithontriptic properties owing, no doubt, to its habitat in rocky crevices which it was supposed to have broken. This best fits a cryptogamous plant, e.g. *Adiantum capillus veneris* L. or maiden hair fern, a plant identified with Pliny's *adiantum* (XXII.30, 64) and which is stated to be useful for breaking up bladder stones. Another plant which sometimes has been identified with

saxifraga is *Lithospermum officinale* L., which Pliny also notes is useful for stones (XXVII.74, 99).

Selinum, or celery (*Apium graveolens* L.). An ingredient in many medications, cf. v. 124 where, however, it is called *apium.*

172. Rabbit brain was a part of traditional medicine, cf. Hippocrates, Diseases of Woman 185 (VIII,366 Littré). Prescribed with wine for urinary disorders by QS 589. For Benedictus' feeling of repugnance, cf. the parallels in QS 541 and 922.

173. *Hyppuris,* or horsetail. The description "from marshy ground" (*humidis ex campis*) suggests the widely distributed genus *Equisetum,* perhaps *E. Telmateia* Ehr. Reportedly useful as a stypic in the *Pseudo-Dioscorides de herbis femininis* (n. 4), p. 610.

179. *Cuminum,* or cummin (*Cuminum cyminum* L.). Applied in the form of pastilles it is effective for nosebleed (Pliny XX.57, 160) and used internally for bloody urine (*ibid.,* XX.43, 111). QS mentions it six times but not in this connection.

180. *Boletus* designates several genera of fungus including mushrooms (cf. Pliny XXII.46, 92-94) as well as several tree-fungi of the genus *Boletus,* e.g. *B. laricis* Jacq. (cf. Pliny XXV.57, 103). The reference here to *pulvis boleti* probably refers to agaric, the crumbly powder obtained from *B. laricis,* and thus likened to the farina or farra lupini, cf. note to v. 83. Benedictus' use of agaric as haemostatic goes back to Pliny XXVI.19, 34.

181. *Mentha,* or mint, is probably the common mint, *Mentha viridis* L. However, the Labiatae are so numerous that exact identification is difficult. For a discussion of the linguistic and taxonomic problems, cf. A. C. Andrews, The Mints of the Greeks and Romans and their Condimentary Uses. *Osiris,* 1958, *13,* 127-149.

183 *Maurella* (lit. black fruits, referring to the dark purple berries), or black nightshade (*Solanum nigrum* L.). Commonly regarded as poisonous but tests indicate that it is inert, U. S. Dispensatory XXII p. 1588; cf. also Henri Leclerc, La pharmacologie de la Morelle. *Presse méd.,* 1938, *46,* 480. The plant often occurs in Latin medical writings as *strychnon* (Pliny XXI.105, 177; Celsus II.33, 2) and in late Latin and medieval writers as *morella* (Alphita ed De Renzi (n. 37), III, 301). This is the first occurrence of *maurella* which becomes normal after Macer Floridus v. 1918-19: herbam, quam Graeci Strignum [i.e. strychnon] dixere, Latini/maurellam dicunt (n. 57), p. 107. For the earlier history of *Solanum nigrum* L. and other members of the Solanaceae, most of which contain narcotic alkaloids, cf. Jerry Stannard, The Plant Called Moly. *Osiris,* 1962, *14,* 254-307.

186. The haemostatic action of lead salts was well known in antiquity and they were accordingly recommended for bleeding hemorrhoids, cf. Diosc. V. 81. For the identity of the various lead salts, cf. E. R. Caley and J. F. C. Richards, *Theophrastus on Stones.* (Columbus, Ohio 1956), pp. 187-191.

192. *Chelidonia,* or celandine (*Chelidonium maius* L.). Used by Pliny XXVI.87, 141 for skin disorders and by QS for erysipelas. Cf. note on v. 193.

193. This is borrowed from QS 764 who specifies both chelidonia and the white of an egg. Elsewhere QS mentions egg white for burns in 1043 and 1047. For other uses of eggs, cf. Diosc. II.50.

194. *Ruta,* or rue (*Ruta graveolens* L.). Much used in antiquity owing probably to the bitter odor of the crushed leaves. QS recommends it as a poultice for erysipelas (125). Benedictus likewise advises rue in a poultice (125) and inhaled for nosebleed (177). QS 646 also advises it in case of uterine hemorrhage.

202. *Savina,* or savine (*Juniperus sabina* L.) by reason of its pungent resin was employed in fumigations and in topical applications, as an ingredient in poultices and plasters, for localized infections and skin diseases, cf. Pliny XXIV.61, 102 and Diosc. I.76. In v. 6 Benedictus recommends *sal savina,* probably the ashes of savine shavings (cf. note to v. 11), for headache. For other uses, cf. L. C. MacKinney, Oleum Savininum: An Early Medieval Synthesis of Medical Prescriptions. *Bull. Hist. Med.,* 1944, *16,* 276-88.

208. *Rapus,* or turnip (*Brassica rapa* L.). The seeds are recommended *carboni tollendo* by QS 723 to which passage Benedictus seems to refer in v. 209.

211. QS advises the application of calcium oxide or quicklime for bruises (1046) and cauterization for gout (774). Quicklime is included in a list of caustics, Celsus v.8, cf. note to Benedictus v. 102. For the widespread practice of cautery, cf. E. Kind, Kauterisation, *Pauly-Wissowa Realencyclopädie* XI/1 (Stuttgart 1921), columns 94-99.

219. *Polypodium,* or polypody (*Polypodium vulgare* L.). Benedictus' synonymy is confirmed by Pliny XXVI.37, 58: polypodi, quam nostri filiculam vocant, similis filicis. In v. 233 it is called simply *filix.* Pliny XXVI.66, 105 recommends it for the sequelae of gout while QS 511 mentions it only in connection with dropsy.

220. *Cariophilus,* or cloves (*Eugenia caryophyllata* Thunb.). Cloves were unknown to Dioscorides and earlier writers and were imperfectly known to Pliny XII.15, 30 for he assigns to them no medicinal properties. QS does not mention them but their condimentary

use was well established, cf. Anthimus, chapters III and XIII. Apparently they enter the materia medica with Galen, *Opera* ed. C. G. Kühn, 20 vols. (Leipzig 1821-33) XIV, 462. After Paul of Aegina, VII.3 ed. J. L. Heiberg *(Corp. Med. Graec. IX/2*, Leipzig 1924) II, 221 they become common. Cf. E. Orth, Gewürznelke. *Pauly-Wissowa Realencyclopädie* VII/1 (Stuttgart 1910) cols. 1353-54.who, however, fails to mention Galen.

221. *Dactylus,* or date *(Phoenix dactylifera* L.). like figs, dates were used in poultices and Pliny XXIII.51, 97 recommends such a preparation for intestinal disorders and bruises. A cataplasm containing dates and aloe is mentioned in an early gynaecological MS, cf. Egert, (n. 9) p. 30. Internally, a potion containing dates was considered to be a restorative, Pliny *l.c.;* Diosc. I.109; cf. Dold (n. 9) p. 211.

Cinamomum, or cinnamon. It can no longer be determined which of several related species furnished the bark to the ancients. It seems likely that it includes *Cinnamomum Cassia* Bl. and *C. Ceylanicum* Nees. It was imperfectly known to Pliny for he speaks of a cinnamon shrub (ipse frutex duum cubitorum altitudine amplissimus. . . XII.42, 89) and its uses were confined to flavoring unguents, wines, oils etc. *(ibid.,* XIII.2, 12; XIV.19; 107; XV.7, 30). A drink containing pepper and *folium* is mentioned by QS 396 for pains in the side. *Folium* or *f. indicum* was the common designation in Latin literature, cf. Anthimus, XIII, p. 10 Rose and was identical with the *malabathron* of Pliny XII.59, 129. Cf. Horace, *Odes* II 7, 8; Diosc. I 12 and B. Laufer, Malabathron. *Journal Asiatique,* 1918, *12,* 5-49.

223. *Ulmus,* or elm *(Ulmus glabra* L.). The mucilagineous innerbark was applied to open wounds, burns, etc. owing perhaps to its astringent action, cf. QS 817; Pliny, XXIV.33, 49; Diosc. I. 84. Its use for arthritis appears to be rare. Benedictus, poetical term *liber,* translated as "inner bark" is borrowed from QS 411, 771.

224. *Rubia,* or madder *(Rubia tinctorum* L.). Used by QS 700 and Pliny XXIV.56, 94 for arthritic conditions of the hip.

225. *Urtica,* or nettle *(Urtica dioica* L.). The burning sensation caused by the application of nettle leaves to the skin possibly determined its use as a counter-irritant. Cf. Pliny XXII.15, 31-36 for its many reputed uses. QS 115, 530 mentions nettles but not for arthritic conditions.

232. A paraphrase of QS 450.

238. *Rucca (rutta,* Cod. Vindob.) is probably the sting-ray *(Trygon pastinaca* Cuv.). The name is not recorded elsewhere; Du Cange, *Glossarium ad scriptores mediae et infimae Latinitatis,* vol. VII (Niort 1886) cites this verse and glosses "piscis species". *Rucca* is probably a local name or dialectical variant for the sting-ray whose barbed "spike or spine [is] 5 or 6 inches long at the root of the tail", D'Arcy W. Thompson, *Glossary of Greek Fishes* (Oxford 1947) p. 270 s.v. *τρύγων*. While Thompson does not record any local names etymologically connected with *rucca,* the evidence from the older naturalists makes the identification probable. Guillaume Rondelet, *L'Histoire entiere des Poissons* (Lion (sic.): Bonhome, 1563) p. 265 lists *bruccho* as a synonym while Conrad Gesner, *Historiae animalium liber IIII. . . de piscium et aquatilium animantium natura* (Tiguri, Chr. Froschover, 1568), p. 802 states "Giftrocche, id est Raia venanata, vel Stahelrocch. . . ab aculeo sive radio in caudo venenato." Finally, Pietro Artedi, *Ichthyologia. . .* (Lugd. Batav., Conr. Wishoff, 1738), p. 100 under Pastinaca states "hodie Bruccus dicta" and mentions two Italian synonyms, "Bruco vel Bucho." Among the ancients the sting-ray was used medicinally, cf. Galen XII, 355 Kühn; Philumenus, *De venenatis animalibus eorumque remediis,* ed. Max Wellmann XXXVII *(Corp. Med. Graec.* X 1, 1; Leipzig 1908), p. 39 and especially Marcellus XV, 61, p. 146 Helmreich: "acu ossea, id est spicula trygonis, quae pastinaca dicitur, struman saepius punge, statim arescit." Both Gesner, *op. cit.,* pp. 799, 801 and Rondelet, *op. cit.,* p. 267 follow the account from Diosc. II. 20. The latest occurrence of the medicinal uses of sting-ray which I have noted is R. James, *Pharmacopoeia universalis* (London, for J. Hodges. . . and J. Wood, 1747), p. 524 s.v. Pastinaca marina.

239. *Prunus,* or plum *(Prunus domestica* L.). Pliny does not mention plums in connection with gout and QS 517 restricts their use to a potion for intestinal disorders.

241. The word "dreadful" is a translation of *dira;* other MSS read *dura.* Brunhölzl and De Renzi present conflicting evidence in their apparatus concerning the reading of the several MSS. Whatever the reading, the verse is an echo of QS 151 omne malum. . . fugabis.

XI

MEDIEVAL ITALIAN MEDICAL BOTANY

The origins of medical botany are lost in the shadows of pre-history. As an organized body of knowledge it first appears in the West in the Hippocratic treatises. With the passage of time, new medical theories were introduced, new social forces changed the tone and direction of medical practice, and new drugs were added to the older armamentarium. Despite such developments as these, medieval European medicine remained basically Greco-Roman. This is particularly evident in medical botany where it was not uncommon to find that for over a millenium the same drug plants were recommended for the same complaints and illnesses (1).

Side by side with the conservative nature of medieval medicine, new forms of phytotherapy were slowly being introduced and the classical tradition of medical botany simultaneously was being modified by new influences. It seems appropriate, therefore, on the present occasion to call attention to some of the highlights of medieval Italian medical botany. As the inheritors of a classical tradition, but also among the first to feel the impact of Islamic contributions, the Italian medical botanists of the Middle Ages richly deserve to be commemorated.

Quite clearly it is impossible to do justice to a subject for which many of the essential texts are inaccessible and about which a vast secondary literature has been assembled. By the same token, space prevents a detailed bio-bibliographical study of even the outstanding medieval authors whose writings enriched medicine in subsequent centuries. Rather, it is our purpose here to call attention to some of the principal themes of medieval Italian medical botany by citing a few representative texts.

As suggested above, medieval botany in general was based on the classics of antiquity. Italian medical botany was no exception. In fact, except for vernacular synonyms, there is little evidence of anything distinctively or pecularly Italian in the early texts, either written in Italy or by one trained in Italy. Benedictus Crispus (*d.* 725/735 A. D.), Archbishop of Milan, may be regarded as representative of this early period. His *Poema Medicinale* is a compilation of classical sources and added little to current medical or botanical knowledge. Still, it remains a valuable document and serves as a reference point for two matters: the lack of any first-hand knowledge of Greek texts on medical botany and secondly, a *terminus post quem* for the introduction of certain Eastern drugs and plant products (2). Moreover, by his allusions to Christianity, Benedictus pointed ahead to a time when successful therapy was believed to depend on the drugs actually used as well as divine grace (3).

Throughout medieval Italian medical botany, the same classical authorities tend to appear in text after text. Among the most popular were Dioscorides, Pliny, and Galen, though others were occasionally cited as well. In texts of the later Middle Ages, references are found to Isaac the Jew, Constantine the African, and various Arabic writers among whom Mesue, Avicenna, and Serapion were the most frequently cited.

One of the most distinctive characteristics of medieval botanical writers is their concern with plant names rather than with the plants themselves. While this is a proper procedure for Peter of Abano's glosses on the *Antidotarium Nicolai* or the lexicographical studies of Simon of Genoa and Matthaeus Silvaticus, the majority of the writers who professed to describe plants or their virtues were following previous authorities. About the beginning of the thirteenth century, this form of traditionalism had reached a point that Platearius could describe the grades and medicinal uses of a plant without supplying any description of the plant or even its synonyms. This method was followed by many later writers and survived even into the incunable editions of the *Hortus sanitatis*. Copying the standard authorities or paraphrasing them was the common literary practice of the time. Another literary technique was the adoption of an etymological method, though it must be admitted that it was not as common in Italian writings as in the French and English encyclopedists who followed the fanciful derivations of Isidore and Hrabanus Maurus. The practical outlook which characterized much of Italian medical botany did not prevent, however, an occasional digression on the reason for an unusual plant name. Platearius, for example, after noting that *corona regia* was a synonym of *melilotum* (*Melilotus* sp.) explained the name on the basis of its semi-circular seed (4). Veneration for authority even extended to the point of attributing spurious texts to well-known authors. The pseudo-Galen *De centaureo* is a good example of attaching the name of a recognized authority to a modest work which otherwise might not have proven popular.

Plant descriptions of the Middle Ages bear little resemblance to the precisely formulated descriptions of modern systematists. Already by Galen's time, the descriptions, most of which were actually borrowed from Dioscorides, had become abbreviated and often essential information was omitted. Only by the presence of the Dioscoridean name or synonyms can one be at all confident in the identification of a plant mentioned or described by later writers. Not infrequently, medieval writers explicitly cited their authorities, with the result that it is a simple matter to trace their sources. In the lexica and the various didactic poems devoted to medicinal plants, it is more difficult to determine sources. Although the anonymous author of the *Regimen sanitatis Salerni* sometimes cited his source, more often than not, the requirements of meter and verse precluded such information. For much the same reason, the descriptions themselves were tailored to fit the verse structure (5).

Names and synonyms are closely linked with the descriptions of plants. In the absence of first-hand knowledge of the plant, its name and synonyms were the only means of keeping alive its traditional therapeutic uses. Both the description of a plant and its names were designed for the common purpose of locating and identifying a plant believed to be useful for a previously specified medical need. That such knowledge was presumed on the part of the reader may be inferred by noting that Saladino of Ascoli devoted a chapter to the proper time for collecting herbs but provided no descriptions (6). In other cases, the description was intended to prevent confusion between two plants, one of which was believed to be poisonous or ineffective in therapy.

Since many of the plants indigenous to the Mediterranean Basin served a therapeutic as well as an alimentary function, there developed a body of knowledge comprised of folklore, empiricism, and the written tradition. Accordingly, it was felt by many of the Italian medical botanists that a common plant or tree required no description. Plants which were presumably known to everyone would thus be instantly recognized when found, though, of course, synonyms might be required in those cases where the plant name differed in different regions. It only remained for the writer to specify its therapeutic uses. Naturally, armchair botanizing of this sort sometimes ended in fruitless polemics having little to do with the plant itself or its virtues (7).

The alimentary uses of those plants or plant products which were also employed as simples or in compound drugs were discussed in many treatises dealing with dietetics and regimen. This is especially evident in texts inspired by the Salernitan teachings. The *Liber* of John of St. Paul, for example, dealt not only with drugs but also with the qualities and actions of different foods, many of which were of vegetable origin (8). Doubtless dietary habits changed over the centuries yet it is instructive to note that many of the same fruits recommended for gastrointestinal complaints in the *Regimen* occur also in later Italian texts (9). Some of the common plants which were used in medicine and in cooking appear to have been cultivated in gardens, as Manlius de Boscho remarked (10). Crescentius devoted one section to garden plants, many of which were intended to serve a medicinal purpose as is indicated by the tell-tale reference to their grades (11).

Dependence upon Greek and Roman writers for the description of plants and their names was necessary in the absence of personal knowledge of the plants themselves. With certain exceptions to be noted below, most medieval medical botanists were content to repeat what they found in their authorities without subjecting them to empirical confirmation. When they disagreed, it usually was in regard to orthography or to the proper synonyms. Philological wrangling of this kind is not surprising when it is recalled that many of the plants known in antiquity and still cited in the Middle Ages were indigenous to areas for which little exact knowledge was then available. Many of the plants from the Near East or portions of Africa were probably known in the Middle Ages only in the form of the dried products of commerce. The bark, seeds, roots, and gums of such plants and trees were extensively used as spices, incense, and, of course, as flavoring agents in foods, beverages, and drugs. That many of these products were plentiful in Italy in the early fourteenth century is proven by Pegelotti's list of prices and tariffs of *spezierie.* Approximately eighty different plant products are mentioned in his list, many of which were used in drugs, e. g. *agiafedita* (*Ferula Narthex* Boiss. and *Scorodosma foetidum* Bunge), *belorigi* (*Terminalia Bellerica* Roxb), *gherofani* (*Eugenia caryophyllata* Thunb.), *istorace* (*Styrax officinale* L.), *serrapino* (*Ferula persica* Willd.) and the celebrated *reopotico* or rhubarb (*Rheum* spp.) (12).

Frequently included in medieval descriptions is a statement concerning the homeland of a certain plant. But such commonplace phrases as *nascitur in India* or *crescit circa Babiloniam* must be accepted with the same caution as the more dubious claim that *herba Sancte Marie* (?*Tanacetum balsamita* L.) *nascitur in paradiso* (13). Statements concerning the nativity of plants were either based on the ancient classics or they represented mere guesswork. The fantastic illustrations in medieval manuscript herbals only strengthen the belief that medical botanists relied on their imagination when no other information was available (14). More confidence can be placed in the somewhat infrequent claims that the plant grew in Italy. It was reported by Platearius, for instance, that *alcanna* (*Lawsonia inermis* L.) grows well in Sicily and that *capparus* (*Capparis spinosa* L.) is found in Apuleia (15). Reliable information, it is true, often came from travelers but it was not until the end of the sixteenth century that Prospero Alpini succeeded in putting to rest some of the fables connected with plants whose products had been used for centuries in medicine (16).

Uncertainty about the geographical origin of plants and the tendency to accept authority uncritically, often led to confusion and error (17). For example, the juice of *cicuta* or poison hemlock (*Conium maculatum* L.) mixed with mother's milk was recommended for podagra (18). Apparently the confusion over this plant and its properties was deep seated for many centuries later Quirico de Augusti identified *cicuta* with *harmel* (19). In point of fact, *harmel* denoted quite a different plant — *Peganum Harmala* L. — which occurred in Italian writings under several different names (20). On the other hand, there were writers who claimed that they had actually seen or handled the plant under discussion. A noteworthy example of personal observation is Peter of Abano's claim to have seen a case of poisoning caused by the *fico de Faraone* (*Ficus Sycomorus* L.) in Sardinia (21). Approximately a century earlier, Rufinus wrote about local Italian plants in such a manner as to leave no doubt that he was personally acquainted with some of them and saw the dried products of other in apothecary shops (22).

Other sources of practical information derived from the apothecaries. The practice of adulterating drugs was already sufficiently established in antiquity to evoke the protests of Pliny (23). There is no reason to doubt that this practice continued through the Middle Ages. A variety of techniques were developed to detect fraudulent drugs, for example Simon of Genoa described the method for detecting adulterated *scammonea* (*Convolvulus Scammonia* L.) while a fraudulent preparation of manna was mentioned by Platearius (24). Closely related was the equally practical task of distinguishing the different qualities of the raw ingredients which went into the compound drugs (25). Similar texts dealt with the preservation of plant products and the period of time they could be stored without losing their efficacy (26).

Italian medical botanists of the Middle Ages were no less dependent upon classical authorities for their knowledge of plant names than for the descriptions. As we have noted above, there was little first hand knowledge of exotic plants that reached the medical texts. For this reason, much importance was attached to plant names and synonyms, especially those derived from Greek. Many plants, even common ones, were known under two, three or more different names. The difficulty of determining the identity of a plant by name alone was increased by the profusion of Italian vernacular names plus the fact that the apothecaries had their own, distinctive names (27). With the advent of Arabic medicine, new plants as well as older ones under new names were introduced into the literature.

But long before Arabic plant names became a puzzle to the average physician, lexica had enjoyed great popularity. The early lexica were concerned primarily with supplying the Latin equivalents of Greek names of plants and drugs (28). As Arabic plant names became more numerous, they were noted in lexica with Latin and sometimes their Greek equivalents (29). As time progressed, the multilingual lexicon appeared in which Italian (φράγγικον) names were added to those in Arabic, classical Greek, and *Koinë* (30).

Largely as a consequence of the synonyms preserved in the lexica and other medico-botanical texts, the majority of the plants cited by Italian medical botanists of the Middle Ages can be reasonably identified. This task has been facilitated by important studies on herbaria in which Latin and vernacular names are associated with the dried plants (31). Some names, of course, pose difficulties but this is partially explained by the abundance and diversity of late Italian names for plants (32). Recent studies have thrown much light on the dialectal peculiarities of the plant names from certain regions in Italy (33).

Just as many of the plant names occurring in medieval Italian texts had a classical origin, so too the majority of plants described in such texts were also known in antiquity. There were, however, new introductions. Examples of medicinal plants, unknown to the ancients, which occur in medieval Italian texts include spinach (*Spinacia oleracea* L.) (34), the fruits of the markingnut tree (*Semecarpus Anacardium* L. fil.) (35), nutmeg (*Myristica fragrans* Houtt.) (36) and some form of camphor (37).

Travelers, too, brought back information regarding unknown plants and some of their uses. Marco Polo appears to have been the first Italian to have observed the practice of chewing the betel nut (*Areca Catechu* L.) (38). Despite his claim that it was a healthy habit, there was little incentive to add it to the list of useful drugs. Polo's near contemporary, John of Marignolli, was enthusiastic in his praise of spoons made from the shell of the coconut (*Cocos nucifera* L.) for the purpose of averting poison (39). Although the coconut, or *nux indica* as it was often termed, had been introduced into both Arabic and European medicine at an earlier time, it appears that John was not acquainted with it in Italy prior to his voyage to the East (40).

Throughout the long tradition of medical botany stretching back to classical antiquity, the vascular plants had received more attention than the lower taxa. Because the reproductive mechanism of the cryptogams was not understood, they were often described as lacking flowers and fruit and that they were spontaneously generated out of mud or debris (41). It is perhaps in virtue of the fact that the cryptogams differed from the better known higher vascular plants, that poisonous and other undesirable properties were ascribed to them (42). Despite this bad publicity, ferns, mosses, and fungi found their place in materia medica and at the table too (43).

It is a short step from attributing undesirable, and sometimes undeserved properties to the lower plants, to the acceptance of magic in medical botany. Although magical incantations appear in Italian texts, they cannot be said to be any more pronounced in medieval Italy than elsewhere (44). As Pitré has shown, the folk tradition of southern Italy contained many magical beliefs, but usually they were confined to an oral tradition and seldom appeared in published medical texts (45). The same aplies to the botanico-astrological manuscripts, recently studied by Professor Pazzini, in which the therapeutic usefulness of various medical herbs are associated with their particular planets (46) Like their contemporaries, Italian writers of the Middle Ages sometimes subscribed to the so-called Doctrine of Signatures. Probably a belief in this underlay certain therapeutic counsels for which it is now difficult to find a reasonable explanation (47). This is not to say, however, that symbolism was ignored for there are many passages which combine the medicinal uses of plants with Christian symbolism. A good example is provided by John of Sancto Geminiano in his attempt to demonstrate that *zinziber* or ginger (*Zingiber officinale* Rosc.) signifies charity, that is, the root of justice. Because it was hot, it signified spiritual fire hile its moisture signified tears (48). The basis for this thinly disguised allegory is the theory of grades or degrees of the four humors which provided medieval physicians with a theoretical justification for drug therapy.

REFERENCES

(1) Rue (*Ruta graveolens L.*) was believed to be effective in correcting dimness of vision and in removing a film from the eyes: Dioscorides III 45 Wellm., Pliny, *Nat. Hist.* XX 51, 124; Piero Giacosa, « Un Ricettario del Secolo XI ». *Atti d. R. Accad. Torino* (Ser. II) 37 (1886) 653; Baptista Fiera, *Coena ... de herbarum virtutibus* (in) Eobanus Hessus, *De tuenda bona valedtudine*. Frankfort, 1564. Fol. 176r.

(2) Cf. Jerry Stannard, « Benedictus Crispus, An Eighth Century Medical Poet ». *Journal of the History of Medicine* 21 (1966) 24-46.

(3) At the end of one of his consilia, Ugo Benzi added, *Et hec puto cum auxilio Dei omnipotentis sibi sufficere*. D. P. Lockwood, *Ugo Benzi*. (Chicago, 1951), p. 295.

(4) Nicolaus praepositus, *Dispensarium ... ad aromatarios ...* Platearius *vulgo Circa instans nuncupatus de simplici medicina*. Lyons: J. Crespin for S. de Gabiano, 1537. Fol. 86v. For another example, cf. Lynn Thorndike, *ed. The Herbal of Rufinus*. Chicago, 1949 p. *278 s. v. saxifraga*.

(5) This is especially noticeable in Otto of Cremona (*fl. ca.* 1200), *De electione et viribus medicamentorum simplicium et compositum*. ed. Lud. Choulant. Leipzig, 1832.

(6) Saladini ... *compendium aromatariorum* (in) Ioh. Mesue, *Opera*. Venetiis apud Iuntas, 1623. Fols. 257r-v.

(7) Cf. Joh. Jac. Manlius de Boscho, *Interpretatio simplicium secundum ritum officinarum* (in) *Vera de herbarum*. Strassburg: Schott, 1532 pp. 167-182.

(8) G. H. Kroemer, *ed. Johanns von Sancto Paulo: Liber de simplicium medicinarum virtutibus*. diss. Borna-Leipzig, 1920.

(9) Compare, e. g., *Regimen sanitatis Salerni* ed. Salv. de Renzi, *Collectio Salernitana* Vol. V (Naples, 1859) verses 814-855 with Baptista Massa, *De fructibus vescendis*. Venice [pr. of Ausonius *ca.* 1472] [ap. Klebs 672.1].

(10) Manlius de Boscho, *ed. cit.* p. 171.

(11) Pietro Crescentio, *D'Agricoltura*. Venice: Bernardino Bindoni, 1542. Libro Sesto « De li horti della natura et utilita cosi d'herbe ... ».

(12) Francesco Balducci Pegolotti, *La Pratica della Mercatura*. ed. Allan Evans. Cambridge, Massachusetts: The Medieval Academy of America, 1936. pp. 293-297.

(13) Tullia Gasparrini *et alii*, edd. *Un inedito erbario farmaceutico medioevale*. Firenze, 1952 p. 101.

(14) Cf. Guido Battelli, « Il prezioso codice erbario medievale della Gambalunghiana di Rimini ». *La Bibliofilia* 38 (1936) 48-53.

(15) Platearius, *ed. cit.* Fols. 74v, 77v.

(16) Prospero Alpini, *De plantis aegypti liber*. Venice: Francisco de Franciscis, 1592.

(17) Cf. Jerry Stannard, « Medieval Reception of Classical Plant Names ». *Revue de Synthèse* IIIe ser. Nos. 49-51 (1968) 153-162.

(18) R. Simonini, « Herbolarium et materia medica Codice Ms. n. 296 della Biblioteca Governativa di Lucca ». *Rassegna di clinica, terapia e scienze affini* 35 (1936) p. 284.

(19) Quiricus de Augustis, *Lumen apothecariorum*. [Lyons: Matthias Huss, *ca.* 1498] [ap. Klebs 123.5] Fol. 16[va].

(20) Cf. Jerri Stannard, « The Plant Called Moly ». *Osiris* 14 (1962), 254-307.

(21) Pietro d'Abano, *Il Trattato De venenis ed.* Alberico Benedicenti. Firenze, 1959, p. 61.

(22) Cf. Rufinus *ed. cit.* pp. 19 *s. v. alleluia* and 196 *s. v. mirasolem.*

(23) Pliny, *Naturalis Historia* XXXIV 25, 108.

(24) Simon Januensis, *Liber servitoris* (in) Joh. Mesue, *Cum expositione Mondini* ... Venetiis apud Iuntas 1502. Foll. 338[ra]; Platearius *ed. cit.* Fol. 86[rb].

(25) Cf. the anonymous *De signis bonitatis medicamentorum* (in) S. de Renzi, *Collectio Salernitana* II (Naples, 1853) pp. 402-406.

(26) Cf. Saladini, *ed. cit.* Fol. 258[r].

(27) Cf. Manlius de Boscho, *ed. cit. passim* who supplies the apothecaries' name and those used by *nostri rustici.*

(28) As examples, cf. *Dynamidiorum libri duo* ed. A. Mai, *Classicorum Auctorum e Vaticanis Codicibus Editorum.* Vol. VII (Rome, 1835) pp. 399-458; Joh. Schmidt, « Das medicinisch-botanische Glossar von Siena ». *Hermes* 18 (1883) 521-545, and the *Synonyma* of Peter of Abano reprinted in, J. L. G. Mowat, *Alphita* (*Anecdota Oxoniensia* Vol. I, Part ii, Oxford, 1887) Appendix pp. 201-235.

(29) Simon of Genoa, *Synonyma Medicinae.* Venice: Anima Mia, 1486 [Klebs 920.4]. Arabic synonyms also occur in Matthaeus Silvaticus, *Opus pandectarum medicinae;* cf. the notes in Mowat's *apparatus* (supra note 28). I have not personally examined a copy of Silvaticus' *Pandectae.*

(30) Cf. Armand Delatte, *Anecdota Atheniensia et Alia.* Vol. II (Liège, 1939) especially nos. XII (pp. 393-417) and XIV (pp. 427-450).

(31) Cf. Ettore De Toni, « Un Codice-Erbario Anonimo », *Mem. d. Pontificia Accad. Romana dei Nuovi Lincei* 22 (1904) [separately paged *estratto* of 44 pp.]; ibid. « Un codice-erbario medico del secolo XV ». *Atti d. Pontificia Accad. Romana dei Nuovi Lincei* 63 (1909) [separately paged *estratto* of 6 pages.].

(32) A strange orthography sometimes disguises Italian vernacular names, cf. Giovanni Carbonelli, « Frammento medico del sec. VII. Cod. Vat. Urb. lat. 293 ». *Rassegna di clinica, terapia e scienze affini* 20 (1921) 5-21.

(33) For example, cf. G. Ineichen, « Die paduanische Mundart am Ende des 14. Jahrhunderts auf Grund des Erbario Carrarese ». *Zeitschrift für romanische Philologie* 73 (1957) 38-123.

(34) *Regimen sanitatis Salerni* verses 755, 785-86. Rufinus *ed. cit.* p. 299 apparently did not have any personal knowledge of it.

(35) Platearius, *ed. cit.* fol. 73vb; Otto of Cremona p. 159 ed. Choulant; Rufinus p. 24 cites Dioscorides but this was an error on his part; Simon of Genoa, *Synonyma ed. cit.* fol. 10^{r} *s. v. anacardus*. The fruits were declared to be poisonous by Peter of Abano, *De venenis ed. cit.* p. 55. An imaginary illustration of the tree is reproduced in Elena Berti Toesca, « Un erbolario del '300 ». *La Bibliofilia* 39 (1937) p. 346.

(36) Simon of Genoa, *Synonyma* fol 65ra *s. v. miristica;* fol. 67rb *s. v. nux miristica.* Cf. Otto of Cremona ed. Choulant p. 164. For the medieval introduction of the nutmeg, cf. Otto Warburg, *Die Muskatnuss* (Leipzig, 1897) pp. 32-48.

(37) The camphor known to Rufinus pp. 63-64; Platearius fol. 76va; Otto of Cremona p. 161; *Regimen sanitatis Salerni* verses 913 and 1205 was *Cinnamomum Camphora* Nees et Eberm. Marco Polo's report (Book III Chap. xi ed. Yule) concerned Borneo camphor from *Dryobalanops aromatica* Gärtn., not available in Europe until later.

(38) *The Book of Ser Marco Polo* ... translated and edited by Henry Yule, Third Edition. (New York, 1903) Vol. II p. 371.

(39) Henry Yule *ed. Cathay and the Way Thither.* Vol. III (London, 1914) p. 236.

(40) The coconut is glossed in the *Alphita,* ed. Mowat p. 126 *nux magna que venit de yndia.* It is thus surprising that the Salernitan *Alphita* ed. de Renzi, *Coll. Salern.* III (Naples, 1854) p. 303 should confuse the matter by stating *nux vomica, nux indica idem.* In point of fact, *nux vomica* refers to *Strychnos nux-vomica* L., a poisonous plant as Peter of Abano, *De venenis* p. 59 and other glossators recognized, cf. Jerry Stannard, « A Fifteenth Century Botanical Glossary (Huntington Library MS HM 64) ». *Isis* 55 (1965) 367. The error made by the Salernitan glossator escaped the attention of Rudolf Creutz, « Das mittelalterliche medizinisch-botanische Vokabularium Alphita ». *Studien und Quellen z. Gesch. d. Naturw. u. d. Med.* 7 (1940) p. 486. Cf. Simon of Genoa, *Synonyma* fol. 66vb *s. v. neregil.*

(41) Cf. Leporace, *ed. cit.* p. 109 and Simon of Genoa, *Synonyma* fol. 43ra *s. v.* fungi.

(42) Peter of Abano regarded *succus usnae* (*Usnea* sp.) as poisonous, *De venenis* p. 35. Elsewhere (p. 58) he recommended the bezoar stone as protection against *malos fungos.* Cf. also, *Regimen sanitatis Salerni* verses 792-93.

(43) Cf. Otto of Cremona pp. 159 (*s. v. agaricus*) and 164 (*s. v. muscus*); Simon of Genoa, *Synonyma* fol. 94vb *s. v. tubera;* fol. 74rb *s. v. polipodion.*

(44) For an example of a magical incantation, cf. Carlo Castellani, *ed. Secreti Medicinali di Magistro Guasparino da Vienexia.* (Milano, 1952) p. 4.

(45) Giuseppe Pitrè, *Medicina popolare Siciliana*. Firenze, 1949.

(46) Adalberto Pazzini, *Virtù delle erbe secondo i sette pianeti*. Milano, 1959.

(47) The use of *satirion* to help a woman conceive (Leporace, *ed. cit.* p. 125) rests on the fact that the swollen, tuberous roots of one of several species of wild orchids (*Orchis* sp. *Ophrys* sp. etc.) were thought to resemble testicles. Cf. Dioscorides III 126 Wellm. For its alleged aphrodisiac powers, cf. Pliny, *Nat. Hist.* XXVI 63, 98.

(48) Johannes de Sancto Geminiano, *Summa de exemplis, et rerum similitudinibus locupletissima* ... Venice: Damiani Zenari, 1576. Fol. 113r.

XII

GRECO-ROMAN MATERIA MEDICA IN MEDIEVAL GERMANY

Precisely when and why the ancient world began to decline are questions which have fascinated historians ever since Gibbon's famous reply to St. Augustine. Less controversial, perhaps, but no less challenging is the attempt to reconstruct the intellectual life of Romano-Germanic culture. The descendents of the earlier, warlike Germanic tribes—people who originally owed little either to Rome or to Athens—were not slow to adopt such portions of another culture as suited their needs. Among those needs, medicine and the healing arts occupied, for obvious reasons, a conspicuous place.[1]

Even a cursory glance at early medieval German medicine reveals the presence of several different and sometimes antagonistic influences. The three which particularly concern us here are Latin Christianity, Germanic folk custom and, of course, Greco-Roman medicine itself. The dissemination of ancient medical teachings, and their assimilation by what was originally an alien culture, owed much to Christianity. The spread of Latin as a language went hand in hand with the spread of Christianity and whatever may have been the actual content of a Latin text, it was an essential part of Romanization. Already by the eighth century, monasticism—one of the most indelible contributions of medieval Christianity—found expression at St. Gall (630 A. D.), Reichenau (724), and Fulda (744). The establishment of monasteries at those places provided a framework for the maintenance of the classical tradition and a means for its transmission. But equally as important, those cultural microcosms provided a meeting place for the two traditions and their respective cultures—a classical tradition, expressed in varying degrees of latinity, and a native Germanic tradition, expressed, originally at least, in the vernacular.

Accordingly, I propose to concentrate on one class of medical writings in which the merger of the two traditions may be most clearly seen, viz., materia medica. Other classes of medieval German medical writings which had classical antecedents could easily be added, for example, pulse

Read at the 44th annual meeting of the American Association for the History of Medicine, Colorado Springs, Col., April 29, 1971.

[1] Cf. Max Höfler, "Ueber germanische Heilkunde," *Janus*, 1897, *2*: 10-22, 137-152.

tracts,[2] the inspection of urine,[3] on fevers,[4] bloodletting,[5] bathing,[6] dietary regimens,[7] surgery,[8] anatomy,[9] veterinary medicine,[10] as well as lapi-

[2] Luzie Drach und Gundolf Keil, "Der 'Altdeutsche Pulstraktat' des Londoner Kodex Wellcome 49," *Janus*, 1967, *54*: 287-296. Among the texts enumerated in the catalogue of the library at Reichenau in 841 was Galen's *liber perisfegmonis* [i. e., Περὶ σφυγμῶν], Karl Baas, "Zur Geschichte der mittelalterlichen Heilkunst in Bodenseegebiet," *Arch. f. Kulturgesch.*, 1906, *4*: 132.

[3] Hans Pohl, *Ein Pseudo-Galen-Text aus dem frühen Mittelalter, betitelt: "De pulsis et urinis omnium causarum" aus der Handschrift Nr. 44 der Stiftsbibliothek zu St. Gallen.* Inaug.-Diss., Leipzig [1922] and Willy Braekman und Gundolf Keil, "Die 'Vlaamsche leringe van orinen' in einer niederfränkischen Fassung des 14. Jahrhunderts," *Niederdeutsche Mitt.*, 1968, *24*: 75-125.

[4] Franz Pfeiffer, "Zwei deutsche Arzneibücher aus dem 12. und 13. Jahrhundert," *Stzb., Akad. d. Wiss., Wien, Phil.-hist. Kl.*, 1863, *42*: 160. A fourteenth-century *Regimen in febribus* was edited by Karl Sudhoff, "Meister Sulko (Sulken)," *Sudhoffs Arch.*, 1908, *2*: 48.

[5] A portion of a ninth-century text was published by Gerhard Eis, "Meister Alexanders Monatsregeln," *Lychnos*, 1950/51, 105. Other texts have been published by Karl Sudhoff, "Eine Aderlassinstruktion aus dem Anfang des 15. Jahrhunderts," *Sudhoffs Arch.*, 1907, *1*: 157-159; "Noch eine ausführliche Aderlassinstruktion auf einem Einblattkalender, gedruckt aufs Jahr 1490 bei Johann Sensenschmidt in Bamberg," *ibid.*, 1908, *1*: 390-391; and "Monatsregeln für den Aderlass aus Cod. monac. lat. 13076 vom Jahre 1356," *ibid.*, 1908, *1*: 447-448.

[6] Carl Ehrle, "Dr. Heinrich Stainhöwels regimen sanitatis," *Deutsches Arch. f. Gesch. d. Med. u. med. Geographie*, 1881, *4*: 328-332. [Ehrle's ascription of the *regimen* to Steinhöwel (!) has now been proven incorrect; cf. G. Eis (n. 5 above), p. 111.], and Josef Hofmann, "Ein fränkisches Arzneibuch von 1398 mit Ortolfs von Bayerland 'Mark aller Erzneien,'" *Mainfränkisches Jahrb. f. Gesch. u. Kunst*, 1955, *1*: 128 ("Badervorschriften").

[7] Manfred Peter Koch, *Das 'Erfurter Kartäuserregimen.' Studien zur diätetischen Literatur des Mittelalters.* Med. Diss. Bonn, 1969. Cf. the bibliography pp. 95-100 for previously published texts.

[8] For texts and documents, see Karl Sudhoff, *Beiträge zur Geschichte der Chirurgie im Mittelalter.* 2 vols. (Leipzig, 1914-1918) [*Studien z. Gesch. d. Med.* 10, 11/12.]. 10, 11/12].

[9] Christoph Ferckel, "Ein deutscher anatomischer Vindiciantext," *Sudhoffs Arch.*, 1914, *7*: 306-318.

[10] Gerhard Eis, "Deutsche Heilmittel für Haustiere aus dem Jahre 1321," *Deutsche Tierärztliche Wchnschr.*, 1958, *65*: 115-116. Other texts are described by Karl Sudhoff, "Deutsche Rossarzneibücher des Mittelalters." *Sudhoffs Arch.*, 1912, *6*: 223-230; 1914, *7*: 335-346. Cf. Christoph Ferckel, "Medizinische Marginalien aus dem Cod. Trevirens Nr. 40," *Sudhoffs Arch.*, 1913, *7*: 138 for a tenth-century incantation (with recipe) for lameness in a horse.

[11] Anton Birlinger, "Kleinere deutsche Sprachdenkmäler des XI. XII. Jahrhunderts," *Germania*, 1863, *8*: 298-303 ("Von den Steinen" pp. 301-303). Cf. Hans Lambel, *Das Steinbuch* (Heibronn, 1877), (cited from the review by Ernst Martin, *Anzeiger für deutsches Alterthum und deutsche Literatur*, 1879, *5*: 224-225, but not available to me); a similar text was edited by J. G. Büsching, "Die Kraefte der Edelgesteine, nach dem Glauben des Mittelalters," *Museum f. altdeutsche Literatur u. Kunst*, 1811, *2*: 52-145.

[12] Cf. Hermann Menhardt, "Wanderungen des ältesten deutschen Physiologus," *Ztschr. f. deutsches Altertum*, 1937 *74*: 37-39.

daries,[11] bestiaries,[12] encyclopedias,[13] and philosophical,[14] and theological [15] texts with medical sections. Even the well-known *Pestregimina* of the later fourteenth century occasionally alluded to classical authors in a desperate, if anachronistic attempt to find an effective remedy.[16] But for the following two reasons, I have concentrated upon materia medica and have drawn my illustrations either from texts written in the vernacular or from Latin texts known to have been read or written in medieval Germany.

First, texts on materia medica and collections of recipes were two of the most popular literary genres in late antiquity and continued to be popular throughout the entire Middle Ages.[17] Consequently, it is sometimes possible to trace a recipe— its name, ingredients, specific use, etc.— back to its original classical form and to compare the two with respect to additions or modifications.[18]

Second, the choice of simples for purposes of healing was an activity which historically required neither pharmacological nor physiological theory. Consequently, Germanic folk custom plus the availability of the

[13] Hildegard of Bingen, "Physica" in J. P. Migne, *Patrologiae Latinae,* vol. 197 (Paris, 1882), cols. 1125 sqq.; Konrad of Megenberg, *Das Buch der Natur,* Franz Pfeiffer, ed. (Hildesheim: Georg Olms, 1962 [facs. repr. of 1861 Stuttgart edition]). Almost contemporaneous with Konrad was Everhard of Wampen (near Greifswald) whose versified encyclopedia exists now only in a fragment, cf. Erik Björkmann, "Everhards von Wampen Spiegel der Natur," *Upsala Universitets Årsskrift,* 1902, XXV, 57 pp. For another side of the encyclopedic genre, see Wolfgang Hirth, "Zu den deutschen Bearbeitungen der Secreta Secretorum des Mittelalters," *Leuvense Bijdragen,* 1966, *55*: 40-70.

[14] Albertus Magnus, *De Vegetabilibus Libri VII, Historiae Naturalis pars XVIII.* edd. E. Meyer et C. Jessen (Berlin: Reimer, 1867). Book VI (pp. 339-588) describes the medical virtues of some 300 plants in alphabetical order.

[15] Hrabanus Maurus, "De Universo" in J. P. Migne, *Patrologiae Latinae,* vol. 111 (Paris, 1864), cols. 9-614. For the influence of medical ideas in theological writings, see Robert Herrlinger, "Die sechs res non naturales in der Predigten Bertholds von Regensburg," *Sudhoffs Arch.,* 1958, *42*: 27-38.

[16] See, for example, Karl Sudhoff, "Pestschriften aus den ersten 150 Jahren nach der Epidemie des 'schwarzen Todes' 1348," *Sudhoffs Arch.,* 1910, *4*: 196, Nr. 11 [anno 1371]: Nunc vestrae magnificentiae conscripsi breve et compilatum de melioribus dictis Gallieni, Avicennae et Ypocratis grecorum medicorum.

[17] Henry E. Sigerist, *Studien und Texte zur frühmittelalterlichen Rezeptliteratur* (Leipzig; Barth, 1923) [*Studien zur Gesch. d. Medizin,* 13].

[18] For example, see Sigerist (n. 17), p. 24 f. (Nr. xiii). Parallels to Celsus' remedy for *vermes* in the ear (*De medicina* VI vii, 5) are examined by Gundolf Keil, "Die Bekämpfung des Ohrwurms nach Anweisungen spätmittelalterlicher und frühneuzeitlicher deutscher Arzneibücher," *Ztschr, f. deutsche Philologie,* 1960, *79*: 176-200. The Galenic sources for recipes are noted in Ludwig Englert, "Die medizinhistorische Bedeutung des Fragmentum Donaueschingense," *Sudhoffs Arch.,* 1931, *24*: 220-244.

required ingredients or admissible substitutes, made possible a fusion of the two traditions at a practical level.

It is significant that the oldest published German medical document, preserved in a Basel manuscript of the eighth or ninth century, contains two short recipes, written in a combination of Latin and Old High German.[19] In the former of these *Basler Rezepte*, as they are commonly called, three of the plant products included as ingredients were exotica, viz., myrrh, frankincense, and pepper. Such foreign drugs as those were not available in Europe at that time except in the form of dried products obtained from an apothecary.[20] The remaining plants mentioned in the recipe were indigenes, some of which were widely distributed in Europe and presumably available locally.

Whether the two recipes from Basel were copied from a now-unknown Latin original and, if so, whether this was done at a monastery, it is impossible to say. There is no doubt, however, that the German monasteries, modeled on earlier Benedictine monasteries, served as focal points for medical aid.[21] An *infirmarium*, for example, is described in the plans of the St. Gall monastery drawn up in 820.[22] In the same manuscript, a *herbularius* or physic garden is also depicted, with sixteen plots, each neatly labeled with the name of the plant to be grown therein.[23] At almost the same time, Walahfrid Strabus (808/09-849) was describing the virtues of the plants growing in the monastery garden at Reichenau.[24] For the present purpose, it is sufficient to note that the plants growing in these gardens—and others like them for which records survive—were identical to those used in Greco-Roman materia medica.[25] Finally, it may

[19] Karl Sudhoff, " Die gedruckten mittelalterlichen medizinischen Texte in germanischen Sprachen," *Sudhoffs Arch.*, 1909, *3*: 274, Nr. 1. The same recipe, but with variant readings, appears in Wilhelm Wackernagel, *Altdeutsches Lesebuch.* Fünfte Aufl. (Basel: Schweighauser, 1873), cols. 233-234. Sudhoff's transcription has been followed here.

[20] See below, n. 54.

[21] Cf. the interesting statement on the need of medicine written in a Benedictine monastery (*ca.* 750-850), edited by Karl Sudhoff, " Eine Verteidigung der Heilkunde aus den Zeiten der Mönchsmedizin," *Sudhoffs Arch.*, 1913, *7*: 223-237.

[22] Reproduced in Henry E. Sigerist, " Aus der Geschichte des Krankenhauswesens im früheren Mittelalter in Morgenland und Abendland." *Sudhoffs Arch.*, 1929, *21*: 188.

[23] Sigerist, *op. cit.* (n. 22), p. 189.

[24] Twenty-nine different plants are mentioned by Walahfrid, *Hortulus,* G. H. M. Lawrence, ed. (Pittsburgh: Hunt Botanical Library, 1966). In a manuscript of the ninth or tenth century, six of Walahfrid's latin plant names are glossed in Old High German, E. Sievers, " Glossen zu Walahfrid Strabus," *Ztschr. f. deutsches Alterthum,* 1872, *15*: 532-534.

[25] Charlemagne, *Capitulare de Villis,* A. Boretius, ed. (in) *Capitularia Regum Francorum* vol. I (Hannover, 1883), pp. 82-91 [*Monumenta Germaniae Historica,* Legum

be noted that there is some evidence of an apothecary, either connected or associated in some loose manner with the early monasteries.[26] Whether or not those *Klosterapotheken* were modeled on the earlier depots which served the Roman colonists or soldiers, the fact remains that by such means some medical supplies would be available when required.[27]

As important as these physical facilities may have been, it was the medical texts of antiquity which provided the basis of medieval German materia medica. From those texts and now-lost epitomes, information was obtained concerning the names, synonyms, descriptions, and uses of the simples employed in ancient medicine. Because of its potential value, this information was gradually assimilated in numerous texts of varying length, purpose, and date, some representatives of which will be examined below.

The beginnings of German materia medica are discernible in the attempt to translate Latin recipes into the vernacular. Returning once again to the *Basler Rezepte, murra* is a simple transliteration of Latin *myrr(h)a* while *peffur* is derived from Latin *piper.* On the other hand, the term *uuiroh* (nhd. *Weyrouch*), translated above as "frankincense," is a translatin of Latin *tus*, itself borrowed from Greek θύος or incense. This meant that by the time the above recipe was written some synonomies already had been established. Perhaps the anonymous compiler had access to a glossed text.[28] or to a bilingual glossary.[29] The latter, however, need not have

Sectio II]; Hermann Fischer, "Mittelhochdeutsche Receptare aus bayerischen Klöstern und ihre Heilpflanzen," *Mitteilungen der bayerischen botanischen Gesellschaft,* 1925, *4*: 69-75; Alois Müller, "Quomodo crocus coli debeat," *Germania,* 1885, *30*: 381-382. For analysis of these and other contemporary documents, see Rudolph von Fischer-Benzon, *Altdeutsche Gartenflora* (Kiel/Leipzig: Lipsius & Tischer, 1894) and H. Steinvorth, "Die fränkischen Kaisergärten, die Bauerngärten der Niedersachsen und die Fensterflora derselben," *Jahreshefte des naturwissenschaftlichen Vereins für das Fürstentum Lüneburg,* 1890, *11*: 33-66.

[26] Julius Berendes, "Ueber Klosterapotheken und Klostergärten im Mittelalter," *Deutsche Apotheker-Zeitung,* 1909, *24*: 113-116; and Ernst Becker, "Die Geschichte der Medicin in Hildesheim während des Mittelalters," *Ztschr. f. klin. Med.,* 1899, *38*: 308-311, 322.

[27] For evidence of a Roman *valetudinarium* and some of the simples stored therein, see Karl-Heinz Knörzer, "Römerzeitliche Heilkräuter aus Novaesium (Neuss Rh.)," *Sudhoffs Arch.,* 1965, *49*: 416-422.

[28] Cf. Hans Butzmann, "Althochdeutsche Glossen aus dem Kloster Tholey im Saarland," *Beiträge z. Gesch. d. deutschen Sprache u. Literatur* [Halle], 1960, *81*: 428-435.

[29] Many glossaries have been published. The following is a selection of those glossaries which contain the names of simples or references to medical matters. L. Bethmann, "Alte Glossen." *Ztsch. f. deutsches Alterthum,* 1845, *5*: 193-211; Renward Brandstetter, "Der Ebingersche Vokabularius," *Arch. f. d Studium d. neueren Sprachen u. Literaturen,* 1885, *73*: 99-105; H. Hoffmann von Fallersleben, "Vocabularius latino-

been a medical glossary. Since *myrr(h)a* and *tus* also occur in the Scriptures,[30] a collection of biblical glosses with German synonyms would have been of use.[31] *Piper*, on the other hand, does not occur in the Scriptures, but it was so often mentioned in classical writings,[32] that it is difficult to believe that it was not known to some,[33] especially to members of those monasteries which possessed a well-stocked library.[34]

An example of a glossed text, similar to that hypothesized above, is known in the form of a ninth-century Latin recipe entitled, *Remedium contra febrim.* But, as the opening words make clear, it was more than a febrifuge. It was, " A powder against all fevers and against all poisons, and the bite of all serpents, and against all difficulties of heart and body." [35] Then follow the names and amounts of fifty-four botanicals, concluding with directions for their preparation. Of the Latin plant names in the

teutonicus," *Ztschr. f. deutsches Alterthum*, 1843, *3*: 368-381; *ibid.*, " Prager Glossen," *Ztschr. f. deutsches Alterthum*, 1843, *3*: 468-477; Leo Jordan, " Ein mittelniederdeutsches Pflanzenglossar," *Ztschr. f. deutsche Wortforschung*, 1902, *3*: 353-356; F. Liebermann, " Hd. Glossen des zwölften Jhs " *Ztschr. f. deutsches Alterthum*, 1887, *31*: 362-363. In addition to the glossaries here noted, see below, notes 31 and 52.

[30] Frankincense and myrrh occur together in two passages: Matthew II, 11 and Song of Songs III, 6. Myrrh and cassia are mentioned in Psalms XLIV, 9 and frankincense and cinnamon in Revelations XVIII, 13. (All references are to the Vulgate). The verse from Matthew is translated in the Old High German *Tatian* (*ca.* 825): Brâhtun imo gold inti uuîhrouch inti myrrûn (quoted by Hans Eggers, *Deutsche Sprachgeschichte* I: *Das Althochdeutsche* [Reinbek: Rowohlt, 1963], p. 101).

[31] Latin plant names, some of them denoting medicinal plants and glossed in German, appear in Adolf Holtzmann, " Die alten Glossare," *Germania*, 1863, *8*: 385-414 and Wilhelm Wackernagel, " Biblische Glossen zu Engelberg und Rheinau," *Ztschr. f. deutsches Alterthum*, 1843, *3*: 123-128.

[32] Hippocrates, *Diseases of Women* I 81 (VIII 202 ed. Littré); Theophrastus, *Historia Plantarum* IX 20, 1; Dioscorides II 159 Wellm.; Pliny, *Hist. Nat.* XII xiv, 26 sq.; Celsus, *De Medicina* II 27; Galen, *De simpl. med.* VIII 11 (XII p. 97 ed. Kühn); Juvenal, *Satires* XIV 293; Persius, *Satires* V 55; Jerome, *Letters* XXXI 2; LII 12; Isidore, *Etymologiae* XVII 8, 8. For further references see H. Steier, " Pfeffer " (in) Pauly-Wissowa, *Realencyclopädie der classischen Altertumswissenschaft* XIX/2 (Stuttgart, 1938) cols. 1421-1425, and Pentti Aalto, " Wo der Pfeffer wächst." *Neophilologische Mitt.*, 1949, *50*: 13-21.

[33] Cf. Walahfrid, *Hortulus (ed. cit.)*, vv. 301-304: Hoc [sc. pulegium] apud Indorum tanti constare peritos/fertur, apud Gallos quanti valet Indica nigri/congeries piperis.

From correspondence dating from the middle of the eighth century, W. Heyd notes that cinnamon, pepper, and perfumes were sent by Roman ecclesiastics to their friends in Germany, *Histoire de Commerce du Levant au Moyen-Age* (Leipzig: Harrassowitz, 1886), II, p. 595, n. 8.

[34] Karl Sudhoff, " Die medizinischen Schriften, welche Bischof Bruno von Hildesheim 1161 in seiner Bibliothek besass . . . ," *Sudhoffs Arch.*, 1916, *9*: 348-356. A catalogue, containing many classical texts, Greek as well as Latin, dated 1155, was published by Anton Birlinger, " Bücherverzeichnis des Klosters Pfäffers," *Ztschr. f. deutsches Altertum*, 1872, *15*: 512-513.

[35] Sudhoff, *op. cit.* (n. 19 above), p. 275, Nr. 2.

main portion of the recipe, twenty-eight of them are glossed in superscript with Old High German synonyms.[36] Though the source of that recipe be unknown, it bears a resemblance to late Latin recipes in which polypharmacy has taken the place of the older and simpler recipes of Dioscorides and Galen. In addition to determining a *terminus post quem* for the establishment of vernacular synonyms—albeit in a sometimes baffling form[37]—the recipe is important for another reason. The sentence immediately preceding the directions for its preparation begins with the phrase " De pigmentis," that is, " Concerning spices." Then follow the names of eight plant products, all but one of which were exotica, viz., zedoary, cinnamon, ginger, costus, rhubarb, pepper, gentian, and cloves.[38] Apart from the question whether such exotica as these were actually available, they were assured a place in the medieval armamentarium.

The random copying of recipes led, in time, to the compilation of receptaria or recipe collections. Several such receptaria and antidotaria, or antidote collections, are extant, dating from the ninth and tenth centuries.[39] By the mid-twelfth century, the more ambitious *Arzneibücher* or general medical texts, and *Kräuterbücher* or herbals, were beginning to take shape.[40]

Following the method of ancient medicine, medieval recipe collections frequently were arranged *a capite ad calcem*—for diseases from the head to the foot.[41] Thus five receptaria, dating from the ninth to the twelfth centuries, begin with recipes entitled *ad capitis dolorem*.[42] But from that

[36] The plants are identified (with a facsimile of the manuscript), by Georg Sticker, " Die gebräuchlichen Heilkräuter in Deutschland zur Zeit Karls des Grossen," *Janus*, 1924, *28*: 21-41.

[37] The origin of the vernacular plant names in a thirteenth-century glossary as well as the dialect, *muss vorerst ein räthsel bleiben*, according to Wilhelm Grimm, " Wiesbader Glossen," *Ztschr. f. deutsches Alterthum*, 1848, *6*: 321-340 (the eighty plant names occur on pp. 323-324).

[38] Sudhoff, *op. cit.* (n. 19 above), p. 275, Nr. 2.

[39] Sigerist, *op. cit.* (n. 17 above), *passim*, and Julius Jörimann, *Frühmittelalterliche Rezeptarien*. (Zürich/Leipzig, 1925) [*Beiträge zur Geschichte der Medizin*, 1].

[40] For the development of medieval German medicine and a characterization of the stages it passed through see Gundolf Keil, " Die deutsche medizinische Literatur in Mittelalter," *Verhandlungen des XX. Internationalen Kongresses für Geschichte der Medizin*. Berlin, 22-27. August, 1966 (*Hildesheim*, 1968), pp. 647-654.

[41] Cassius Felix, *De medicina*, ed. Val. Rose (Leipzig: Teubner, 1879); Quintus Serenus, *Liber medicinalis*, ed. R. Pépin (Paris: Presses Universitaires de France, 1950); Plinius Secundus, *De medicina libri tres*, ed. Alf Önnerfors (Berlin, 1964) [Corpus Medicorum Latinorum, 3], lib. I (pp. 7-33). Cf. Pliny, *Historia Naturalis* XXV lxxxiii, 132 sq., and Isidore, *Etymologiae*, ed. W. M. Lindsay (Oxford, 1911), IV vii 2 sq.

[42] Jörimann, *op. cit.* (n. 39), pp. 10, 40, 63; Friedrich Wilhelm, *Denkmäler deutscher Prosa des 11. und 12. Jahrhunderts* (München: M. Hueber, 1960), Abt. I, 39, 53.

point onwards, there is little agreement as to the ingredients and not one of the five sections dealing with pain of the head agrees very closely with the corresponding recipes in ancient texts.[43]

There is a greater agreement between classical sources and medieval texts regarding the simples believed to have been of therapeutic value. As in ancient materia medica, simples of plant origin predominate, though products of animal and mineral origin are not wanting in medieval recipes. Many of the plants employed in those recipes were indigenous to and widely distributed over central Europe and a large number were cultivated in gardens, either for their medicinal or alimentary uses. Because of the practical purposes served by the recipes, information which would facilitate the recognition and collection of plants growing wild is seldom included. There are, to be certain, occasional references to some distinguishing characteristic[44] or to the location where the simple might be obtained.[45] But in the main, even in the early *Kraüterbücher*, the information consists in listing the complaints for which the simple was allegedly useful. In a twelfth-century herbal which enumerates the virtues of nineteen herbs, *liquaricia* or liquorice (*Glycyrrhiza glabra* L.) is dismissed in these words, " It is good for the cough and for chest pain." [46] Judging from references specifying the use of fresh materials, green leaves, and the juice of ripe fruits, we may suppose that some plants were intended to be gathered locally.[47] In addition to the above data, sometimes one also finds a reference to the precise location where the desired plant can be found growing.[48] By including such information, the medieval author

[43] Cassius Felix, *op. cit.* (n. 41), cap. 1 (pp. 2-10, Rose); Pliny, *Historia Naturalis* XXV lxxxiii, 134; Caelius Aurelianus, *On Acute Diseases and On Chronic Diseases*, ed. I. E. Drabkin. (Chicago: Univ. of Chicago Press, 1950), p. 444.

[44] Erba splenion [*Ceterach officinarum* Willd.] . . . nascitur secus rigum aquarum longa habens folia et subincisa. Jörimann, *op. cit.* (n. 39), p. 69.

[45] " Schabiosa [*Scabiosa columbaria* L.] dryackir cruedt, daz vindt man an deme berge in deme sande," J. Zacher, " Zwei medicinische Recepte," *Ztschr. f. deutsches Alterthum*, 1867, *13*: 382; " nim . . . polei [*Mentha pulegium* L.], daz man an den wisen vindet," Pfeiffer, *op. cit.* (n. 4 above), p. 137; " Nym wedewynden [*Convolvulus* sp.] van den bomen," Karl Regel, " Aus dem Gothaischen Arzneibuche," *Jahrbuch des Vereins für niederdeutsche Sprachforschung*, 1879/80, *5*: 65.

[46] Wilhelm, *op. cit.* (n. 42), pp. 44-45.

[47] " . . . vnde stote denne bathonyen [*Betonica officinalis* L.] und lorberen [*Laurus nobilis* L.] vnde wringk dat sap vth." Regel, *op. cit.* (n. 45), p. 75; " nim i lb. sarbachboum [*Populus alba* L.], die vff den vordresten spitzen der esten stont . . . ," G. Jenny, " Alte Recepte und Hausmittel," *Alemannia*, 1891, *19*: 34. The phrase, wir suechen ein chraut daz zu der wunden guet sei, implies that the plant was to be gathered locally; Anton Schönbach, " Segen aus Grazer HSS," *Ztschr. f. deutsches Alterthum*, 1875, *18*: 80.

[48] " Contra lapidem. Item eadem decana [sc. in Vilike, i. e. Vilich] monstravit michi

was only following the lead of ancient texts on materia medica in which an abundance of similar data supported their practical function.

More frequent in medieval texts is the listing of one or more synonyms of the simple in question. Vernacular synonyms have been mentioned above, but Latin synonyms also occur occasionally.[49] While it may appear strange to find such matters discussed in a medical text, it is a reminder of the difficulties of writing in, and for, a bilingual culture with its imperfect command of technical terms in either language. This is especially true for Latin plant names which were based upon Greek plant names, for a reading knowledge of Greek was a rare achievement in the early middle ages.[50] The situation was somewhat eased by the compilation of glossaries in which Greek plant names gradually became recognized.[51] In addition to that service, medieval glossaries frequently contained data regarding contemporary medicine and materia medica.[52]

Despite denunciations by the Roman poets and satirists concerning the

in cymiterio Vilike duas herbas quarum quamlibet divisim bulivit cum vino et dixit maxime valere in dolore renum. Una herba fuit disposita sicud pes columbinus et fuit in radice rubea," Ernest Wickersheimer, "Faits cliniques observés à Strasbourg et à Haslach en 1362 et suivis de formules de remèdes," *Bull. Soc. franç. d'Hist. Méd.*, 1939, *33*:90.

[49] Jörimann, *op. cit.* (n. 39 above), pp. 57, 65, etc.; "(T)olle de radicibus salvatoris i. de regia herba, quod dicitur alant [*Inula helenium* L.] . . . ," I. V. Zingerle, "Recepte aus dem XII. Jahrhundert," *Germania*, 1867, *12*: 468. For further examples, see Zacher, *op. cit.* (n. 45), pp. 382-383.

[50] Jerry Stannard, "Benedictus Crispus, An eighth century medical poet." *J. Hist. Med.*, 1966, *21*: 28-29. Cf. Max Manitius, "Die Kenntnis des Griechischen im frühen Mittelalter," *Beilage, Münchener allgemeine Zeitung*, Nr. 193 (1905), pp. 355-357.

[51] The abbreviation "g'" (*i. e. graece*) appears in superscript following the latinized Greek plant names which are in synonymy with the Middle High German glosses, F. L. K. Weigand, "Nomina lignorum avium piscium herbarum mit deutschen Glossem aus der Frankfurter Hs," *Ztschr. f. deutsches Alterthum*, 1853, *9*: 394-398.

[52] The miscellaneous material contained in glossaries cannot even be summarized here. Suffice it to say, representative entries like the following indicate something of the nature of the medical and scientific information transmitted by the glossary.

Apotecā.cellariū (Conrad Hofmann, "Über Bruchstücke einer Handschrift mit althochdeutschen Glossen," *Germania*, 1863, *8*: 12; *cucumerarius.curbizgarto* (Hoffmann v. Fallersleben, "Prager Glossen" [note 29] 470a); *glicida.pionie vel confectio amara; kimina lepra. gepulverte bonen; musa. fructus in quo Adam peccavit, alii dicunt, quod in ficu peccavit* (F. J. Mone, "Botanisches Glossar," *Anzeiger f. Kunde d. teutschen Vorzeit*, 1835, *4*: cols. 245, 246, 247); *dyagridium. scamonea cocta* (M. Kleemann, "Ein mittelniederdeutsches Pflanzenglosser." *Ztschr. f. deutsche Philologie*, 1878, *9*: 202a); *Thus.wirochboum.Cuius gutta etiam dicitur wirouch* (Max Rieger, "Altmitteldeutsche Glossen zu Heinrici Summarium," *Germania*, 1864, *9*: 22; also a section on anatomical terms, p. 27). In addition to the scattered information contained in the glossaries noted above (note 29), see Hartmut Broszinski, "Das Heilmittelglossar das Hans Suff von Göppingen," *Centaurus*, 1967, *12*: 114-131.

use of Eastern spices and perfumes,[53] exotica commonly were included in medieval texts on materia medica. It is sometimes difficult to estimate the availability of exotica as opposed to mere literary references thereto. For, in most early medieval texts, there is little evidence of a personal knowledge of the drugs, although supplies from an apothecary are occasionally mentioned.[54] This is especially true for texts written or compiled in central Europe, where commercial supplies in bulk were less available, and hence more expensive, than in the port cities to the south. Nevertheless, exotica are mentioned as early as the ninth century and continued to play a major role in medicine and dietetics throughout the Middle Ages, as medieval drug inventories amply testify.[55]

The effectiveness of medieval drug therapy may well be questioned in the light of such advice as the following recipe for pleurisy: "Take the stone which the swallow bears and hemp seed and cabbage seed. Mix with salvia and place under the tongue." [56] But, effective or not by contemporary standards, medieval writers often reinforced their drug therapy with such paramedical techniques as charms, prayers, and incantations of various kinds.[57] While those techniques are not unknown in ancient materia medica, they are comparatively less frequent.[58]

Prayers against a variety of complaints appear very early in German medical literature.[59] Often their pagan origin is thinly disguised by a Christian veneer, abetted by words and phrases that vaguely recall pas-

[53] Pliny, *Historia Naturalis* XXII lvi, 118; Horce, *Odes* I, 38, and Juvenal, *Satires* VI, 465-66; XIV, 187-88.

[54] ". . . nim iiii lott wisses wachses vss der appoteggen," Jenny, *op. cit.* (n. 47 above), p. 35; ". . . hale se vth vnde ghut dar ynne muskelenolye, de ys in der apoteken . . . ," Regel, *op. cit.* (n. 45 above), p. 73; "Ergo non intromisi me de cura ita difficili, quia non potuit solvere res in apoteca, sed ego laborassem gratis amore Dei.," Wickersheimer, *op. cit.* (n. 48 above), p. 80; "er [sc. gebranntes Wasser or brandy] hat auch die tugennd wenn allen würtzen behält er die krafft in allen Appodegken," Wilh. Wackernagel, "Kochbuch von Maister Hannsen des von Wirtenberg Koch," *Ztschr. f. deutsches Alterthum,* 1853, *9*: 367; ". . . köffe in der apdecken ein gummin, heisset aloe, . . . ," Anton Birlinger, "Aus einem elsaessischen Arzneibuche des XIV Jahrhunderts," *Alemannia,* 1882, *10*: 220.

[55] F. A. Flückiger, "Die Frankfurter Liste." *Arch. d. Pharmacie,* 1872, *201*: 433-464, 508-526, and *idem,* "Das Nördlinger Register," *Arch. d. Pharmacie,* 1877, *211*: 97-115.

[56] Wilhelm, *op. cit.* (n. 42 above), I, p. 59. Cf. Pliny, XXX xxvii, 91.

[57] Oskar Ebermann, *Blut-und Wundsegen in ihrer Entwickelung dargestellt.* (Berlin: Mayer & Müller, 1903), (*Palaestra,* XXIV).

[58] See Ricardus Heim, "Incantamenta Magica Graeca Latina," *Jahrbücher f. klassische Philologie,* Supplm-Bd. 19 (1893) pp. 465 *et passim.*

[59] F. W. E. Roth, "Althochdeutsches aus Trier," *Ztschr. f. deutsches Altertum,* 1910, *52*: 169-182 (a ninth/tenth century fragment from Trier with prayers and the uses of amulets).

sages from the New Testament.[60] There are also some charms which are more directly related to materia medica in virtue of their ability to increase the efficacy of a simple. For example, several charms are associated with verbena (*Verbena officinalis* L.), a plant to which many curative properties were early assigned.[61] Another plant invested with almost miraculous powers was *vet(t)onica* (*Betonica officinalis* L.).[62] The choice of that particular plant, out of dozens of equally inefficacious plants, may have been influenced by the *Liber de vetonica*, a short, late-classical text written by Antonius Musa.[63] Since that tract is often associated with the popular and influential *Herbarius* of Pseudo-Apuleus, the magic-laden recipes of the latter may have drawn attention to its more modest companion piece.

Several other medicinal plants were the subject of specialized tracts, perhaps in the analogy of the now-lost tracts on cabbage and its manifold virtues attributed by Pliny both to Pythagoras and Chrysippus.[64] The virtues of oak,[65] mistletoe,[66] and juniper [67] were enumerated in a series

[60] Cf. F. Dietrich, "Drei altheidnische Segensformeln," *Ztschr. f. deutsches Alterthum*, 1867, *13*: 193-217.

[61] Anton Birlinger, "Bairische Besegnungen," *Germania*, 1879, *24*: 75; F. Jecklin, "Proben aus einem Arzneibuch des 15. Jahrhunderts," *Schweizerisches Arch. f. Volkskunde*, 1927, *27*: 85; Weigand, "Kräuterbeschwörungen und Kräuteraberglaube," *Ztschr. f. deutsche Mythologie u. Sittenkunde*, 1855, *2*: 171-172; K. Sudhoff, "Der Abschnitt über die Kräfte der Verbena (Eisenkraut) aus dem Münchener Cod. lat. 614 (13-14 Jahrh.)," *Sudhoffs Arch.*, 1920, *12*: 83; Oswald Zingerle, "Segen und Heilmittel aus einer Wolfsthurner Handschrift des XV. Jahrhunderts," *Ztschr. f. Volkskunde*, 1891, *1*: 322. For the etymology of *Eisenkraut* (or *Eisenhart*) see H. Hoffmann von Fallersleben, "Volkswörter," *Arch. f. d. Gesch. deutscher Sprache u. Dichtung*, 1874, *1*: 245.

[62] Roth, *op. cit.* (n. 59), p. 175; Joseph Haupt, "Ueber das md. Arzneibuch des Meisters Bartholomaeus," *Stzb. d. Akad. d Wiss., Wien, Philos.-hist. Cl.*, 1872, *71*: 528 n. Betony and verbena were the principle ingredients in *Valkensteins Salbe*, cf. Siegfried Sudhof, "Ein Salzburger Sammelcodex mittelalterlicher Medizin und Naturwissenschaft," *Med. Monatsschr.*, 1954, *8*: 191.

[63] Antonius Musa, *De Herba Vetonica Liber*, Pseudo-Apuleius, *Herbarius*, anonymous, *De Taxone Liber*, Sextus Placitus, *Liber Medicinae ex Animalibus*, E. Howald et H. E. Sigerist eds. (Leipzig/Berlin: Teubner, 1927) [*Corpus Medicorum Latinorum*, IV].

[64] Pliny, *Historia Naturalis* XX xxxiii, 78. Cf. Jerry Stannard, "Lost botanical writings of antiquity," *Actes de XIe Congrès International d'Histoire des Sciences, Varsovie-Cracovie 24-31 Août 1965* (Varsovie: Maison d'Edition de l'Academie Polonaise des Sciences, 1968), vol. V, pp. 319-322.

[65] J. Telle, "Altdeutsche Eichentraktate aus medizinischen Handschriften," *Centaurus*, 1968, *13*: 37-61.

[66] Heinrich Marzell, "Zur Geschichte der Mistel als Heilmittel," *Deutsche med. Wchnschr.*, 1937, *63*: 835-837; Willem Daems, "Der Misteltraktat des Wiener Kodex 3811," *Sudhoffs Arch.*, 1965, *49*: 90-93; K. Sudhoff, "Zwei deutsche Reklamezettel zur Empfehlung von Arzneimitteln—Petroleum und Eichenmistel—gedruckt um 1500," *Sudhoffs Arch.*, 1910, *3*: 397-402; *ibid.*, "Handschriftliches vom Eichenmistel," *Sudhoffs Arch.*, 1910, *4*: 313-314.

of late tracts. The weedy little plantain (*Plantago* sp.) was, and remains even in recent German folk medicine,[68] a plant with extraordinary virtues, especially when carried or worn as an amulet.[69] The peony (*Paeonia* sp.)[70] and bay leaves (*Laurus nobilis* L.)[71] likewise fulfilled quasi-magical roles in ancient and medieval materia medica alike.[72]

It may be useful at this point to review briefly some of the classical medical sources known to and utilized by the authors of medieval German medical texts. Bearing in mind that a citation of an author's name does not necessarily imply that that author actually was read, the most frequently cited authorities were Galen[73] and Hippocrates.[74] Doubtless Pliny's influence was almost as great, especially insofar as he was one of the major sources for Isidore,[75] though explicit references to Pliny

[67] Josef Werlin, "Drei deutsche Wacholder-Traktate des Spätmittelalters," *Sudhoffs Arch.*, 1965, *49*: 250-254, and Rudolph Zaunick, "Ein alter deutscher Traktat über das Wacholderbeerwasser," *Sudhoffs Arch.*, 1916, *9*: 183. For later uses, cf. Heinrich Marzell, "Der Wacholder im altbayerischen und oberpfälzischen Volksglauben," *Das Bayerland*, 1915/16, *27*: 203-205.

[68] Vagn J. Bröndegaard, "Wegerich als Wundheilmittel in der Volks—und Schulmedizin," *Sudhoffs Arch.*, 1963, *47*: 127-151. For current folk uses, see Hugo Hertwig, *Knaurs Heilpflanzenbuch* (München: Knaur, 1969), pp. 253, 264.

[69] Pseudo Apuleius, *op. cit.* (n. 63 above), I, 1 (p. 22). Cf. Max Höfler, *Volksmedizinische Botanik der Germanen* (Wien: R. Ludwig, 1908) [*Quellen und Forschungen zur deutschen Volkskunde*, 5] p. 12.

[70] Regel, *op. cit.* (n. 45 above), p. 105; cf. Ferckel, *op. cit.* (n. 10 above), p. 131. For classical references see Galen *De simplicium medicamentorum temperamentis ac facultatibus* VI, 3 (*Opera Omnia*, ed. Kühn, vol. XI (Leipzig, 1826), p. 859); Pseudo-Apuleius, *op. cit.* (n. 63 above), LXV (p. 120), and Alexander of Tralles I, 571, ed. Puschmann (Wien, 1878). Peony was extensively used in medieval medicine for epilepsy, cf. Alexander of Tralles, *loc. cit.*, and Owsei Temkin, *The Falling Sickness*, 2nd ed., rev. (Baltimore: Johns Hopkins Press, 1971), pp. 25, 112.

[71] Jörimann, *op. cit.* (n. 39 above), p. 13.

[72] Carolus Wessely, "Synopsis Florae Magicae"; *Bull. de l'Institut français d'Archéologie Orientale*, 1930, *30*: 17-26, and Gundolf Keil, "Zauberpflanzen und Wunderdrogentraktate," *Leuvense Bijdragen*, 1968, *57*: 165-175.

[73] Galen is mentioned in the following texts cited above: Koch (n. 7), p. 30; Telle (n. 65), p. 40; Pfeiffer (n. 4), p. 127; Eis (n. 5), p. 130; Haupt (n. 62), p. 495; Birlinger (n. 54), p. 229; Hofmann (n. 6), p. 141.

In addition, see W. L. Wardale, "A low german-latin miscellany of the early fourteenth century," *Niederdeutsche Mitteilungen*, 1952, *8*: 25 and J. Follan, *op. cit.* (below n. 90), p. 65.

[74] Hippocrates (or Ypocras) is cited in the following texts cited above: Roth (n. 59), p. 173; Regel (n. 45), p. 87; Telle (n. 65), p. 40; Pfeiffer (n. 4), p. 127; Eis (n. 5), p. 131; Hofmann (n. 6), p. 128; Ferckel (n. 10), p. 137.

In addition, see Karl Regel, "Zwei mnd. Arzneibücher, Cod. Chart. Goth. 980 und Cod. Wolfenb. 23, 3," *Jahrbuch d. Vereins f. niederdeutsche Sprachforschung*, 1878/1879, *4*: 18.

[75] Though not explicitly cited in any of the texts examined for this paper, Isidore's definitions and fantastic etymologies, are the basis of Hrabanus' definitions (note 13)

are rare.[76] There is in addition, evidence that the *Herbarius* of Apuleius Platonicus (now attributed to Pseudo-Apuleius) was known [77] along with the writings of Cassius Felix,[78] Sextus Placitus,[79] and possibly Marcellus of Bordeaux,[80] and Alexander of Tralles.[81]

Hippocrates, or Ypocras as it is often spelled in medieval texts, is a good example of an author to whom writings were attributed that bear little relation to the critical editions we know today. The popular *Arzneibuch Ipocratis* and its descendents are examples of texts whose titles were chosen in order to bestow upon them the *auctoritas* necessary for a wide hearing.[82] But Hippocrates was not the only one to whom spurious works were attributed. Aristotle, for example, is cited in the *praefatio* of a ninth-century receptarium,[83] while a charm in the Bamberg Antidotarium ends with the phrase "this was used by Plato and me." [84] Dioscorides, the fountainhead of ancient materia medica, was comparatively unknown to early German writers, and become popular only later.[85]

and his influence is also evident in later glossaries; see Rieger, *op. cit.* (n. 52 above), and Ferckel, *op. cit.* (n. 10 above). Some of Isidore's theological writings were translated into Old High German by the eighth century, see the selections in Wackernagel, *op. cit.* (n. 19 above), cols. 219-230.

[76] Regel, *op. cit.* (n. 74), p. 6.

[77] Erhard Landgraf, "Ein frümittelalterlicher Botanicus." *Kyklos*, 1928, *1*: 114-146.

[78] Alois Fauser, "Ein Dillinger Fragment einer therapeutischen Schrift aus einer Unzialhandschrift des 8. Jahrhunderts," *Sudhoffs Arch.*, 1956, *40*: 156-167; cf. Josef Heeg, "Das Münchener Unzialfragment des Cassius Felix (clm. 29136)," *Stzb. kgl. preuss. Akad. d. Wiss.*, 1910, *13*: 284-291.

[79] Cf. Gundolf Keil, "Die mittellateinische Übersetzung vom Harntraktat des 'Bartholomäus,'" *Sudhoffs Arch.*, 1963, *47*: 435, 436 n., and Karl Sudhoff, "Codex medicus Hertensis (Nr. 192)," *Sudhoffs Arch.*, 1917, *10*: 266.

[80] Some parallels to Marcellus' recipes are noted by Fischer, *op. cit.* (n. 25 above), p. 71, and Ferckel, *op. cit.* (n. 10 above), p. 130. It is possible that the phrase "Von demselbm kraut [*sc. Verbena*] schreybt uns Marco der hoch arzt . . . ," Birlinger, *op. cit.* (n. 61 above), p. 75, refers to Marcellus.

[81] Cf. Eis, *op. cit.* (n. 5 above), p. 112.

[82] Walter L. Wardale, "The excerpta Ipocratis vnde Bartholomei of Göttingen MS. Hist. Nat. 51." *Niederdeutsche Mitteilungen*, 1954, *10*: 5-23. Cf. Pfeiffer, *op. cit.* (n. 4 above), pp. 118-127, for the text of the *arzinbuoch Ypocratis.*

[83] "In Christi nomine incipiunt dicta quem Aristotelis archieter dictauit," Jörimann, *op. cit.* (n. 39 above), p. 10. For other references to Pseudo-Aristotle, see Hirth, *op. cit.* (n. 13 above), Haupt, *op. cit.* (n. 62 above), p. 480, and the amusing dialogue between Plato and Aristotle in Sven Norrbom, *Das Gothaer mittelniederdeutsche Arzneibuch und seine Sippe* (Hamburg: Hartung, 1921) [*Mittelniederdeutsche Arzneibücher*, 1], pp. 193-194.

[84] Sigerist, *op. cit.* (n. 17 above), p. 30. In a fourteenth-century calendaric regimen, the phrase "Der meister Plotonice sprichet . . ." presumably refers to Plato, Hofmann, *op. cit.* (n. 6 above), p. 126.

[85] There are isolated passages in which Dioscorides is cited. Cf. Eis, *op. cit.* (n. 5 above), p. 128; Jecklin, *op. cit.* (n. 61 above), p. 90; Pfeiffer, *op. cit.* (n. 4 above), p. 137;

The empirical and eclectic stage of German materia medica gradually was replaced by the growing influence of Salernitan medicine. It is beyond the scope of this paper to dwell at length upon the recipients of the new learning from the South. Suffice it to say, authors as different as Albertus Magnus, Konrad of Megenberg, and the so-called *Deutscher Macer,*[86] the anonymous translator of the *De virtutibus herbarum* of Macer Floridus, all profited from a knowledge of the theoretical and philosophical bases of Greek and Arabic materia medica. The culmination of this new period is the *Arzneibuch* of Meister Bartholomaus.[87] Based upon the *Practica* of Bartholomew of Salerno,[88] that text was the precursor of the many Middle Low German *Arzneibücher* of the fourteenth century.[89] Even in the later, more comprehensive texts of Ortolf of Bavaria,[90] the *Cirurgia* of Peter of Ulm,[91] the *Hortus Sanitatis,*[92] or the *Feldtbuch* of Hans von Gersdorff,[93] the Greco-Roman foundations are supplemented, not supplanted. It is not, in fact, until the writings of Paracelsus, that German materia medica considered the present, not the past, to be its teacher.

Sudhoff, *op. cit.* (n. 79 above), p. 267. It is possible that Dioscorides' influence was established via the anonymous *De herbis femininis* (H. E. Sigerist, "Materia Medica in the Middle Ages," *Bull. Hist. Med.*, 1939, *7*: 419-420), but the matter requires a closer study of the manuscript tradition than can be attempted here.

[86] Cf. Cyrill Resak, *Odo Magdunensis, der Verfasser des "Macer Floridus" und der deutsche Leipziger Macer Text.* Med.-Diss. Leipzig, 1917. For selections from the *deutscher Macer* see Rudolf Blum, "Urform und Quelle des deutschen Macer," *Mitt. z. Gesch. d Med. u. d. Naturwiss.*, 1935, *34*: 1-14; Hofmann, *op. cit.* (n. 6 above), pp. 133-137; Haupt, *op. cit.* (n. 62 above), pp. 529-542; J. Zacher, "Macer Floridus und die Entstehung der deutschen Botanik," *Ztschr. f. deutsche Philologie*, 1881, *12*: 215, and Karl Sudhoff, "Ein deutscher Text von Graden und Qualitäten der Naturdinge, speziell der Arzneistoffe. (Aus dem 'Deutschen Macer')," *Sudhoffs Arch.*, 1914, *8*: 223-224. My former pupil, Professor Bruce Flood, is now preparing an English translation with commentary of Macer Floridus' *De virtutibus herbarum.*

[87] Versions of the *Arzneibuch* occur in Pfeiffer, *op. cit.* (n. 4 above), pp. 127-158; Norrbom, *op. cit.* (n. 83 above), pp. 177-202; Regel, *op. cit.* (n. 74 above), pp. 12-16; Robert Priebsch, "Deutsche Prosa-Fragmente des XII. Jahrhunderts, II," *Modern Language Rev.*, 1916, *11*: 321-334; Hofmann, *op. cit.* (n. 6 above), pp. 137-142; and Haupt, *op. cit.* (n. 62 above) *passim.*

[88] Salvatore de Renzi, *Collectio Salernitana* IV (Naples, 1856), pp. 321-406.

[89] Conrad Borchling, "Die mittelniederdeutschen Arzneibücher," *Janus*, 1902, *7*: 131 *et passim.*

[90] James Follan, *Das Arzneibuch Ortolfs von Baierland nach der ältesten Handschrift (14. Jhdt.)* (*Stadtarchiv Köln, W 4° 24**) (Stuttgart, 1963) [*Veröffentlichungen d. Internationalen Gesellschaft f. Gesch. d. Pharmazie*, N. F., 23].

[91] Gundolf Keil, *Peter von Ulm. Untersuchungen zu einem Denkmal altdeutscher Fachprosa mit kritischer Ausgabe des Textes.* Diss. Heidelberg, 1960.

[92] See Jerry Stannard, "The Herbal as a medical document," *Bull. Hist. Med.*, 1969, *43*: 212-220, for references to the literature.

[93] See Jerry Stannard, "Hans von Gersdorff and some anonymous Strassburg apothecaries." *Pharmacy in History*, 1971, *13*: 55-65.

XIII

Albertus Magnus and Medieval Herbalism

At first glance, any connection between Albertus Magnus and herbalism may seem unlikely. Although he is universally recognized as one of the most learned and distinguished scholars of the High Middle Ages for his numerous contributions to philosophy, science, and theology, Albert is not, in any obvious manner, associated today with herbalism or the ancillary disciplines of medicine and pharmacy. Yet, on closer examination, the conjunction of the *Doctor Universalis* and the *herbolarii* may not appear so unusual. He was, even in his own day, famous for the encyclopedic breadth of his knowledge which, in no small measure, was the result of his commentaries on Aristotle.[1] That knowledge covered, inter alia, the biological sciences and mineralogy, portions of which were closely allied to herbalism. Moreover, a substantial segment of one of his major

[1] Throughout the Middle Ages, Aristotle was regarded as the author of *De plantis*, now attributed to Nicholas of Damascus. For the complex history of its MS tradition and the several translations, cf. M. Bouyges, "Sur le De plantis d' Aristote-Nicholas à propos d'un manuscript arabe de Constantinople," *Mélanges de l'Université Saint-Joseph, Beyrouth* 9 (1924), 71-89; L. Labowsky, "Aristoteles De plantis and Bessarion," *Medieval and Renaissance Studies* 5 (1961), 132-154; B. Hemmerdinger, "Le De plantis, de Nicholas de Damas à Planude," *Philologus* 111 (1967), 56-65; H. J. Drossaart Lulofs, "Aristoteles Περὶ φυτῶν ," *Journal of Hellenic Studies* 77 (1957), 75-80. For references to the supposed Aristotelian authorship, cf. Thomas of Cantimpré, *De natura rerum*, ed. H. Boese (Berlin: De Gruyter, 1973), p. 317 (Dicit Aristotiles in libro primo De vegetabilibus) and p. 322.

scientific writings covered the same ground, used the same methods and sources, and employed the same terminology, as did contemporary herbals and certain kinds of medical writings.

Book VI of Albert's *De vegetabilibus libri septem* may, in fact, almost be considered to be a herbal.[2] To be sure, there are some differences for, after all, Book VI is but one portion of a much larger work that deals with plant life and plant form in a broad, philosophical sense. Other, more detailed differences will be noted below. But, and this is the major claim here advanced, in form and content alike, Book VI taken independently, may be regarded as a herbal. In partial support of this claim, it may be noted, that later writers treated it as if it were a herbal and sometimes cited Albert by name as a source for various kinds of data traditionally found in herbals.[3]

Before we proceed further, it will be convenient to define the basic terms. A herbal is a series of descriptions of plants (sometimes including animal and mineral substances), usually in alphabetical order, which are regarded as medicinal, accompanied by medical, pharmacological, and scientific data concerning their names, properties, uses, habitat, and related forms of practical information.[4]

A herbalist, hence, is one who practices the art of herbalism and/or who records such findings in a herbal or related document. The medieval herbalist combined the roles of physician, apothecary, plant collector, dietician, and student of natural history. Leaving aside differences in background, training, clientele, and the like, the herbalist's art centered upon a belief, shared by most non-herbalists as well, in the healing properties of plants and plant products.

---

[2] Albertus Magnus, *De vegetabilibus libri septem*, ed. Ernst Meyer and Carl Jessen (Berlin: Georg Reimer, 1867). Book VI (of 492 sections) occupies pp. 339-588. References to *De veg.* are by page and section; the number to the left of the diagonal indicates page, to the right, the section. Similar techniques are used in citing other works, when applicable.

[3] Albert is twice cited on therapeutic matters by Mayster Albrecht: cf. W. L. Wardale, ed., *Albrecht van Borgunnien's Treatise on Medicine*, (Oxford: for St. Andrews University, 1936), 26/47 and 67/237. See also Erich Schmidt, "Die Bedeutung Wilhelms von Brescia als Verfasser von Konsilien," diss., Leipzig, 1922, p. 30: "De hac facit Alberto causam in suo libro De plantis. . . ." "Albre Major" is cited as authority in Clovis Brunel, ed., *Recettes médicales alchimiques et astrologiques* (Toulouse: Privat, 1956), 48/405 and, by A. Feyl, ed., "Das Kochbuch des Eberhard von Landshut," *Ostbairische Grenzmarken* 5 (1961), 352-366, cf.360/81: ". . .als do spricht Albertus." Finally, in the late medieval *Hortus sanitatis germanice* (Mainz: Peter Schöffer, 28 March 1485), Albert is cited in eight chapters: 9 cinoglossa, 204 gagates, 222 Karabe, 242 lapis magnes, 243 lapis margarite, 337 rosa, 410 urtica, 412 verbena. It is possible that the abbreviation "Alber." may also refer to Albertus in T. G. Leporace et alios, ed., *Un inedito erbario farmaceutico medioevale* (Firenze: Olschki, 1952), p. 90.

[4] See Jerry Stannard, "Medieval Herbals and their Development," *Clio Medica* 9 (1974), 23-33.

Most of the claims made by medieval herbalists concerning the therapeutic usefulness of plants cannot be substantiated today, either clinically or pharmacologically. This does not mean, however, that the plants were physiologically inert or that their administration possessed no therapeutic value. It is rather the case that in the absence of other forms of therapy, the majority of persons in the Middle Ages were dependent upon medicaments of plant origin and a supervised diet, mainly of plant substances, as the main lines of defense in times of illness.[5] For this reason, a large body of information accumulated and gradually was systematized in the form of herbals and other kinds of medical writings such as leechbooks, receptaria, and antidotaria.[6] The material found in these writings was organized and explained by the categories available to, and accepted by, their authors or compilers, for example, the theory of the four elements or, alternatively, the theory of the four humors. The writers of these texts were often physicians, although there was nothing to prevent others from compiling their favorite recipes or rearranging traditional material to suit their particular needs.[7] Because much of the information comprising a typical medieval herbal was common knowledge, it was widely diffused and easily modified in accordance with the specific occasion.[8] This has direct bearing upon the present

---

[5] See Gerhard Eis, "Meister Alexanders Monatsregeln," *Lychnos* (1950/51), 104-136; Manfred Koch, "Das Erfurter Kartäuserregimen," diss., Bonn, 1969. For the close connection between plants as foodstuffs and condiments and plants as spices and medicaments, see Hermann Fischer, "Mittelhochdeutsche Receptare aus bayerischen Klöstern und ihre Heilpflanzen," *Mitt d. bayerischen botanischen Gesellsch.* 4 (1925), 69-75 and Jerry Stannard, "The Botanico-Medical Background of Baptista Fiera's *Coena de herbarum virtutibus*" in *Civilità dell' Umanesimo*, Atti del VIII Convegno del Centro di Studi Umanistici, Montepulciano (Firenze: Olschki, 1972), pp. 327-344.

[6] See Jerry Stannard, "Greco-Roman Materia Medica in Medieval Germany," *Bulletin of the History of Medicine* 46 (1972), 455-468.

[7] This is illustrated by comparing specific entries in medieval herbals. There exist three herbals written within either Albert's lifetime or a decade of his death:

(1) Henrick Harpestraeng (ob. 1244), *Liber de herbarum*, ed. Poul Hauberg (Copehhagen: Bogtrykkeriet Hafnia, 1936).

(2) Rufinus (*fl.* 1280), *De virtutibus herbarum*, ed. Lynn Thorndike (Chicago: Univ. of Chicago Press, 1949).

(3) Hermann of Heiligenhafen (anno 1284), ed. H. Ebel, *Der Herbarius communis des Hermannus de Sancto Portu* (Würzburg: Triltsch, 1940).

All of these writers, like Albert, depend heavily on Matthew Platearius (ob. 1161), *Circa instans.*, ed. Hans Wölfel, "Das Arzneidrogenbuch Circa instans in einer Fassung des XIII. Jahrhunderts aus der Universitätsbibliothek Erlangen," diss., Berlin, 1939.

[8] This is especially clear in the case of two compilers of encyclopedias, both of whose books on plants are indirectly connected with Albert: Thomas of Cantimpré, *Liber de natura rerum*, ed. Boese, (libri X-XII = pp. 312-350) and Konrad of Megenberg, *Das Buch der Natur*, ed. F. Pfeiffer (Hildesheim: Olms, 1962; facsimile reprint of 1861 ed.)

study because, as we shall observe, Albert availed himself of much traditional material, used many of the major sources available in his time, and arranged his material in accordance with his particular purpose.

With this as a brief background of the herbal, the herbalist, and the latter's role, let us return to the herbal, as defined above, and examine the extent to which Book VI fits that pattern.

It is an unprofitable question to ask what is the most important kind of information in a herbal for the simple reason that the various kinds of information are mutually supportive. Despite the difficulty of neatly separating all the components of a medieval herbal, the descriptions of the plants are among the most obvious. This is the case, not only in the few illustrated herbals, but in those which lack any illustrative material.

In herbals, the descriptions have commonly been regarded as subserving a practical function, viz. the identification and recognition of therapeutically useful plants such that they can be collected and used appropriately. To what extent the physician, apothecary, or herbalist actually relied on written description in order to recognize the plants sought, is somewhat unclear. An equally good case can be made for the well-known fact that herbalists, to this day, possess a fund of empirical information on the basis of which they recognize and hence collect the desired plant. The herbal, thus, served other purposes as well, for the description of the plant was only propadeutic to the preparation and administration of medicaments, the composition of which usually included substances of plant origin. Regardless of which way the descriptions be interpreted, it is clear that Albert was not describing plants solely for the purpose of recognizing potentially useful plants for a therapeutic end.

In the opening sentences of Book VI, Albert states, almost apologetically, that he will deal with specific plants and their properties.[9] It

[9] "In hoc sexto libro vegetabilium nostrorum magis satisfacimus curiositati studentium quam philosophiae. De particularibus enim philosophia esse non poterit. Nos autem in hoc sexto libro proprietates quasdam intendimus ponere, quae particularibus plantis convenire videntur" (338/1). Concluding his account of *thus* (the oleo-gum resin derived from *Boswellia sacra* Flueck.; frankincense), Albert states, "Est autem stypticum restrictivum, et memoriae confert glutitum, et ad alias multas medicorum praeparatur operationes, de quibus hic non intendimus, nec est per singula dicendum" (457/235).

is significant that his ultimate purpose is not medical, but rather to provide a brief description of those plants whose *formae* and *naturae* were discussed from a philosophical point of view in the preceding five books.[10] It is even more significant that among the *proprietates* chosen to illustrate the specific plants, he includes their medicinal uses and pharmacological action, that is, the *virtutes* and *operationes* whose enumeration is one of the normal features of medieval herbals.

The fact that his descriptions of some four hundred different species were intended to subserve a philosophic or scientific purpose, rather than a medical purpose, permitted Albert to utilize a wide range of descriptive data. The only major limitation was the lack of information concerning exotica, i.e. herbs and trees neither indigenous to nor naturalized in Western Europe in Albert's time. For like the great majority of his contemporaries, Albert had no personal knowledge of the parent plant of Near Eastern exotica. As a consequence, he was restricted to making inferences, based on the dried commercial forms (e.g., cloves, pepper, cinnamon), and repeating, or paraphrasing, descriptions of those plants which he found in his sources, especially Avicenna and Constantine the African, two of the principal literary sources for Book VI.[11]

Taken together, the different kinds of information assembled by Albert in his descriptions are impressive. No doubt it helped the botanically innocent reader to visualize some of the herbs and trees discussed earlier by being able to relate their knowledge of common, everyday plants to Albert's compressed and sometimes technical descriptions.

Leaving aside temporarily the problem of nomenclature — synonyms, varietal names, etymologies and the like — many of Albert's descriptions begin with a simple, yet revealing remark to the effect that the plant in question is well known or common. *Fenugraecum* (*Trigonella foenumgraecum* L.; fenugreek), for example, *est herba*

10 "Sicut enim in animalium scientia non scimus naturam eorum, nisi cognitis cibis et operibus animalium et partibus eorum: ita etiam in scientia plantarum nequaquam cognoscitur natura ipsarum, nisi sciantur et partes earum et qualitates et effectus. Haec igitur in quibusdam plantarum investigata describemus, ut in aliis per similem modum natura plantarum inveniatur" (472/263).

11 *Liber canonis Avicenne*, revisus et ab omni errore mendaque purgatus summaque cum diligentia impressus (Venice: P. de Paganinis, 1507), (the *simplicia* occupy Lib. II, Tract. ii = folios 88rb-162rb); Constantine the African, "De gradibus simplicium," in his *Opera medica* (Basel: Henr. Petri, 1536), pp. 342-387.

nota, while the description of *diptannus* (*Dictamnus alba* L.; dittany) begins *est herba communis satis*.[12] Like many phrases in herbals, the statement that a plant is *valde nota* or *bene nota* might be regarded as nothing more than a typical herbalistic formula, on a par with the ubiquitous *probatum est*. But in Albert's case it is not a hollow formula; it is a way of setting the stage for subsequent information. By designating a plant as "common" or "known to all," Albert, in effect, is advising his reader to look about or make a simple inquiry if more information is required about the plant or the appropriateness of his analysis.

Immediately following the opening statement that the plant is common or well known, Albert, like most medieval herbalists, turns to a physical description of the plant itself. Ever since the pioneer studies of Ernst Meyer, historians of botany have recognized Albert's attention to details and, equally important, his ability to record them with a degree of precision unmatched in his day.[13] Not only did he observe attentively the plant or some portion thereof, in true scientific fashion he smelled, tasted, or felt some of the plants which he described. Perhaps without completely realizing it, he was engaged in making a differential diagnosis (see below, p. 363). For example, *ruta* (*Ruta graveoleus* L.; rue), described first as *herba nota*, is further characterized as "markedly bitter" (*amara valde*), an apt phrase for one of the most popular of herbs of traditional materia medica.[14] Its dominant characteristics are, in fact, its bluish-grey foliage which, when slightly bruised, emits a pronounced, bitter and pungent odor. Elsewhere, in describing *marmacora* (? *Melissa officinalis* L.; lemon balm), he notes, *est boni odoris, aromaticus*.[15] This too, is appropriate for another old garden favorite whose crinkled leaves when bruised immediately give off an unmistakable but pleasant, lemon-like scent.

In addition to carefully observing, Albert was able to communicate that information in a succinct fashion. His vocabulary was so well chosen, in fact, that the great majority of his plants can be identified readily to the generic level and many even to the specific level.[16]

---

[12] *De veg*. 515/342 (*fenugraecum*); 506/327 (*diptannus*).

[13] Ernst Meyer, "Albertus Magnus. Ein Beitrag zur Geschichte der Botanik im dreizehnten Jahrhundert," *Linnaea* 10 (1836), 641-741; idem, "Albertus Magnus, Zweiter Beitrag zur erneuerten Kenntniss seiner botanischen Leistungen," *Linnaea* 11 (1837), 545-595.

[14] *De veg*. 559/428.

[15] *De veg*. 536/382.

[16] See Jerry Stannard, "Identification of the Plants Described by Albertus Magnus, *De vegetabilibus*, lib. VI," *Res Publica Litterarum* 2 (1979), forthcoming.

Because of the studies by Meyer, Sprague and others, it is unnecessary here to deal at length with the technical aspects of Albert's descriptions or with the various organs denoted by his unusually large botanical vocabulary.[17] Suffice it to say, his descriptions of the various plant organs — root, stem, leaf, flower, fruit and seed — are the sorts of details to which herbalists paid close attention, though for quite a different purpose than Albert. For therapeutic purposes, it is important to know, and to record, which portion of the plant is to be used, as it is well known that some portions of the same plant are more physiologically active than others. It speaks well for Albert's utilization of his sources, written and oral alike,[18] when he remarks that it is the root of *nenufar alba* (*Nymphaea alba* L., white water-lily) which physicians use in the treatment of ulcers.[19] Elsewhere, in concluding his description of *humulus* (*Humulus lupulus* L., hop vine) he states that it is the flower alone which is used.[20] This is an interesting observation and one which may have been based on the use of the dried strobili for flavoring beer. It is possible, moreover, that Albert had seen hops trained along poles during his many travels in southern Germany, for Bavaria is to this day one of the centers for the production of hops in Germany. Like many of the other traditional garden plants, hops may have been grown on lands farmed or owned by monastic houses visited by Albert in the course of his duties as provincial of the Dominican Order *in germania*.[21]

Several other kinds of descriptive information are also included in Book VI, each of which is represented in contemporary herbals. Albert, for example, frequently likens one plant to another though, on occasion, he mentions some character by means of which two plants can be distinguished from one another. In the latter case, the plants may in some respects be similar, but with reference to a stipulated character they are distinguishable.

The use of the former of the two techniques, based on the recogni-

---

[17] T. A. Sprague, "Plant Morphology in Albertus Magnus," *Kew Bulletin* 9 (1933), 431-440; id. "Botanical Terms in Albertus Magnus," *Kew Bulletin* 9 (1933), 440-459. A more detailed study is in preparation and will appear as "Die Botanik des hl. Albert," in *Albertus-Magnus-Festschrift* (Köln, 1980).

[18] In his discussion of the hop vine, Albert states that the dried flower is long lasting and concludes, "ita quod vulgaris opinio est, quod numquam putrescit" (525/361).

[19] *De veg*. 542/395.

[20] *De veg*. 526/361.

[21] It is significant that hops are mentioned in a document (undated) from St. Emeran's in Regensburg, where Albert briefly served as bishop; see C. O. Cech, "Über die geographische Verbreitung des Hopfens in Alterthume," *Bulletin de la Société Imperiale des Naturalistes de Moscou* 57 (1882), 57.

tion of the concept of similarity as an irreducible philosophic category, is an effective device for purposes of emphasizing some noteworthy or conspicuous character. Such a technique, however, assumes that the comparandum be a well-known plant, for otherwise it loses its effectiveness.[22] Albert's use of this technique thus tells us something about the general knowledge of plants in his time. For example, by likening *anetum* (*Anethum graveolens* L.; dill) to *feniculum* (*Foeniculum vulgare* Mill.; fennel) or *cepa* (*Allium cepa* L.; onion) to *porrum* (*Allium porrum* L.; leek) he tells us indirectly that some knowledge of fennel and leek can be expected on the part of nearly everyone.[23] But care must be taken not to misread Albert's use of *similis est* for when describing exotica, any similarity between the parent plant and another plant clearly derives from his sources.

The establishment of certain plants or, as is often the case, certain organs of those plants, as norms for the purpose of comparing and describing less well known plants, is most useful in connection with the description of those plants which lack a conspicuous characteristic. Such plants, often small or weedy, are easily overlooked and, in fact, many such European indigenes are not recorded in medieval herbals. For Albert's purpose, however, the fact that they played no role in medicine is irrelevant. Indeed, with respect to *gauda* (*Reseda luteola* L.; weld), Albert explicitly states that it has no medicinal use or, at least, that it is unknown.[24] Although medieval herbals occasionally devoted a section to a plant of which it is stated that it had no medicinal use, a tell-tale sign of how descriptive botany will eventually outgrow its herbalistic origins, it was not a common practice.[25] It was far more common simply not to mention any medicinal uses, a practice which Albert understandably used frequently.[26]

---

[22] The statement that *gariofilus* (*Syzygium aromaticum* [L.] Merr. et. Perr.; cloves) "est sicut id, quod vocatur *sambacus*" (396/115) is a case in point. Few if any Europeans in Albert's time had seen the clove-bearing tree and second, *sambacus* probably denotes *Jasminum sambac* (L.) Ait.; Arabian jasmine, itself an exoticum, not available as a living plant in Western Europe until the late eighteenth century.

[23] *De veg.* 481/282 (*anetum*); 487/295 (*cepa*).

[24] *De veg.* 521/352.

[25] Rufinus, *teste* Isaac Judaeus, states of *jujube* (*Zizyphus jujuba* Mill.), "nec sanos custodiunt nec sanitatem egris restituunt" (Rufinus, ed. Thorndike, p. 161). Referring to *cinoglossa* (*Cynoglossum offinale* L.; houndstongue) Konrad writes: "daz kraut ist guot für der viertagleichen riten, und sagt daz puoch ze latein niht mêr dâ von. sô vinde ich auch in andern meinen püechern, diu von den kräutern sagent, niht mêr dâ von" (ed. Pfeiffer, 390/20).

[26] For example, of *crocus* (*Carthamus tinctorius* L.; safflower), Albert states, "hic non est bonus, neque bene tingit, neque condit cibaria, et facit nauseam" (488/297). The crocus of commerce or saffron (*Crocus sativus* L.) is termed *crocus hortensis* (488/297) to distinguish it from safflower.

The second of the two methods mentioned above, viz. where two species of plants are distinguished from one another, was readily used by Albert. The basis for this differential diagnosis, as it is termed by some field botanists, is the recognition of a single but typical character by whose presence or absence two species, otherwise regarded as similar, can be differentiated. This method was intuitively employed by herbalists and plant collectors for the practical purpose of distinguishing between two species of similar appearance, one of which was believed to be inert or, alternatively, one of which was believed to be toxic. Referring to *centinodia* (*Polygonum aviculare* L.; knotgrass), Albert first states that its habitat is similar to that of *cicorea* (*Cichorium intybus* L.; chickory) and then proceeds to describe its habit of growth: "it creeps over the ground, but does not entwine itself around those plants which grow nearby. In this, it differs from *volubilis* [*Calystegia sepium* (L.) R. Br.; bindweed] which entwines itself around nearby plants."[27] Another example is interesting for an additional reason. *Esula* (*Euphorbia esula* L.; leafy spurge), he states, has leaves like those of *linaria* (*Linaria vulgaris* Mill.; toadflax) but they differ in that *esula* produces a milky latex but *linaria* does not.[28] The reason why this passage is of double interest is that it indicates an Albertian source not otherwise formally acknowledged. The phrase "esula produces. . .linaria does not" is simply a rewording of a passage from the *Regimen sanitatis salernitanum* though it is possible that Albert derived the passage from a presently unidentified or intermediate source.[29]

It is a tribute to Albert's profound understanding of the diversity of plant life that when he distinguished between two species or compared them, he selected stable characters which were typical and invariant for the species in question. Thus, while he, like all herbalists, often noted the color of the flowers, he recognized that color is not a constant character and that in some species there is a consider-

---

[27] *De veg.* 504/322. Another example concerns the differentiation of *rizum* (*Oryza sativa* L.; rice) and *ordeum* (*Hordeum vulgare* L.; barley), 558/427. The description of the rice grain may have been based on personal observation, but whether Albert saw the rice plant growing is doubtful.

[28] *De veg.* 512/336: "esula lactescit, et non facit linaria."

[29] The Salernitan passage, which later served as a mnemonic aid, appears in the *Collectio Salernitana*, ed. Salvatore de Renzi (Naples: Sebizio, 1859), 5: 33, verse 1173: "esula lactescit, linaria lac dare nescit." The verse does not occur in the *Circa instans*, Hermann of Heiligenhafen, or Henrik Harpestraeng. It appears in Rufinus, ed. Thorndike, p. 173 as "esula lactescit; sine lacte linaria crescit" and, in a still later form as "esula in ir hait milch und linaria keyn milch" (*Hortus sanitatis* [1485 ed.], cap. 235).

able range of normal color forms.[30] The color of the flowers was not, for Albert, the only criterion for either field recognition or taxonomic discrimination. For such purposes, morphological characters were better criteria and these Albert often noted in his descriptions.

In summing up the descriptive data assembled by Albert, it must be emphasized that although he took over some details from herbals and other medical writings, he applied those details to his own ends. This may be illustrated in his descriptions of several different species of trees.

Along with their more obvious features — general appearance, relative size, shape of leaf, nature of their fruit, etc. — several times he described the texture of the wood, insofar as he could without any optical aids, and noted some of the economic uses of the wood. For example, of *platanus* (*Acer pseudoplatonus* L.; plane tree maple), he noted that it is commonly known and that in parts of Germany and elsewhere it is used for domestic wares.[31] Again, he noted that *fagus* (*Fagus silvatica* L.; beech), on the contrary, is useless for construction purposes because of the susceptibility of the wood to insect attack.[32] Both of these statements are the result of personal observation and they remain true even today, as may be determined by walking through provincial farming communities in central Germany.

Closely related to the foregoing kinds of descriptive data is the specification of the habitat of the various plants described in Book VI. For obvious reasons, it was as important for the herbalist to specify as it was for the collector to know the habitat of the desired plant.[33] There is, of course, no guarantee that a written specification of the habitat of a given plant will enable the collector to find it. But Albert, like Theophrastus before him, observed that some species are only found growing in marshes, while other species prefer a dry and rocky soil while still others thrive by the side of heavily travelled

---

[30] For example, the range of colors of *lens* (*Lens culinaris* Medik.; lentil) is described as "flos eius rubeus declinans ad albedinem aliquantulum" (531/372). On the other hand, by describing the flower of *mellilotum* (537/385) as "est aliquantulum flos eius declinans ad albedinem, cum tamen sit croceus," he has probably combined two species under one name, viz. *Melilotus albus* Medik. and *M. officinalis* (L.) Pall., the latter being yellow flowered.

[31] *De veg*. 430/183.

[32] *De veg*. 390/105.

[33] Already by the end of the thirteenth century, this information was assembled in the form of several small, anonymous tracts. Several were edited by Ernest Wickersheimer, "Nouveaux textes médiévaux sur le temps de cueillette des simples," *Archives internationales d'histoire des sciences* series 3, vol. 29 (1950), 342-355.

roadways. Although the early botanists lacked the conceptual apparatus to describe an ecological niche or to isolate the relevant edaphic factors, they empirically recognized the optimum habitat for many species.[34]

Since it was not Albert's object to describe plants in order that they might be collected, it is all the more significant that he included references to their habitat: by the seashore, in moist places, as weeds in grain fields and in domestic gardens, near old walls, even on the roofs of houses.[35] These references to habitat are not only another indication of the care Albert took in recording details, they also illustrate some of the ways he used his sources. As noted above, he relied on herbals and the herbalistic sections of medical writings because they were the major and in some cases the only sources for descriptive data regarding a particular species.[36] Albert often excerpted the accounts of his sources and sometimes supplemented them with data obtained through personal observation.[37] But in other instances, he seems to have recorded the information much as he found it on the grounds, presumably, that it helped to provide a more complete description. This is especially true in the case of plants for which Albert had little or no personal knowledge.[38] But it is also evident when one considers those descriptions in which Albert refers to the actual practice of collecting. In medieval herbals, data regarding the proper place and time to collect are often included for practical purposes, as are instructions on the preparation and administration of medicaments composed of the plants so collected. It is noteworthy that

---

[34] This is well represented by Albert's description of *capparis* (*Capparis spinosa* L.; caper bush): "nascitur autem,. . .in petrosis, et ex loco contrahit caliditatem et siccitatem; in frigido autem et humido dissipatur et perit" (371/68).

[35] "In littore maris" (408/138); "in humidis crescens" (534/376); "nascitur in frumento" (543/396); "visus in viridariis" (524/358); "nascitur. . .in siccis locis iuxta rimas murorum" (510/334); "in tectis plantatur" (484/288).

[36] For example, the discussion of distinct sexes of *palma* (*Phoenix dactilifera* L.; date palm) (425/172) goes back, ultimately to Theophrastus, *Historia plantarum* II, 8, 4; it is unlikely that Albert knew the remarks of Herodotus, *Historia* I, 193.

[37] He notes how the spiny seeds of *lappa* (*Arctium lappa* L.; great burdock) "adhaerent vestibus" (534/376) and how the branches broken from *salix* (*Salix alba* L.; white willow) "cito recrescunt" (449/218). Clearly, his description of the use of torches by the Germans, prepared from the resinous wood of *picea* (*Pinus silvestris* L.; scotch pine), was based on personal observation (432/187).

[38] Of *lignum aloes* (*Aquilaria agallocha* Roxb.; aloes wood), Albert writes: "cuius originem nullus hominum apud nos habitantium sufficienter usque hodie cognovit" (344/11). Elsewhere, in describing *blitus* (*Beta vulgaris* L.; beet) he qualifies the phrase "ut quidam dicunt" by adding, "quia hoc non sum ego expertus" (463/246).

Albert's references to collecting for the most part avoid mention of the various incantations or magical rites which sometimes accompanied the actual collection. But from notices elsewhere in Book VI, it is clear that Albert was aware of such practices (see below pp. 374-375).

Another kind of data that is commonly found in herbals concerns a means of recording the different kinds of relations which exist between different kinds of plants. That is, admittedly, a clumsy way of expressing what taxonomists today would clearly recognize as an early attempt at plant classification. But only at the risk of committing an anachronism can one say that medieval herbalists were studying the relations between different species or different genera; that cannot be said because our modern concepts of species and genus had not been formulated.[39] It was empirically recognized by all herbalists, indeed probably by all farmers and gardeners, that individual plants differ and that some differences are more marked than others. But it was not the herbalists' object to study those empirically recognized differences as ends in themselves. Rather, a recognition of some of the differences was another type of information which was worthy of communicating or recording. In turn, that information was considered to be of practical value for the collection and ultimate utilization of plants for therapeutic purposes.

Throughout Book VI, Albert employed a variety of quasi-taxonomic techniques in order to describe the similarities and differences between plants. Several of these techniques have been noted above. But, like all herbalists, Albert also employed another, more obviously taxonomic device in those cases where the nature of the plants in question required finer discrimination. This technique was essentially the careful application of a linguistic convention to natural objects. That is, by incorporating as part of a plant's biverbal name, some descriptive term that demarcated that plant, not only from many, quite different plants, but from plants which appeared to be similar and thus closely related, a provisional taxonomy was created. By this technique, which was not systematically applied to all plants until much later, the world of plants was, in theory at least, capable of being subdivided into smaller, discrete units about which it was comparatively simple to communicate information. That informa-

[39] In medieval nomenclature, *genus, species, varietas*, and *forma* were used interchangeably. Sometimes Albert uses *genus* to denote what we would accept as a *genus* (533/375) but in other places his *genus* is practically equivalent to our *species* (531/372; 547/405; 550/409).

tion, of course, may be practical, as was the intention in herbals, or it may be philosphical, as it was in Albert's case.

A good example of Albert's technique concerns the taxonomy, and inevitably the nomenclature, of *apium*.[40] Normally the term *apium*, without any modifier, denoted *Apium graveoleus* L. or celery and it was used in that fashion from Pliny's time to the Renaissance.[41] But for a variety of reasons, some of which are quite unclear, the term *apium*, plus modifier, was used to designate other species. Hence, Albert's section devoted to *apium* actually refers to four different species, each of which was described by Albert as a kind (*genus*) of *apium*. It is unnecessary here to paraphrase the descriptions of the four taxa recognized by Albert, beyond noting that three of the four taxa represent three different genera of the *Umbelliferae* or carrot family, to which celery belongs; the fourth taxon of *apium*, and the most difficult to identify, is almost certainly not an umbelliferous plant, and its only claim to be called an *apium* is that its many-cleft leaf is typical of many umbellifers.[42]

There is no need to discuss other examples of the taxonomic distinctions made by Albert, e.g., *maior/minor, masculus/femina, aestivum/hiemale*, etc.[43] In all instances where taxa are so distinguished, his purpose remains the same, to describe the plants and illustrate some of their characteristics for the benefit of his readers.

It is only a short step from taxonomic data to nomenclature. As indicated above, nomenclatural matters cannot easily be separated from taxonomy. But there are other aspects of medieval herbalistic nomenclature which deserve comment. Frequently the individual chapters or sections of medieval herbals begin by listing one or more synonyms of the plant in question. In addition, there are sometimes

---

40 *De veg.* 480/281.

41 Pliny, *Historia naturalis* XX, 46, 112 seq.; Isidore, *Etymologiae* XVII, 11, 1; Constantine, "De gradibus," p. 379; Matthew Platearius, *Circa instans*, ed. Wölfel, p. 8; Hermann of Heiligenhafen, ed. Ebel, p. 4; Rufinus, ed. Thorndike, 28/49; Konrad of Megenberg, ed. Pfeiffer, 382/3; Thomas of Cantimpré, ed. Boese, 343/4; *Hortus sanitatis* (1485 ed.), cap. 6 etc.

42 In summary, Albert's description of *apium* is as follows:

i. apium domesticum = *Apium graveolens* L.; celery.

ii. apium montanum = *Athamanta macedonica* (L.) Spr.; no common English name; in French, persil de macedoine.

iii. apium silvestre = *Smyrnium olusatrum* L.; alexanders

iv. apium aquaticum = *Ranunculus* sp.; probably including *R. flammula* L. and *R. aquatilis* L.

43 "Maior/minor": 529/368, 574/461; "masculus/femina": 361/49, 378/81, 396/115 etc.; "aestivum/hiemale": 547/405; "silvestris/hortulana": 507/329, 508/331, 509/332 etc.

etymological notices and differences of opinion concerning the application of the name or synonyms by previous writers. All of these nomenclatural matters are represented in Book VI, though Albert's purpose was certainly not lexicographical nor was the enumeration of synonyms regarded by him as an end in itself.[44] Rather, because of the absence of a universally accepted nomenclature and a precise taxonomic system, the transmission of information regarding a particular plant was difficult and complex.[45]

Many plants were known under a plurality of latin names and there is sufficient evidence to demonstrate that the use of vernacular names and synonyms was widespread outside of academic circles.[46] Moreover, there are instances where one name designated two or more different species.[47] Under these circumstances, and as part of the fund of practical information contained in a herbal, it was essential to describe and name a plant in such a manner that it could be recognized and hence collected or otherwise obtained.[48] Since it is not always possible to recognize a plant solely on the basis of a written description, especially in the absence of any other information regarding that plant, for example, alimentary or economic uses, one or more of the most common synonyms was added. Usually they fol-

---

[44] Over twenty etymologies are provided by Albert, e.g. *agnus castus* 349/24; *ficus* 385/48; *semperviva* 568/447 etc. More interesting are the attempts to explain the origin and meaning of synonyms, e.g. *juniperus* (*Juniperus communis* L.; juniper): "cum cypresso multam habet similitudinem; propter quod etiam cypressus silvestris vocatur" (398/121).

[45] This is well summed up by Rufinus in his discussion of *iacea nigra* (*Centaurea scabiosa* L.; greater knapweed): "item licet non scribantur virtutes istarum herbarum et aliarum multarum, tamen tota die occurrent nomina earum tam in libris medicine quam in libris cyrugie, et ibi videbitur ad quid valent, nam sapientes quos sequor de huiusmodi non locuntur" (ed. Thorndike, p. 155).

[46] Rufinus concludes his discussion of *gratia dei* (*Gratiola officinalis* L.; gratiola) by stating, "alio nomine dicitur portulaca silvatica a laycis" (149/13). Albert uses *carvi* <OHG *garba* to denote *Achillea millefolium* L.; yarrow (503/318) which is thus etymologically unconnected with lat. *carvi*, the normal medieval name of *Carum carvi* L. or caraway; the latter is called *cimunum domesticum* by Albert (493/303). Another vernacular name used by Albert is *fibex* (*Betula pendula* Roth; white birch, 390/107). For the synonymy cf. *Hortus sanitatis* (1485 ed.), cap. 421: "vibex-byrck"; and Leo Jordan, "Ein mittelniederdeutsches Pflanzenglossar," *Ztschr. f. deutsche Wortforschung* 3 (1902), 354.55: "vibex vel pinnosa. berke".

[47] For example, *sponsa solis* designated two different species in *De vegetabilibus*:
(i) "sponsa solis sive solsequium" (570/451) is *Calendula officinalis* L.; calendula.
(ii) "cicorea, quae et sponsa solis vocatur" (504/321) is *Cichorium intybus* L.; chickory.

[48] Plants were obtained not only from the *herbolarii* who collected them (cf. Rufinus, ed. Thorndike, pp. 70, 115, 151 etc.); many simples were also available *in apoteca* or *ex apotecario*. For examples, see Jerry Stannard, "Hans von Gersdorff and some Anonymous Strassburg Apothecaries," *Pharmacy in History* 13 (1971), 55-65.

lowed immediately upon the name chosen as the chapter or sectional heading.

Another reason why synonyms played such an important role is that many of the most commonly used medicinal plants were described under different names in different herbals. For example, *Althaea officinalis* L. or marsh mallow, described by Albert as *altea*[49] was listed in contemporary herbals sometimes as *bismalva*, sometimes as *malvaviscum*, and sometimes as *malva domestica*.[50] But short of knowing that all these, and other vernacular names, denoted the same species, one would be hard pressed to identify the plant solely on the basis of the descriptive data provided by the several herbals. Without such knowledge that *altea, bismalva*, and *malvaviscum* denoted the same species, the reader of a herbal would not be able to coordinate the data found in different herbals under different names nor be able to use effectively the many recipes which required a specific ingredient.[51]

In Albert's day, plant nomenclature was further complicated by the introduction of Arabic plant names in herbals and pharmacological writings. Although the transliteration of Greek still exhibited some inconsistencies, at least a pattern was emerging by means of which transliterations and Latin-Greek translations were becoming standardized. This cannot be said, however, of Arabic or for indivi-

[49] "Altea — vocatur etiam bismalva, ea quod habet folia sicut malva, sed est major ea, habens crura longa plurima ex radice una. Vocant autem quidam eandem malvaviscum" (*De veg.* 483/285).

[50] A selection of the descriptions and names of *Althaea officinalis* L. in medieval herbals and leechbooks follows: Avicenna (1507 ed.), 96/76 (altea); *Circa instans*, ed. Wolfel, p. 72 (malva domestica); Herman of Heiligenhafen, ed. Ebel, p. 4 (althea); *Hortus sanitatis* (1485 ed.), cap. 12 (altea-ybisch); Konrad of Megenberg, ed. Pfeiffer, 385/10 (alcea-weizpapel); *Regimen sanitatis salernitanum*, ed. De Renzi, p. 24 (altea. . .malvae species). Rufinus has four entries for *Althaea officinalis*, in all of which but the last, the plant is described and its virtues enumerated: "aeviscus" (p. 9); "altea" (p. 18); "bismalva sive herba ungarica alba" (p. 60); "malvaviscus" (p. 178). In addition, see Macer Floridus, De *viribus herbarum*, ed. Lud. Choulant. (Leipzig: Voss, 1832), p. 43 (althaea); Serapion, "De simplicibus ex plantis," in his *Practica* (Lyon: Jac. Myt, 1525), fol. 133 vb (altea); *Boec van Medicinen in Dietsche*, ed. William Daems (Leiden: Brill, 1967), p. 135 (malva. . .groetpappel).

[51] Clearly a knowledge of the names and synonyms underlies the following statement by an anonymous author "Ut perfecte operetur medicus oportet eum cognoscere complexiones cum gradibus, virtutes cum operationibus, ut sciat que herbe vel species et cetera humanum corpus immutantia sint calida, que frigida, que humida, que sicca et in quo gradu" (Léopold Delisle, "Note sur un manuscrit de Tours renfermant des gloses françaises du XIIe siècle," *Bibliothèque de l'Ecole des Chartes* 6e ser., 5 (1869), p. 322.

dual words which were thought to be Arabic.[52] In the absence of a standardized method for the transcription of Arabic, it was difficult to apply many of the plant names which appeared in herbals and other genres of medical writings.[53] To be certain, Latin-Greek-Arabic synonyms helped, but often the synonymy or the inclusion of the wrong ingredient in a compositum, might make a significant difference. Although this was not Albert's concern, his reliance on Gerard of Cremona's translation of Avicenna's *Liber canonis* occasionally led him onto uncertain ground.[54] In defense, it may be noted, that Albert used the best sources available to him in an effort both to describe and to name, in a meaningful fashion, those plants whose properties were analyzed in Books I-V. He was, moreover, sufficiently critical to query some of the reported claims.[55]

---

[52] Some, but not all, of the Arabic names appearing in *De vegetabilibus*, most of which occur in Book VI, have been collected in Meyer and Jessen's edition (p. 693, "Plantae Albertinae: Dubiae Restant" and pp. 700 seq., "Index Rerum"). A careful study of such names, on the model of F. Schühlein's *Index nominum Arabicorum* (in Stadler's edition of *De animalibus*, 2: 1655-1663) is badly needed.

[53] Inconsistencies in orthography and uncertainties of synonyms can still be dectected much later, e.g., M. Kleemann, "Ein mittelniederdeutsches Pflanzenglossar," *Ztschr. f. deutsche Philologie* 9 (1878), 196-209 (saec. XIV); *Alphita*, ed. J. L. G. Mowat (Oxford: Clarendon Press, 1887), (saec. XV).

[54] A case in point is provided by Albert's discussion of *manna*, which occurs at the end of his description of *zucarum* (*Saccharum officinarum* L.; sugar): "est autem quiddam simile zucaro, quod vocatur etiam zucarum, et cadit in terra, quae vocatur alhusar de aere, et est species mannatis et roris, et est sicut frusta salis" (471/262). Earlier, in his discussion of *ladanum*, a resinous exudation from several related species of *Cistus*, he notes, "non esse aliquid plantae vel arboris, sed potius esse de genere roris, sicut mel et manna et id, quod vocatur tereniabin" (394/113). Albert's account derives, in part, from Avicenna (136/493, s.v. manna) and *Circa instans* (72, s.v. manna), and that another of his sources was of Arabic provenience is indicated by the name *tereniabin* (cf. Pierre Guigues, "Les noms arabes dans Serapion, Liber de simplici medicina," *Journal Asiatique* 6 (1905), 58/360 (*men*) and 84/497 (*tereniabin*). It appears that Albert has confused two substances, similar in appearance and in their properties but distinct in their origin: the exudate of the common manna ash, *Fraxinus ornus* L., which he may easily have known while in Italy or Southern France and *tereniabin*, possibly an exudate produced by an insect; cf. Max Meyerhof, "The earliest mention of a manniparous insect," *Isis* 37 (1947), 32-36.

[55] Cf. the devastating criticism in 430/183 concerning "arbores parvas in insulis Germaniae crescentes," the falsity of which is clear; Albert begins, "quoniam nec Germania insula est. . . ." Elsewhere, in regard to the *arbor peredixion*, Albert queries the description and the association of its fruit with doves (437/198). It is problematic, however, whether Albert realized that this fabulous entity derived from the Φυσιολόγος , c.34: Περὶ δένδρου περιδεξίου , ed. Fr. Lauchert, *Geschichte der Physiologus* (Strassburg: Trübner 1889), pp. 264-265.

At this point, it will be convenient to pause momentarily and to consider the kinds of data discussed so far. One of the functions of the herbal, it will be recalled, was to provide practical information. A large portion of that information, but by no means all of it, has been discussed above as descriptive data. Originally, those data had, as one of their functions, the identification of medicinal plants such that they could be collected or otherwise procured. But there is another kind of practical information contained in medieval herbals which is obviously and thematically connected with the descriptive data. This latter class is generally termed "therapeutic data." It is a fund of practical advice comprised of recipes, dietary counsels, domestic or household recommendations, and explanations regarding the nature of a medicinal plant and the preparation, administration, dosage, and storage of the medicaments made from or containing plant products.

The class of therapeutic data is no less important than the descriptive data, though its application is more restricted and its relations to developments in the history of science less obvious. But in the typical medieval herbal, the therapeutic data supplement the descriptive data in such a fashion that in the absence of either, the value of the herbal is proportionately lessened.

One might expect, *a priori*, to find little therapeutic data in Book VI, especially when it is recalled that *De vegetabilibus* was designed to be a commentary on the Pseudo-Aristotelian *De plantis*. But, as always, Albert was encyclopedic in his coverage and, as if to compensate for the lacunose descriptions of plants and their properties in *De plantis*, he attempted to remedy the situation by bringing together the salient information. Once the decision was made to use herbalistic data, it was difficult to draw a line between the descriptive and the therapeutic data. It may also be noted that in his earlier *De animalibus*, a wide range of therapeutic data is also included, though in no sense could that work be considered a medical text.[56]

Thus, it is not surprising that Book VI contains a selection of therapeutic data. Admittedly, they are neither as systematized nor as complete, as one might expect in a herbal or a leechbook. But whatever the deficiencies, it is more than compensated by Albert's efforts to explain on philosophic grounds the reasons for the therapeutic effec-

[56] Extensive portions of Books XXII and XXIII, deal respectively with diseases of horses and falcons, and their remedies. Cf. Albertus Magnus, *De animalibus Libri XXVI*, ed. Hermann Stadler (Münster: Aschendorff, 1916-1920).

tiveness of a given plant. The explanations generally were formulated in humoralistic terms. Both the composition of sound and sick bodies as well as the plant substance itself and its properties were analyzed in terms of qualitative changes. In turn, they were explicated by the theory of the four elemental qualities of heat, coldness, dryness, and moisture. The properties of a plant substance, and hence its physiological action, were thus explained in terms of its elemental nature; for example, *ysopus* (*Hyssopus officinalis* L.; hyssop) is stated to be hot and dry while *portulaca* (*Portulaca oleracea* L.; purslane) is cool and moist.[57] On the basis of its nature, the therapeutic employment of a plant followed as a matter of course. A dry plant dried etc. or a cool plant was beneficial for conditions caused by too much heat, etc.[58]

But this kind of explanation of a plant's therapeutic usefulness did not exhaust the store of therapeutic data found in Book VI. Like the descriptive data, however, the therapeutic data cannot be sharply separated into distinct classes. Because many of the plants and plant products belonging to the traditional materia medica served a plurality of functions, they can be discussed from an equal number of different points of view. Since that is not feasible here, it may merely be noted that Albert's references to the therapeutic uses of plants and their products were, in all essential matters, those of the contemporary herbalist. Depending principally upon the mode of administration, a plant or some portion thereof might serve as part of a normal diet in times of health or in a restorative diet in times of illness.[59] Moreover, those same plants might serve as a relish or condiment,[60]

---

[57] *De veg.* 581/477 (ysopus); 548/406 (portulaca).

[58] Of *lingua arietis* (*Plantago* sp., plantain), Albert states, "et siccitas eius, ut diximus, non est mordicativa, et ideo consolidat ulcera optime, et ad haec nihil est melius ea" (529/369). A cooling and moistening plant such as *orpinum* (*Sedum telephium* L.; orpine) "valet autem calefactioni hepatis, et infrigidat vehementer" (546/402).

[59] Prior to enumerating the *virtutes* of *petroselinum* (*Petroselinum crispum* (Mill.) Nym.; parsley) Albert states, "est autem calidum et siccum, plus medicina quam cibus" (551/413). Nearly the same statement is made in connection with *lupinus* (*Lupinus* spp.; lupine) in 533/375.

[60] Beans (*Vicia faba* L.) are made more acceptable, "cum. . .comedantur cum pipere et asa et sale et oleo et origano et similibus" (513/338). Each of the three plant substances here mentioned, *piper* (*Piper nigrum* L.; black pepper); *asa* (the gum resin from *Ferula assa foetida* L.; asafedita) and *origanum* (*Origanum vulgare* L.; oregano) is described by Albert and its virtues enumerated.

as ingredients in composita[61] or even be taken or applied singly for specified complaints.[62]

On the whole, the rationale underlying the therapeutic data assembled by Albert was humoralistic and was based on the traditional peripatetic theory of the four elements or elemental qualities. But there are traces here and there of a more primitive rationale, though they are partially obscured by the humoralistic interpretation. It is entirely in keeping with his utilization of herbalistic sources that Albert would adopt on occasion references to therapeutic practices based upon the belief in a plant's signature. This belief rests on a supposed likeness between some characteristic of a plant and either some pathological condition or the agent reputed to produce that pathological condition or some other form of bodily harm. For example, *Dracunculus vulgaris*, Schott or dragon arum, known to Albert as *basilicus, dracontea*, or *colubrina* was, somewhat fancifully, likened to a serpent. To what extent one can see any resemblance is irrelevant for once the resemblance is asserted, and bolstered by its set of synonyms, it is no surprise to find that its principal virtue is to cure the bite of a serpent.[63]

As in all medieval herbals and, indeed many later ones as well, there is a further class of data included in Book VI. By its very nature, this class of data is neither easily summarized nor readily organized into sharply defined categories. Yet no account of

[61] Many of the sections devoted to the individual plants, conclude with a reference to "praeparationes medicorum" or "operationes in medicorum praeparatum," e.g., 448/216; 457/235 502/316; 534/376 etc. In addition, Albert sometimes provides simple recipes which include, as one of their ingredients, the plant discussed in that section, e.g., 555/421; 564/436. Under *enula campana* (*Inula helenium* L.; elecampane), a detailed recipe is included for the preparation of *vinum enulatum* (509/332). A recipe, bearing the following title "Pillole de Alberto: e credesi di Alberto magno: Ma stimiamo sieno di Alberto Bolognese" occurs in *Nuovo Receptario. . .della inclita cipta di Firenze* (Firenze: Ad instantia delli Signori Chonsoli, 21 January 1498), sig. gv[rb].

[62] Thus Albert devotes as much space to the *virtutes* as to the alimentary or nutritive properties of such plants as *beta* (*Beta vulgaris* L.; beet) (485/292); *caulis* (*Brassica oleracea* L.; cabbage) (493/304); *ordeum* (*Hordeum vulgare* L.; barley) (544/399); and *rizum* (*Oryza sativa* L.; rice) (558/427).

[63] "Basilicus [codd. basiliscus] est herba, quae dracontea vel serpentaria dicitur. . . .Primum autem in siliqua profert ea, quae est sicut posterior pars serpentis, in fine sicut cauda serpentis, et habet in stipite suo varietatem serpentis. Et valent contra serpentis morsum succus eius, et etiam dicitur, quod portata tutat a serpentibus omnibus" (484/290).

medieval herbalism would be complete without at least mentioning a stratum of beliefs loosely but popularly termed "magic and folk belief."

Quite apart from the legends associating Albert with magic[64] and the strange assortment of "secret" writings spuriously attributed to him,[65] it cannot be denied that there are over a dozen references to the magical arts in Book VI. In these passages, however, Albert is reporting the uses of plants by magicians and enchanters, not subscribing to those uses. Such references are historically important because they are a reminder that Albert's enlightened, philosophic approach to the therapeutic uses of plants was not representative of all ranks of herbalists and other medical writers. It is no surprise that Albert noticed some of the darker practices since the employment of herbs and portions of trees for a variety of magical purposes was widespread in his time.[66] Not only are these practices well attested in medical literature; they are also mentioned, and sometimes roundly denounced, in various genres of nonmedical literature.[67] The detailed analysis, required to explain the origin of the magical practices reported by Albert, cannot be undertaken here. But a few observations bearing on the relations between herbalism and magic will help to round out an analysis of Book VI.

One very common practice, alluded to in nearly all medieval herbals, concerns the use of plants or plant products as apotropaic devices. A good example is provided by Albert in his discussion of *lingua arietis* (*Plantago* sp.; plantain). "The enchanter says," states Albert, "that its root, suspended from the neck of a child, cures scrophulous sores."[68] This passage, chosen from several similar references, illustrates some of the interrelated aspects of herbalism and

---

[64] For a summary, see Karl Helm, "Albertus Magnus" in *Handwörterbuch des deutschen Aberglaubens*, ed. Hanns Bächtold-Stäubli (Berlin: De Gruyter, 1927), vol. 1, cols. 241-243.

[65] Cf. G. Meersseman, *Introductio in opera omnia B. Alberti Magni O.P.* (Bruges: Beyaert, 1931), pp. 145-148, and Lynn Thorndike, "Further Considerations of the Experimenta, Speculum Astronomiae and De secretis mulierum ascribed to Albertus Magnus," *Speculum* 30 (1955), 413-443.

[66] See the passage pertaining to plants referred to by Berthold of Regensburg, a contemporary of Albert, in Anton Schönbach, "Zeugnisse Bertholds von Regensburg zur Volkskunde," *Stzb. Akad. d. Wissenschaften zu Wein, Phil.-hist. Cl.* 142/7 (1900), esp. pp. 35-50, 138-148.

[67] See Oskar Ebermann, "Zur Aberglaubensliste in Vintlers Pluemen der Tugent (v. 7694-7997)," *Ztschr. f. Volkskunde* 23 (1913), 1-18, 113-136, and Theod. Zachariae, "Abergläubische Meinungen und Gebräuche des Mittelatters in den Predigten Bernardinos von Siena," *Ztschr. f. Volkskunde* 22 (1912) 113-134, 225-244.

[68] *De veg.* 529/369.

magic. By the term "enchanter" (*incantator*), Albert is probably referring to Avicenna whose chapter on *lingua arietis* is almost certainly Albert's source.[69] The passage in question, however, is Albert's paraphrase, not a verbatim quotation.

The use of natural products, whether of plant, animal, or mineral origin, as apotropaia, was not, in itself, regarded as a form of black magic. Not only were various kinds of pendants and medallions tacitly accepted, but various plants, when subjected to ecclesiastically approved rites, were invested with extraordinary properties and hence useful for a wide range of medical and nonmedical purposes alike. Plants which had been consecrated in such a manner can be termed magiferous plants.[70] Unlike imaginary or fabulous plants, they were real plants. But as a result of their exposure to various rituals, their natural properties or *virtutes* were temporarily reinforced with greatly enhanced properties, useful in many forms of healing. Several of the plants described by Albert fall into this category, for example, *betonica, ruta, salvia*, and *verbena*, though the descriptions touch only lightly upon such matters.

There are, finally, other references where the linkage between herbalism and magic is more obvious and potentially more dangerous. Despite the efforts of the authorities to stamp out such practices, references to them in herbals and medical texts seem to suggest that these practices were widely adopted. Albert again provides some examples, not because he advocated them but rather, since they were so well known and commonly accepted parts of the folk scene, they added another dimension to the herbalistic descriptions of plants. For example, the use of oak galls by the *aeromantici* for purposes of divination,[71] the use of *iusquiamus* (*Hyoscyamus niger* L.; henbane) by necromancers intent upon invoking demons,[72] or the use of a drink prepared from the leaves of *populus* (*Populus* sp.; poplar) to be given to women, *non concipiat, sed sterilis efficiatur*.[73] This last example is doubly interesting because Albert attributes to the enchanters

[69] Avicenna (1507 ed.), fol. 128va, cap. 433 "De lingua arietis." A different, but no less unusual property of plantain is reported by Henrik Harpestraeng, ed. Hauberg, pp. 104-106.

[70] See Jerry Stannard, "Magiferous Plants and Magic in Medieval Medical Botany," *Maryland Historian* 8 (1977), 33-46.

[71] *De veg.* 441/206. For a summary of the problem, see the fifteenth-century text edited by Gerhard Eis as "Prophezeiung aus dem Gallapfel," in his *Wahrsagetexte des Spätmittelalters* (Berlin/Munich: Erich Schmidt, 1956), p. 69.

[72] *De veg.* 527/363. Cf. the reference to "qui in nigromantia student" (458/235).

[73] *De veg.* 432/185.

(*incantatores*) a typical herbalistic practice. Independent of the question whether there is any valid evidence of its effectiveness, this is a practice which is no more magical than the use of certain plant substances, mentioned by Albert without comment, as abortifacients.[74]

Only a thin, uncertain line separates magical practices from folk beliefs and, for obvious reasons, there is much overlapping. Folk beliefs, as might be expected, are of such a heterogeneous nature, that only a few comments are possible here. Despite the problems involved in defining and separating the two subjects, one point is reasonably clear. The heart and center of medieval herbalism was its empiricism. Albert was far too perceptive not to recognize the force of custom and the pragmatically justified concept of success which together sustained herbalistic practice. Despite the conceptual elegance of scholasticism or the inherent reasonableness of Aristotelian natural philosophy, herbalism was entrenched among the folk. If a certain plant was effective for a specified condition, as measured by the standards that pevailed among the folk, e.g., squill for dropsy, that was important to the patient, significant for the herbalist, and far outweighed the lack of an explanation from the schools.[75]

Thus, to take a simple example, in his description of *sinapis* (*Sinapis alba* L.; mustard), Albert first lists several properties: it clarifies the face of unsightly blotches, opens blockages, dries the tongue, and is effective for baldness. Then, he continues, "And some say, that if drunk while fasting, it produces a good intellect."[76] Short of controlled, laboratory and clinical tests, how could this be gainsaid? Other examples, for which parallels can easily be found today, concern a field test to determine whether fungi are poisonous or not,[77] and the belief that by eating beforehand *allium* (*Allium sativum* L.; garlic) one is protected from harm when collecting *elleborus niger* (*Helleborus niger* L.; black hellebore.)[78]

---

[74] Cf. 374/73; 393/111; 487/294; 499/311, etc.

[75] For Albert's reference to *squilla* (*Urginea maritima* (L.) Bak.; squill) see 561/431. This passage, along with others, has been discussed by Jerry Stannard, "Squill in ancient and medieval materia medica, with special reference to its employment for dropsy," *Bull. New York Academy of Medicine* 50 (1974), 684-713.

[76] *De veg*. 568/446. Another interesting claim is made by Albert in regard to *raphanus* (*Armoracia rusticana* Gaertn., Mey. et Schreb.; horse radish): "apud nos etiam maniacus, raphano contuso cum succo ligato super caput eius rasum, infra triduum recipit beneficium sanitatis" (558/426).

[77] *De veg*. 517/344.

[78] *De veg*. 510/333.

Whether these and the many other similar beliefs found in *De vegetabilibus* and contemporary herbals alike are always capable of being distinguished from the philosophic beliefs of the schools is problematic. But from the point of view of herbalism, the issue is clear cut: whatever offered any help in times of illness or served to avert harm, was deserving of recognition. The collection and compilation of such data, not their validation, was the task of the herbalist. With this in mind, one can, perhaps, better understand why Albert included a range of data over and above what we would term descriptive. Those other data, no less than the descriptions themselves, pertained to plants in all their diversity. It was Albert's task to organize them within the framework available to him.

XIV

THE BOTANY OF ST. ALBERT THE GREAT

Among the many reasons why Albertus is honored as ›Doctor Universalis‹ is the enormous range of knowledge that he was able to assemble, organize and explain. Following Aristotle, he believed that ›naturalia‹ were capable of being understood to the extent that they could be explained causally. Since ›naturalia‹ were amenable to philosophic analysis, they were within his encyclopedic grasp. With respect to the natural sciences, this included the biological as well as the physical sciences. Albertus' aim, as he stated in a well known passage, was nothing less than to make the fundamental concepts of Aristotle understandable to his brother Dominicans[1]. His method of doing that took the form of commentary or, more accurately, paraphrase. Since it was his task to comment upon the entire Corpus Aristotelicum, he naturally found a place for *De plantis libri duo,* then commonly attributed to Aristotle[2]. The result of Albertus' commentary on *De plantis* was his own *De vegetabilibus libri septem*[3].

Although there are scattered throughout Albertus' writings, many miscellaneous references to plants and to the fundamental principles by means of which plant life can be explained, the present study will concentrate on *De veget.*, for two reasons. First, in order to keep this study within manageable size, it is not possible to examine in detail the relevant passages in Albertus' other writings, many of which are noted in passing in *De veget*[4]. Second, *De veget.* represents Albertus' concentrated efforts to organize a wide range of data, collected over a period of years, into an integrated, self-contained whole. Moreover as a commentary, as opposed to an original treatise, the limits of *De veget.* were to a certain extent determined by *De plantis.* And even though Albertus added much material not covered by *De plantis,* his aim was the completion of one more in a series of commentaries on Aristotle. That he accomplished, and the result was a comprehensive study of plants which cannot easily be compared with any text that came before or after.

---

1 »Intentio nostra in scientia naturali est satisfacere pro nostra possibilitate fratribus Ordinis nostri . . . ex quo libros Aristotelis competenter intelligere possent.« *Phys.* lib. I, tr. 1, c. 1.

2 *Nicolai Damasceni De plantis libri duo Aristoteli vulgo adscripti.* ed.E. H. F. Meyer. Lipsiae, MDCCCXLI.

3 *Alberti Magni ex ordine praedicatorum De vegetabilibus libri VII.* Editionem criticam ab Ernesto Meyero coeptam absolvit Carolus Jessen. Berolini 1867. (All references to *De veget.* are to this edition.)

4 For a convenient summary cf. *De veget.* ›Index personarum‹ 694–695 Jess., s. v. Albertus Magnus . . . »Opera quae ab ipso aut a nobis laudantur«.

Ever since Meyer's pioneer studies, which culminated, though posthumously, in 1867 with the appearance of the first critical edition of *De veget.*, scholars have been nearly unanimous in regarding it as a masterwork[5]. Except for some unfortunate historical accidents of the sort that are not uncommon in the history of science, *De veget.* might have been a milestone in the historical development of botany. That is, it might have changed the direction of botanical studies by its emphasis upon the need for understanding the basic principles that govern plant life rather than by continuing to assemble references to medicinal plants culled from old authors which was the herbalistic tradition[6]. Albertus recognized the distinction between a philosophical, generalized study of plants and the approach of the herbalists and the compilers of ›receptaria‹[7]. As it turned out, new directions in philosophy and theology, coincided with new developments in the natural sciences and technology. As a result, *De veget.* was virtually forgotten, with the consequence that the first major contribution to general botany since the time of Theophrastus was ignored. When it was rediscovered by Sprengel and Meyer, so to speak, the new scientific botany had already reached the point where a return to *De veget.* would be of little profit to botanists.

The past several hundred years have witnessed an interest in the capabilities of science and in its application unparalleled in the history of mankind. But with respect to the pure intellectual powers of attempting to understand natural phenomena and to explain, in consistent theoretical terms, the processes of life, two earlier periods can be favorably compared, the Golden Age of Greece and the High Middle Ages. It is thus no accident that the present study unites those two periods by means of two texts: *De plantis* representing, albeit in a much debased form, the theoretical inventiveness of Greek philosophy and *De veget.*, representing part of a systematic plan to assimilate and disseminate the teachings of Aristotle. Inevitably, the contributions of both texts, as well as their respective philosophic presuppositions, must be understood in terms of the present.

---

[5] Meyer's earlier works are: *Albertus Magnus. Ein Beitrag zur Geschichte der Botanik im dreizehnten Jahrhundert.* Linnaea 10 (1836), 641–741; idem, *Albertus Magnus. Zweiter Beitrag zur erneuerten Kenntnis seiner botanischen Leistungen.* ibid., 11 (1837), 545–595; idem, *Geschichte der Botanik.* vol. 4 (Königsberg 1857), 9–84. For valuable material on Meyer, his relations with Karl Jessen, and the steps leading up to the publication of *De veget.*, cf. FRANS VERDOORN, *Ernst Meyer and his Geschichte der Botanik.* Utrecht 1965. (Communicationes Biohistoricae Ultrajectinae, IV.)

[6] Cf. JERRY STANNARD, *Albertus Magnus and Medieval Herbalism,* to appear in *Albertus Magnus and the Sciences,* Commemorative Essays ed. James A. Weisheipl O. P. Toronto 1980.

[7] »Aliter autem haec dicenda sunt, quam dicantur a medico. Nos enim herbarum et generaliter plantarum naturam non dicimus, nisi ut ex operationibus earum naturam inveniamus: medicus autem talia inquirit ad receptionem medicinarum.« *De veget.* lib. VI, tr. 2, c. 1, p. 472 Jess.

This does not mean, of course, that the findings of either *De Plantis* or *De veget.* should be interpreted in the light of modern botany. To do so would be anachronistic and would add little to our understanding of Albertus' contributions. Despite the philosophic significance of many of Albertus' claims and his remarkable originality of attempting to explain physiological processes in terms of the limited categories available to him, not all of his findings are acceptable today[8]. Moreover, many of his leading concepts, for example, the plant-animal analogy, have been replaced by concepts more appropriate to experimental inquiry and testing. As a result, Albertus' botanical investigations can be, and have been, interpreted in different ways, depending upon the purpose at hand[9].

What makes the *De veget.* particularly difficult for the modern reader is the presence of several strata of data which must be distinguished if Albertus' contributions are to be fully understood. There is, first of all, a stratum representing *De plantis* which, in itself is a composite work. Albertus adopted many of the concepts and arguments of *De plantis*, and some of its Greek technical terms are still discernible beneath the Arabico-Latin translation used by Albertus[10]. Second, is Albertus' adoption of Aristotelian philosophic principles, not all of which, however, are to be found in *De plantis*. Third, is the wide range of extra-Aristotelian literary material utilized by Albertus for explanatory and illustrative purposes. Some of this material was easily adapted to the language of thirteenth century Aristotelianism, though not all of it. Fourth, are the more specific examples and observations which frequently were added by Albertus to clarify or illustrate the abstruse philosophic discussions. Many of these examples are easily recognizable as deriving from personal experiences gained by Albertus in the course of his travels. But the casual manner in which some of them are mentioned, and without any clue as to geographical locale, has caused some of his most valuable observations to go relatively unnoticed.

---

8 Cf. KAREN REEDS, *Albert on the Natural Philosophy of Plant Life (Books I–V of De vegetabilibus)*, to appear in *Albertus Magnus . . .*, Commemorative Essays, ed. James A. Weisheipl, O. P. Toronto 1980.

9 Among the older literature, the following are important: RUDOLPH VON FISCHER-BENZON, *Altdeutsche Gartenflora*. Kiel u. Leipzig 1894; J. WIMMER, *Deutsches Pflanzenleben nach Albertus Magnus (1193–1280)*. Halle a. S. 1908; STEPHAN FELLNER, *Albertus Magnus als Botaniker*. Wien 1881; T. A. SPRAGUE, *Plant Morphology in Albertus Magnus*. Kew Bulletin 9 (1933), 431–440; idem, *Botanical Terms in Albertus Magnus*. ibid. 9 (1933), 440–459. cf. also, HEINRICH BALSS, *Albertus Magnus als Biologe*. Stuttgart, 1947, 79–187 and FRANZ STRUNZ, *Albertus Magnus. Weisheit and Naturforschung im Mittelalter*. Wien u. Leipzig 1926, 106–119.

10 For example the series ὄπτησις, πέπανσις and στάθευσις (on the latter, cf. *De veget.* lib. V, tr. 2, c. 3, p. 328 Jess.) These terms are discussed by MEYER, Praelibanda, Particula VI (*De veget.* pp. XLII–XLIII ed. Jess.)

It may be said, then, that *De veget.* is not only a commentary on an earlier botanical text or a record of what Albertus thought about plants, it is a ›summa botanica‹ of what was known and what was believed about plants in the mid thirteenth century. For these reasons, plus the fact that Albertus chose to include material that is usually not found in general botanical studies of a later date e.g. folk lore, references to magical practices, etymological excursus, etc., scholars have found it difficult to organize Albertus' findings in such a way as to do justice to him but, at the same time, to make them capable of being understood by the modern reader.
As a result of over a century of scholarship devoted to *De veget.* it has become increasingly evident that it can neither be appreciated as a historical document nor evaluated as a contribution to science solely on the basis of the methods and models current at a given time. While there is no doubt that the experimental method and its demand for testable hypotheses have enriched our understanding of plants, it would be an anachronism to apply today's standards to a text written some seven hundred years ago which, in turn, was based on a text written a millenium earlier. Under these circumstances, the present study is an attempt to explicate the botany of St. Albert in a fashion that does justice both to history and to science. Since this procedure precludes a detailed analysis of the myriad data with which *De veget.* abounds, a consideration of its scope and general plan will provide the context for the subsequent portions of this study.

I. Format

Both the form and the structure of *De veget.* initially were determined by *De plantis*, but modified by Albertus through the addition of new material. The addition of such material, designed to supplement the brief and sometimes nearly incomprehensible exposition of *De plantis*, necessitated a special structure for *De veget.* This was accomplished by the insertion of several ›digressiones‹ containing Albertus' own thinking and only loosely connected with the ›Urformat‹ of *De plantis*.
Because of these modifications, a synopsis of *De veget.* as a whole may be useful. This is especially true because *De veget.* is not a mechanical, line-by-line commentary but rather a combination of paraphrases and digressions through which Albertus was able to assemble information relating to plants not found in *De plantis*. In fact, out of the seven books of *De veget.*, only two (libri I and IV) are commentary in the strict sense and, even then, one half of lib. I is composed by chapter-size ›digressiones‹ representing Albertus' own contributions. Of the remaining five books, three of them (libri II–III, V) are ›digressiones‹ while books VI and VII are independent treatises devoted to specific subjects.

De veget. lib. I, tr. 1, cc. 2–8 are commentary to *De plantis* lib. I where the central issue is whether, and to what extent, it can be maintained that plants possess life. But *De veget.* lib. I is introduced, however, by the first of many ›digressiones‹; in the present instance, by Albertus' statement of his aim and of synopsis of the contents of the remaining portions of lib. I. The second half of the first tractate (cc. 9–14) is composed of a series of six ›digressiones‹ which have as their purpose an extended discussion of the six philosophic problems involved in the main question concerning life in plants. In this manner, Albertus was able to discuss the central issue – »de vita occulta plantarum« – without being confined to the order of presentation of *De plantis* or restricted by the limited evidence adduced therein. Underneath the problem of whether plants have life was the more significant issue of whether the possession of the ›anima vegetabilis‹ by plants – a fact that Albertus never questioned – entailed that plants, like animals, possess similar physiological responses to perceptual stimuli; i.e., whether plants feel pain and joy, whether they sleep and awaken, and finally, whether separate sexes can be predicated of plants as they can of most, but not all, animals. The second tractate of lib. II, »De diversitate et anatomia plantarum« resumes the commentary to *De plantis* lib. I. by stating, in effect, a method that applies to all of Albertus' commentaries on natural science: »Si tamen non Aristotelem, sed nos ipsos sequeremur, pro certo aliter procederemus[11].« Without going into needless detail, the second tractate is built about two themes: the natural diversity of plant life and the plant-animal analogy, best known from Theophrastus' *Historia plantarum* lib. I. Whereas a modern reader might hesitate before accepting that analogy, for Albertus it was a useful explanatory concept provided one did not push it too far and was prepared to view it as a heuristic device[12]. As noted above, Books II–III and V are explicitly stated to be ›digressiones‹ in which Albertus' order of presentation and choice of examples were not rigidly determined by *De plantis*. One of the major reasons for those three ›digressiones‹ is announced in the opening sentence of Book II: »Haec omnia tradita sunt ab antiquis physice de plantis loquentibus, et videntur quandam habere confusionem[13].« This passage, is doubly instructive for it is a clue to the many literary sources utilized by Albertus in the preparation of *De veget.* and to his undoubted command of the relevant literature. But there is also the implication sub rosa that his personal observations do not in all details agree with what he has read. As a result of such discrepancies, plus the fact that *De*

---

11 *De veget.* lib. I, tr. 2, c. 1, p. 57 Jess.

12 Thus, Albertus writes: »Venter etiam plantae metaphorice convenit plantae.« Lib. I, tr. 2, c. 1, p. 58 Jess. and later: »Sed tamen augmentum plantarum non est sicut augmentum animalium«, lib. V, tr. 1, c. 6, p. 308 Jess.

13 Lib. II, tr. 1, c. 1, p. 103 Jess. For the methodological function of a ›digressio‹, cf. G. Meersseman O. P., *Introductio in Opera Omnia B. Alberti Magni O. P.* Brugis 1931, 6.

plantis does not even touch upon certain matters which were important for Albertus, his ›digressiones‹ were justified both philosophically and pragmatically. The ›digressiones‹ which constitute Books II–III and V may thus be regarded as quasi-independent treatises whose overall purpose was to examine those subjects applicable to plants in general.

Book II is divided into two tractates of six and seven chapters respectively. The first, »De origine diversitatum plantae supra enumeratum, et de his quae convenient eisdem, ut ordinatus sciantur causae earum«, raises a series of questions so basic to plant life, that simple empirical data, taken by themselves, can do little more than provide a redefinition of the central problem, viz. the nature of plant qua plant. At issue is the question of how the inner life (vita occulta) of a plant can be satisfactorily explained. Albertus answers by rephrasing the question in such a way that it may be answered within the framework of Aristotelian philosophy: by a knowledge of the several functions of the plant, plus an understanding of the physical or material conditions by which the essential functions can be both understood and modified[14]. Those functions, common to all plants in virtue of their possession of the vegetative soul, are nutrition, growth, and generation[15]. The success by means of which these three vital processes are carried out, depends upon the proper mixture or blendings of the four primal elements in cooperation with a ›virtus coelestis‹[16]. The latter, at a very minimum, includes the heat and light of the sun, as providing the necessary conditions for both germination and subsequent development, terminating in fructification. Plant diversity, an explanation of which is the central motif of the first tractate, can be expressed in other ways as well, e.g., by means of the distinction, as obvious on the perceptual as on the linguistic level, between trees and non-trees. This quasitaxonomic device, complicated by the existence of ›fungi‹ whose life cycle was then unknown, is discussed at greater length later. Diversity again is evident and can be explained by a different, though related, approach. What are commonly called parts or organs of a plant are clearly of different kinds and serve different functions. The ›partes officiales‹, or functional parts, include sap, nodes, root, veins, pith, bark, and ligneous tissue, if present[17]. Finally, a last index of diversity is generalized shape or habit, which is explained by Albertus in

---

14 ». . . et natura communis, qua planta dicitur, est secundum formam quidem actus ad potestatem vitae occultae, secundum materiam autem est susceptibilitas huiusmodi actus vitae«, lib. II, tr. 1, c. 1, p. 104 Jess.

15 »Cum autem sint tres operationes animae vegetabilis in plantis, nutrire scilicet et augere et generare«, lib. I, tr. 1, c. 14, p. 52 Jess.

16 »Et cum complexio tanto sit facilior, quanto est similior, sufficit in productione ejus agens unum commune, quod est virtus coelestis cum materia convenienter commixta«, lib. II, tr. 1, c. 1, p. 105 Jess.

17 Cf. *De veget.* lib. I, tr. 2, c. 1, p. 61 et lib. II, tr. 1, c. 3, pp. 114 sqq.

terms of ›virtus et lumen coelestis‹[18]. Just as difference in shape can be explained causally, so too can color, the subject of tr. 1, c. 6. The second tractate of the second book continues the morphological approach by considering the role of the four organs essential to fructification and generation, viz. leaves, flowers, fruit, and seed[19]. Differences in leaf shape and texture are next explained (tr. 2, cc. 2–3). The last four chapters are devoted to an explanation of the great differences in shape, size, colors and odors of the flower itself.
Book III, tr. 1, »De fructibus et seminibus«, treats of the two remaining organs essential to fructification and generation while the second tractate, »De saporibus succorum et fructuum et seminum planatarum«, identifies the different flavors and explains them in humoralistic terms by the interaction of the four elements. Although Albertus relied heavily on Avicenna and Galen for his explanation of differences in flavor, he was not restricted, here or elsewhere, from making use of his own experiences, as some of his examples testify[20]. In the first tract, when discussing the ripening of fruit, Albertus takes care to note that what applies in one region may not be valid elsewhere[21]. In a similar fashion, he points out that although horticultural practices may drastically affect the bearing-capacity of fruit trees, in the long run nature is more effective than art, »quia opus naturae habet motorem ad esse speciei se moventem; opus autem artis suum extra se habet motorem«[22] – a passage which owes as much to Albertus' habit of observing living plants as it does to his reading.
Book IV, the counterpart to Book I, is a commentary on *De plantis* lib. II. The four tractates which comprise this book, however, go far beyond *De plantis* and point ahead to *De veget.* lib. V. The first tractate, »De virtutibus originalibus« (sc. plantae), demonstrates the explanatory powers of the four-element theory by applying it to a series of problems relating to plant physiology and plant chemistry. In brief, each of the four primal elements – earth, water, fire and air – fulfills a specific function in the development of a plant[23]. The second tractate, »De modo et loco generationis plantarum«, is a

---

18 Following an explanation of columnar shape, Albertus continues: »Cum enim virtus et lumen coelestis circuli pyramidaliter impendeat loco generationis, est virtus ejus fortior in circulo, qui est basis luminis pyramidalis, et in centro ejusdem pyramidis fortissima«, lib. II, tr. 1, c. 5, p. 128 Jess.
19 »Hae autem partes sunt folia, quae sunt propter fructuum cooperimenta, et flores, qui sunt indicia fructuum, et fructus et semina«, lib. II, tr. 2, c. 1, p. 138 Jess.
20 »Et exemplum hujus (sc. differences in taste) hujus est in ferro et aere, quae aut insipida sunt, aut debilem habentia saporem in lingua . . . et ita est de lignis multis«, lib. III, tr. 2, c. 1, p. 192 Jess. Cf. also his account of fermenting wine, tr. 2, c. 7, p. 210 Jess.
21 »Licet autem dixerimus, maturari fructus ad esum, non tamen aliquem fructum de his, qui noti sunt in nostris climatibus . . .«, lib. III, tr. 1, c. 6, p. 184 Jess.
22 Lib. III, tr. 1, c. 7, pp. 187–188 Jess.
23 »Plantae igitur quatuor virtutes habent a quatuor elementis: a terra quidem fixio-

further application of the four-element theory, succinctly summed up in the statement: »herbae et omnes plantarum species non fiunt nisi per compositionem et commixtionem elementorum, et non per naturam alicujus simplicis elementi[24].« The proper blending of the elements or their harmony, a phrase that Albertus used elsewhere[25], is adopted principally to explain why different species have different environmental preferences or ecological niches. The third tractate, »De principiis generationis et fecunditatis plantarum«, examines the five factors – ›semen‹, ›putredo‹, ›humor‹, ›aqua‹, ›plantatio‹ – which affect the growth of all forms of plant life. The concluding tractate, »De accidentibus plantae quoad visum et gustum«, is again a study of the application of the four-element theory, in this case to such perceptible properties as the color of foliage and various characteristics of sap, fruits, and seeds. The full scope of Book V is only partially indicated by its title: »De convenientia et differentia plantarum et de effectibus earum.« But considering the specialized nature of Books VI–VII, the present book can be regarded as Albertus' summary statement concerning various matters raised in, but not settled by, *De plantis*. This is evident, for example, in Ablertus' enumeration of the seven conditions necessary for generation and the maintainance of life, common to all forms of plant life[26]. A departure from the normal or typical pattern of development can be explained, however, in terms of a failure of one or more of the components in the ›elementorum mixtio‹[27]. Book V concludes with several chapters concerning phenomena encountered in agricultural and horticultural practice whose aim is to provide ›per artem‹ if necessary, the conditions responsible for optimal growth. Such practices as grafting and transplanting are likewise explained in terms of the interworkings of the four elements and the derivative humors, the effects of which, are discussed in the concluding chapters.

The final two books of *De veget.* are, in contrast to the philosophically oriented commentaries and digressions, much more in the nature of pragmatic statements with clearly defined, practical purposes. The title of Book VI, »De speciebus quarundam plantarum«, does not fully indicate the contents of its two tractates: But taken together, tr. 1 »De arboribus« and tr. 2 »De herbis specialiter secundum ordinem alphabeti« form a series of short chapters in which the various species of plants mentioned in the preceding books are described and some of their uses are enumerated. Bibliographically speaking,

---

nem, ab aqua conglutinationem sive continuationem, ab igne vero unitionem fixionis et continuationis, ab aere autem raritatem«, lib. IV, tr. 1, c. 1, pp. 218–219 Jess.

[24] Lib. IV, tr. 2, c. 1, p. 232 Jess.

[25] Lib. V, tr. 1, c. 1, p. 292; tr. 2, c. 1, p. 324 Jess.

[26] »In generatione autem sunt septem, sine quibus nulla omnino nascitur plantą ...«, lib. V, tr. 1, c. 1, pp. 290–291 Jess.

[27] »Et causa coloris talis (sc. nigra) est, quia est ex terrestri grosso non bene commixto . . .«, lib. III, tr. 1, c. 5, p. 180 Jess.

Book VI if taken by itself, may be considered to be a herbal and, in fact, it was so regarded by some later writers[28]. Because Albertus utilized herbalistic sources, the individual descriptions often contain pharmacological and medical data of the sort commonly found in contemporary herbals and ›Arzneibüchern‹. It was not, of course, his intention to compile a herbal for therapeutic purposes. But in order to describe succinctly yet accurately some 400 different plant species, not all of which he had personally seen in the living state, Albertus used the best sources available to him. This included not only the afore mentioned herbals but the herbalistic portions of larger, more general works, such as Avicenna, *Liber canonis* lib. II. The purpose of Book VI, therefore, was to provide his readers, few of whom were so well acquainted with the world of nature, with the information necessary to understand the preceding books and particularly the relevancy of the plants singled out as examples. Basically, that information included (i) a description of the living plant in its native habitat, (ii) its name and synonym(s) and (iii) some indications of its uses or properties. The descriptions, in short, were designed to help the reader visualize the plant in question and to associate that name with a locally growing plant or, in the case of the Near Eastern exotica, with some use of that product in medicine or in the preparation of food.
Although Book VI, »De mutatione plantae ex silvestritate in domesticationem«, is explicitly stated to be a digression, it has more in common with Book VI than it does with the other book-length ›digressiones‹ (Libri II–III, V). The reason for this is that Book VII, like Book VI, owes virtually nothing to *De plantis*. If anything, Book VII may be regarded as a commentary on Palladius' *Opus agriculturae*. Like so many other, imperfectly known Graeco-Roman texts, Palladius' work was in process of becoming assimilated in the Latin West, hence commenting on it was a useful service. But more important, Book VII was functionally connected to the preceding six books by virtue of several themes, some of which, e. g. grafting and transplanting, were discussed earlier. But unlike the previous references to those and other matters, Book VII treats of such subjects, not from a philosophical point of view, but from an agricultural or horticultural point of view.

II. Literary Sources

The primary source for *De veget.* was *De plantis* which throughout the Middle Ages was accepted as a genuine work of Aristotle[29]. Already in the Renaissance, doubts were raised concerning its authenticity, but the identity

[28] The data are discussed in STANNARD, op. cit. (vide supra adn. 6).
[29] Roger Bacon, for example, writes: »dicit Aristoteles quod pars anime reperitur in planta, scilicet vegetativa, et hoc dicit in hoc libro et in libro De Anima.« *Questiones supra de plantis,* (in) Opera hactenus inedita Rogeri Baconi, ed. Robert Steele, fasc.

of the author as well as the translators remained a matter of conjecture[30]. The issue became clarified, though by no means completely solved, with the publication in 1841 of a critical edition of the medieval Latin text of *De plantis*. Its editor, Ernst Meyer, who was later responsible for the text of *De veget.*, established that *De plantis* was written by Nicholas of Damascus (64 B.C. – post 4 B.C.), advisor to King Herodes and a personal friend of the Emperor Augustus[31]. Nicholas, a peripatetic polymath, was the author of many writings, which today exist only in the form of fragments[32]. Subsequently, scholars generally have accepted Meyer's attribution of *De plantis* to Nicholas[33]. It is another issue, albeit related, whether Aristotle wrote a Περὶ φυτῶν, traditionally attributed to him in the older doxographic literature[34]. Whatever may be the final verdict on such an Aristotelian text, there appears to be no grounds today for identifying our *De plantis* with Aristotle's (lost?) Περὶ φυτῶν. Subsequent studies devoted to Nicholas and to the establishment of the Aristotelian canon have resulted in a maze of detail, only portions of which are relevant to the present study.

Although there are still many puzzles concerning the textual transmission of *De plantis*, the chronology of the several translations or versions and especially whether there existed a second medieval Latin translation known to or used by Roger Bacon[35], Meyer's text plus a Greek version, late and unreliable as it is, are sufficient for placing *De plantis* within the peripatetic tradition. From even a cursory reading of the Latin and Greek texts, it is apparent that

---

XI (Oxford 1932), 177. For another reference to Roger, cf. AUGUSTE PELZER, *Une source inconnue de Roger Bacon: Alfred de Sareshel, Commentateur des Météorologiques d' Aristote.* Archivum Franciscanum Historicum 12 (1919), 44–67, (ref. 62). Likewise, Thomas of Cantimpré stated: »Dicit Aristotiles in libro primo De vegetabilibus . . .«, THOMAS CANTIMPRATENSIS, *Liber de Natura Rerum* ed. H. Boese. Berlin 1973, X, 10, 8 p. 317. Finally, Geoffrey of Aspall (obit 1287) wrote a commentary, presumably lost, on *De plantis;* cf. ENYA MACRAE, *Geoffrey of Aspall's Commentaries on Aristotle.* Mediaeval and Renaissance Studies 6 (1968), 94–134.

30 Cf. LOTTE LABOWSKY, *Aristoteles De plantis and Bessarion.* Mediaeval and Renaissance Studies 5 (1961), 132–154.

31 Cf. EDW. SCHWARTZ, *Zu Nikolaos von Damaskos Biographie des Kaiser Augustus.* Hermes 33 (1898), 182–184 and KARL PATSCH, *Zu Nicolaus von Damascus.* Wiener Studien 12 (1890), 231–238.

32 Cf. H. J. DROSSAART LULOFS, *Nicolaus Damascenus on the Philosophy of Aristotle.* Leiden 1965.

33 Cf. BERTRAND HEMMERDINGER, *Le De plantis, de Nicolas de Damas à Planude.* Philologus III (1967), 56–65 and H. J. DROSSAART LULOFS, *Aristotele's* ΠΕΡΙ ΦΥΤΩΝ. Journal of Hellenic Studies 77 (1957), 75–80.

34 Cf. DIOGENES LAERTIUS, *De clarorum philosophorum vitis libri decem.* lib. V, c. 1, s. 25 Περὶ φυτῶν α' β'. For the evidence that Aristotle did write such a text, cf. PAUL MORAUX, *Les Listes Anciennes des Ouvrages d'Aristote*. Louvain 1951, 109.

35 The evidence is assembled by S. D. WINGATE, *The Mediaeval Latin Versions of the Aristotelian Scientific Corpus, with Special Reference to the Biological Works.* London 1931, 55–72.

Nicholas was conversant with Aristotle's major writings, but whether he understood them is more difficult to judge. It must be admitted that the compressed style and abbreviated arguments are not always easy to understand. But on the supposition that *De plantis* was not intended to be an original work but rather a compendium of peripatetic botany, hence the inclusion of material from Theophrastus' Περὶ ἱστορίας φυτῶν and Περὶ αἰτιῶν φυτῶν, it should not be judged as if it were a genuine work of Aristotle. At any rate, Nicholas' Greek text was translated into Syriac at some undetermined date by a now unknown translator. Whether the later Arabic version by Ishaq ibn Hunayn circa A. D. 900 was based entirely on the Syriac text or whether it was controlled by some knowledge of the Greek text is still problematic, just as is the fate of the original Greek text which is presumably lost[36]. The Arabic text was then translated into Latin by Alfred of Sareshel (ca. 1180–ca. 1220) or Alfredus Anglicus, as he is sometimes termed[37]. It was Alfred's translation, or one very similar to it, which Albertus used. However, there is no evidence in *De veget.* that Albertus knew of Alfred's dedication to Roger, probably Roger of Hereford (fl. 1176/78) in which it is stated that the Latin text was a translation from the Arabic[38].

De plantis, along with the genuine Aristotelian writings which Albertus already had read and commented upon, supplied the conceptual framework and much of the data of *De veget.* Of the other two sources of data, one, the empirical data resulting from Albertus' personal observations, will be discussed below in section III. Here we must note some of the other literary sources used by Albertus in the preparation of *De veget.*

Next to Aristotle himself, the most important, identifiable literary source is Avicenna (A. D. 980–1037). His *Liber canonis* was available to Albertus in the translation of Gerard of Cremona (ca. 1114–1187). Book II, tr. 2 of the *Liber canonis* was the principal authority for the descriptions in *De veget.* lib. VI, though Albertus made use of that same tractatus elsewhere without citing it by name. Albertus' reliance on Avicenna for source material is important for several reasons. Like the vast majority of Europeans of the thirteenth cen-

---

[36] Cf. MAURICIO BOUYGES, S. J., *Sur de plantis d'Aristote-Nicolas, à propos d'un manuscrit arabe de Constantinople.* Mélanges de l'Université Saint-Joseph Beyrouth (Syrie). t. IX, fasc. 2 (1924), 70–115.

[37] On Alfred and his writings, cf. especially CLEMENS BAEUMKER, *Die Stellung des Alfred von Sareshel (Alfredus Anglicus) u. seiner Schrift De muto cordis in der Wissenschaft des beginnenden XIII. Jahrhunderts.* Stzb. Bayr. Akad. Wiss., philos.-hist. Kl., Abh. 9 (1913); CHARLES H. LOHR S. J., *Medieval Latin Aristotle Commentaries.* Traditio 23 (1967); 355–356 and A. PELZER, op. cit. (vide supra adn. 29). For the wider context cf. MARTIN GRABMANN, *Die wissenschaftliche Mission Alberts des Großen und die Entstehung des christlichen Aristotelismus.* Angelicum 6 (1929), 325–351, bes. 332.

[38] » . librum scilicet Aristotelis de vegetabilibus ex Arabico in Latinum transferens.« NICOLAUS DAMASCENUS, Ed. cit. (n. 33); *Alfredè ad Rogerium Dedicatio,* 4, Ed. Meyer.

tury, Albertus had little knowledge of the parent plants of eastern exotica beyond what he had read in texts or inferred as the basis of examining their dried products, for example, cinnamomun, cloves, ginger, nutmeg and pepper[39]. These, and other products, were widely available, at least in the major cities, in western Europe because of their demand as ingredients in medicaments, for flavoring foods and beverages, and, in the case of some of the aromatic gums and resins, for liturgical purposes as incense. The basis, then, of Albertus' description of these exotica depended on Avicenna or upon other sources, either cited by Avicenna or known to Albertus independently of Avicenna. Of the latter, he may have read Galen, who is cited in connection with the medicinal properties of plants on several occasions. Other writers, however, may have been known to him only through the intermediary of Avicenna or some other literary source, for example, Pliny the Elder, whom he knew through Solinus; or Dioscorides, whom he could have known through Avicenna, Galen, or Constantine the African, whose *De gradibus simplicium* may have been used by Albertus[40]. A second reason for Avicenna's importance concerns the arabic, or arabicized, plant names, that frequently appear in *De veget.* As we shall note later, one of the indications of Albertus' personal knowledge of plants is provided by the examples that he uses for purposes of illustration. Many of them refer to trees and herbs that either were indigenous to western Europe or completely naturalized. But there also occur a few instances where they refer to plants that Albertus could not have seen personally. Those names, therefore, indicate a second reason why Avicenna, and possibly other Arabic authors, provided important data for Albertus.

In addition to the above mentioned authors whose writings played a significant role in the preparation of *De veget.*, a number of others are mentioned occasionally, for example, Alexander of Aphrodisias, Isidore of Seville and Isaac the Jew. Several other Greek and Roman writers are mentioned in connection with specific points, for example, Josephus, Paul of Aegina, and Virgil, but their contributions appear to be negligible.

Finally, Albertus probably consulted other texts that have not yet been identified completely[41]. These may have included epitomes or collections of ex-

---

39 A good example is provided by Albertus' comment on ›lignum aloes‹ (Aquilaria agallocha Roxb.): »Aloes experimento visus scitur lignum esse, cuius originem nullus hominum apud nos habitantium sufficienter usque hodie cognovit.« *De veget.* lib. VI, tr. 1, c. 11, p. 344.

40 Constantine is cited four times in Book VI, and significantly, only in connection with exotica cf. the ›Index Personarum‹, *De veget.* 696.

41 This includes such references as »qui poemata scribunt« (lib. VI, tr. 1, c. 29, p. 429); »ut tradunt veteres« (tr. 1, c. 33, p. 453) and »quae veteres tradiderunt« (tr. 1, c. 36, pp. 467–468).

tracts similar to those used for his *De mineralibus*[42]. At this point it is unnecessary to try to determine more precisely the nature of these postulated sources. It is sufficient to note that they included a large assortment of different kinds of information, for example, doxographical, philosophical and, for Book VI especially, medical and pharmacological.

III. Botanical Sources

The older philosophical literature, which he used wisely and critically[43], was not his only source regarding plants. Unlike the great majority of his contemporaries, he recognized that the natural world of plants also provided valuable information. Although artisans and tradesmen of various sorts had long made use of the local flora in connection with their professions, the schoolmen rarely regarded natural products as a means of confirming or supplementing their literary data. It is to Albertus' credit that he did both.

It is difficult to state at what point in his career Albertus recognized that the flora of *De plantis* differed from that of western Europe. But there are two kinds of evidence which support the claim that by the time he composed *De veget.*, he intuitively recognized the difference. The two kinds of evidence pertain to Albertus' choice of plants as ›exampla‹ and secondly, the more specific references to the seven ›climata‹, each, which contained a different representative flora.

Approximately sixty one different plant names occur in *De plantis* while Book VI of *De veget.* alone contains approximately 479 different plant names[44]. Obviously, the number of names is no measure of a writer's knowledge of plants when taken by itself. But in the present case, there is ample evidence that Albertus had a wide command of the flora of western Europe. This is clearly indicated by the plants chosen by Albertus to serve as ›exempla‹.

The purpose of the ›exempla‹, noted above, derived from *De plantis* but was considerably amplified by Albertus and, in so doing, substantially modified. An examination of the plants cited as ›exempla‹ in *De plantis* reveals first, that many of the plants were not indigenous to western Europe and second, that some of the plant names were so badly corrupted that it is difficult to deter-

[42] Cf. the passages assembled by Georg v. Hertling, *Albertus Magnus. Beiträge zu seiner Würdigung*. 2. Aufl. Münster 1914, 66, adn. 4.

[43] On occasion Albertus challenges the beliefs of others or points out an error in one of his sources. Cf. his comments on the erroneous identification of ›napellus‹ (Aconitum napellus L.) with ›squilla‹ (Urginea maritima [L.] Bak.) lib. VI, tr. 2, c. 17, p. 561 and his remark on Isidore (lib. VI, tr. 1, c. 26, p. 404).

[44] Cf. Jerry Stannard, *Identification of the Plants Described by Albertus Magnus, De vegetabilibus, lib. VI.*, to appear in Res Publica Litterarum 2 (1979).

mine the original Greek form[45]. Under these circumstances, the usefulness of citing a plant name as an ›exemplum‹ was considerably lessened. Hence, while Albertus retained some of the names which he found in *De plantis*, even though they designated plants which one would rarely if ever encounter in western Europe, he added the names of indigenous species to his ›exempla‹. As instances of the latter class, the following are but a few of the plants with which Albertus was well acquainted: ›coctanus‹ (quince; Cydonia oblonga Mill.); ›corilus‹ (haselnut; Corylus avellana L.); ›feniculum‹ (fennel, Foeniculum vulgare Mill.); ›petroselinum‹ (Parsley; Petroselinum crispum [Mill.] Nym.); ›plantago‹ (plantain; Plantago spp.); ›platanus‹ (plane tree maple; Acer pseudoplatanus L.); ›ulmus‹ (elm; Ulmus procera Salisb.); ›viscum‹ (mistletoe; Viscum album[L]). It may be noted that none of these plants occur in *De plantis*.

As the foregoing illustrations indicate, Albertus' choice of ›exempla‹ was not restricted to his literary sources, though admittedly he borrowed from *De plantis* and other sources, especially Avicenna. Before we examine the extent of Albertus' own personal experience with these and other plants, a note on the concept of the ›climata‹ is necessary, for these hypothetical divisions provided the theoretical foundations for Albertus' contributions to plant geography.

Commenting on *De plantis* II, 7, in which the subject is an explanation of the edible properties of plants, Albertus followed Nicholas in asserting that the hot, light, and high regions of the third and fourth climatic zones are especially productive of edible plants[46]. This is a subject, he then notes, that he has discussed independently in another work, *De natura locorum*. It is not necessary here to follow all of the details, beyond noting that ›clima septimum‹ was regarded by Albertus as the northern most vegetational zone. Roughly, this seventh zone covered a west-east band cutting across central Europe, bounded on the north by a line bisecting Rennes, Orleans, Freiburg, München, Vienna and Bratislava to a line southward bisecting Bordeaux, Grenoble, Torino, Venice, Belgrade and on eastward to the Black Sea, where the Danube ends. Significantly, this area includes, from Regensburg eastward, the drainage system of the Danube Basin, and it was there, in Lauingen (Schwaben)

---

[45] Cf. e.g., the following two series: i. ἐν φυτῷ τῷ λεγομένῳ ὀπιγαΐς (Aristoteles quae feruntur *De plantis* . . ., ed. Otto Apelt. Lipsiae 1888, 818 b 38 p. 16) – »in dicitur opigaidum« (*Nicolai De plantis* I, 10, p. 16 Mey.) – »in planta, quae graece vocatur epygradium« (*Alberti De veget.* lib. I, tr. 2, c. 3, pp. 69–70 Jess.) ii. τὰ ἐν Αἰγύπτῳ φυτὰ τὰ λεγόμενα μαργαρῖται (Arist. ed. cit., 819 a 11, p. 16 Apelt) – »ut arbores Aegyptiae, quae dicuntur vargariation« (Nicolai ed. cit. I, 11, p. 16 Mey.) – »sicut . . . arbores Egyptiae, quae vargavariton graece dicuntur« (*Alberti De veget.* lib. I, tr. 2, c. 4, p. 72).

[46] *De veget.* lib. IV, tr. 2, c. 4.

that Albertus was born and spent his early years[47]. In other words, the seventh zone, was the norm by which Albertus evaluated the flora, both what was known to him personally and what he obtained from his literary sources. This is indicated by Albertus in several different ways. Some plants, for example, ›alnus‹ (Alnus glutinosa [L.] Gaertn.) are explicitly said to flourish in the seventh ›clima‹[48] while the pear tree (Pyrus communis L.) grows in all lands[49] and the ›promegranate‹ (Punica granatum L.) is described as »arbor nota in omnibus climatibus usque ad septimum«[50]. Especially noteworthy is Albertus' description of ›mirtus‹ (Myrica gale L.): »Multum enim abundant in littore maris oceani in fine septimi climatis versus Daniam« – an observation that is borne out by modern botanists[51]. Another way of expressing the same thought occurs in reference to »fungi meliores«, that is, the non-poisonous and edible mushrooms, »qui sunt in nostro climate«[52]. Other, plants with which Albertus had personal experience were said to grow »in nostris habitationibus«, for example, the chestnut (Castanea sativa Mill.) and ›epithimum‹ (Ajuga chamaepitys Schreb.)[53] while ›platanus‹ (Acer pseudoplatanus L.) is described as »arbor magnae quantitatis vale, et nota in habitationibus nostris«[54]. Finally, the habitat of numerous plants is indicated by the simple expression, »apud nos«, for example, ›bedegar‹ (Rosa rubiginosa L.), ›buxus‹ (Buxus sempervirens L.), ›scolopendria‹ (Phyllitis scolopendrium [L.] Newm.), ›urtica‹ (Urtica dioica L. and U. urens L.) etc.

---

47 Referring to ›populus‹ (Populus alba L. et P. canescens L.) Albertus wrote: »crescit iuxta aquas, et maxime in insulis aquarum fluentium, et praecipue in insulis Danubii.« (*De veget*. lib. VI, tr. 1, c. 30, p. 431 Jess.). A similar geographical locale is implied elsewhere, e.g. »Et hoc experti sunt aucupes terrae nostrae, quae est superior Germania.« *De animalibus* lib. I, tr. 1, c. 3. In general, cf. J. A. ENDRES, *Albertus Magnus und die bischöfliche Burg Donaustauf*. Historisch-politische Blätter für das Katholische Deutschland, 149 (1912), 829–836.

48 »Alnus est arbor nota in sexto et septimo climatibus.« *De veget*. lib. VI, tr. 1, c. 6, p. 350 Jess.

49 »Pirus est arbor nota valde, quasi in omni terra convalescens.« *De veget*. lib. VI, tr. 1, c. 30, p. 433 Jess.

50 *De veget*. lib. VI, tr. 1, c. 22, p. 397.

51 *De veget*. lib. VI, tr. 1, c. 26, p. 408 Jess. Wagner concludes his account of ›gemeiner Gagel‹ (Myrica gale L.) by describing its geographical range: »In Deutschland vom Niederrhein durch Westfalen, Hannover, Ostfriesland, Holstein . . . bis Westpreußen und der Niederlausitz.« HERMANN WAGNER, *Illustrierte Deutsche Flora*. Stuttgart 1871, 695.

52 *De veget*. lib. VI, tr. 2, c. 7, p. 516. In describing the cool and moist environment favored by many ›fungi‹, Albertus notes what may be either moss or lichens, »sicut videmus in arboribus omnibus septimi climatis, in quo frigus et humor dominantur«. *De veget*. lib. V, tr. 1, c. 3, p. 299.

53 »Castanea autem est arbor apud nostram habitationem multum nota«, *De veget*. lib. VI, tr. 1, c. 10, p. 360; »epithimum . . . in nostris habitationibus nascitur«, ibid. lib. VI, tr. 2, c. 6, pp. 507–508.

54 *De veget*. lib. VI, tr. 1, c. 30, p.430.

Perhaps the most noteworthy characteristic of Albertus' use of those observational data which reflect his personal experience of plants, is their wide range. While some of them concern minute details of the sort that one expects only from a scientifically-trained naturalist, others were drawn from common experiences of the everyday world. Such matters of common knowledge, however, took on special meaning when they could be used to support a philosophical discussion. No detail was too small or too insignificant to escape Albertus' eye and it is much to his credit that he made use of them in the appropriate place. This is well illustrated by the phrase, used in the descriptions of many plants, to the effect that a certain tree or herb is »satis nota, valde nota« or some similar phrase[55]. Those simple phrases must be taken seriously for they are not mere empty formulae, uncritically repeated from some literary source. In Albertus' case, they represent the experiences gained from many years of observing plants and trees in their native habitat. No better illustration can be provided than Albertus' description of the growth pattern in a disturbed forest[56]. The ecological process of regrowth and succession, culminating in a climax vegetation, was not fully elucidated until the present century. Only through repeated observations, undertaken over a period of time, could Albertus have observed that the forest spontaneously recovered, though with a different primary composition. Because of the need to site monastic establishments ›in solitudine‹, forest clearance was extensively practiced in Albertus' day and one may well imagine him observing not only the construction of monastic buildings but the simultaneous destruction of the forest[57]. It was through his observations and travels, therefore, that Albertus was able to state authoritatively that certain plants were well known or common. This included, many trees and herbs that historians have assumed were locally available but for which demonstrative evidence was lacking. A particularly interesting example is furnished by Albertus' remarks concerning the cherry and plum tree which, along with the pear tree, were widely known[58].

---

55 »Morus (Morus nigra L. et M. alba L.) est arbor nota . . . et illorum mororum sunt plurima genera, quae nota sunt omnibus«, lib. VI, tr. 1, c. 26, p. 411; »corilus (Corylus avellana L.) est arbor nota,« lib. VI, tr. 1, c. 27, p. 415; »betonica (Stachys officinalis [L.] Trev.) est herba nota«, lib. VI, tr. 2, c. 3, p. 484; »coriandrum (Coriandrum sativum L.) est herba satis communis«, lib. VI, tr. 2, c. 4, pp. 494–495; »centaurea (Centaurea centaurium L.) est herba valde communis,« lib. VI, tr. 2, c. 4, p. 498.

56 *De veget.* lib. V, tr. 1, c. 7, pp. 313–314 Jess. This passage is specifically noted by Hermann Stadler, *Albertus Magnus als selbständiger Naturforscher*. Forschungen zur Geschichte Bayerns 14 (1906), 95–114 (ref. 1017).

57 Cf. Dieter von der Nahmer, *Über Ideallandschaften und Klostergründungsorte*. Studien u. Mitteilungen zur Gesch. d. Benediktinerordens 84 (1973), 195–270.

58 »Quaedam autem pluries, sicut ficus et quaedam piri, quae ter in anno fructificant, et per totam aestatem habent fructum diversum successive in se florentem; quae abundant in Colonia et in partibus Reni circa Coloniam«, lib. I, tr. 2, c. 11, p. 98. Concer-

Since this is not the place to attempt to classify the different kinds of empirical data of *De veget.*, a few samples must suffice. Several examples of Albertus' ability to note details and to use the results of such observations in his attempt to explain vegetative processes are contained in his discussion of seeds and fruits. In his discussion of the diversity of their shapes, he notes that as the seed capsule of leguminous plants matures and dries, the seeds are ejected with great velocity[59]. Passing on to an explanation of the cause of the shape of seeds and fruit, many of his examples are drawn from species indigenous to or naturalized in western Europe. He observed, for example, that pears are ovoid in shape, except the little ones which are spherical[60]. While this may be a reference to immature fruit, it is also possible that Albertus is referring to a wildling, for many passages in *De veget.* indicate his interest in explaining how wild varieties of fruits and vegetables become domesticated. Turning from shape to odor, he noted the unmistakable odor of a ripe apple that remains on one's hands after handling it[61]. This is a good example of his ability to relate a common everyday experience to the elucidation of a philosophical problem.

The empirical data assembled by Albertus were not, however, restricted to the common fruits and vegetables of the garden. He made use of an unusual datum wherever he found it. Thus he stated that he had often seen that some plants grew spontaneously on roof tops and suggested, quite correctly, that their growth was maintained by deposits of organic matter collected in the roof's texture[62]. In a similar fashion he observed that the slender twigs broken off from a willow easily take root and grow[63], and that a milky sap exudes from some plants, such as esula or endivia when a leaf is torn off.[64]. Although

---

ning the fruit trees, ›cerasus‹ (Prunus cerasus L.) and ›cinus‹ (P. domestica L. ssp. insititia [L.] Schreb.) Albertus writes: »De ceraso autem et cino hic non facimus tractatum, quia nota sunt omnibus, neque oportet nos talibus occupari«, lib. VI, tr. 1, c. 16, p. 381.

59 »Aliae autem semina sua proferunt in testis quibusdam, quae maturato semine arescunt et aperiuntur – et impetu ventositatis ejicitur granum«, lib. III, tr. 1, c. 1, pp. 166–167

60 »Pira enim magis videntur esse pyramidalia secundum genus suum, exceptis paucis, quae sphaerica sunt«, lib. III, tr. 1, c. 4, pp. 177.

61 ». . . quando (sc. poma) premuntur aut revolvuntur manu, ut currat in eis humidum, et spiret in eis odor«, lib. III, tr. 2, c. 6, p. 208. Another example pertaining to the apple is the detailed description of the five, different substances or tissues which can be seen in cross-section, *De veget.* lib. VI, tr. 1, c. 25, p. 402.

62 ». . . plantas saepe videmus nasci in tectis lapideis, quando in superficie parum putrescunt lapides«, lib. IV, tr. 2, c. 4, p. 252.

63 »Salix (Salix alba L.) . . . est arbor in locis humidis melius convalescens . . . sed est laevis in iuventute, ramorum multorum, qui, cum praeciduntur ex ea, cito recrescunt«, lib. VI, tr. 1, c. 33, p. 449.

64 Following a general remark that some trees and herbs have a milky sap, such as ›ficus‹, ›esula‹ (Euphorbia esula L.) and ›endivia‹ (Cichorium endivia L.), Albertus conti-

these two plants belong to widely separated families, many genera of each are characterized by the presence of latex, a useful field mark which, in quite a different context, is mentioned by Albertus[65].

IV. Taxonomy and Nomenclature

Because of the size and diversity of the plant world, at least two principles are required in order to understand it: some technique to organize the diversity of size, structure, shape etc. and some means of referring to those recognized differences. That was accomplished, though not on scientific grounds, by the adoption of various nomenclatural conventions by which class terms and names of smaller, more discrete units (our ›genera‹ and ›species‹) denoted, and thereby distinguished, different groupings of plants, for example, trees and herbs. Naturally, there was little uniformity in the application of such terms, not only because of linguistic differences but also because of the differences in the flora of various regions. This was, roughly, the state of affairs that Albertus found, both in his literary sources and as reflected in popular speech. It was not Albertus' intention to propose a taxonomic system but to comment on *De plantis* (lib. I, c. 12) and to explain its crude division of the plant world. Albertus adopted Nicholas' division of trees, shrubs, and herbs, as well as the strange subdivision of herbs and ›olera‹[66]. This is not a very satisfactory method of distinguishing plants and it would be difficult to understand were it not for Theophrastus' text which formed the basis of Nicholas' text. According to Theophrastus, all or nearly all plants fall into one of four classes: trees, shrubs, undershrubs or herbs[67]. Why Nicholas complicated this simple but

---

nues, »et tunc totus humor sic attractus retinebit colorem lacteum, qui etiam apparenter est in eo, quando decerpitur planta«, lib. IV, tr. 4, c. 2, p. 280.

[65] »Esula . . . habet folia sicut linaria (Harnkraut, Linaria vulgaris Mill.), sed differt in hoc, quod esula lactescit, et non facit linaria«, lib. VI, tr. 2, c. 6, p. 512. This is an interesting passage for it indicates that Albertus had access to some Salernitan medical texts or excerpts which seem not to have been noted. The anonymous *Regimen sanitatis salerni* (Salvatore de Renzi, Ed. Collectio Salernitani, t. V, Napoli 1859, 33 versus 1173) states: »Esula lactescit, linaria lac dare nescit.« This useful piece of information later is incorporated in medical texts as a mnemonic device. E. G. Rufinus, *De virtutibus herbarum*, ed. Lynn Thorndike. Chicago 1949, 173: »Esula lactescit; sine lacte linaria crescit.« Cf. the *Hortus sanitatis germanice* (Mainz: Peter Schöffer, 28 March 1485) cap. 235 Linaria – »eyn krut also genat . . . Auch beschriben uns die meister . . . mit dissem verss / Esula lactescit sine lacte linaria crescit.«

[66] Commenting on *De plantis* I, 12 (pp. 17–18 Ed. Meyer), Albertus writes: »Quaedam inter arbores et herbas ipso nomine suo se medias demonstrant esse, quae graece quidem ambragyon, latine autem arbusta dicuntur communi nomine. Quaedam autem sunt herbae, et quaedam olera virentia. Fere autem omnis planta sub his cadit nominibus.« *De veget.* lib. I, tr. 2, c. 5, p. 75 Jess.

[67] Πρῶτα δέ ἐστι καὶ μέγιστα καὶ σχεδὸν ὑφ' ὧν πάντ' ἢ τὰ πλεῖστα περιέχεται τάδε, δένδρον θάμνος φρύγανον πόα. Theophr. *Hist. Pl.* I, iii, 1.

clear four-fold division based on morphological criteria cannot now be determined. What is important for us is how Albertus responded to the divisions as stated in *De plantis*. One indication is provided by Albertus' attempt to find a place for ›fungi‹. Since their life cycle was unknown, they created a difficult problem[68]. Clearly they were living entities but just as clearly they were not animals. Thus only by an extension of meaning, could they be called plants[69]. As such, they were placed within a modified taxonomy which included, in addition to the above groupings, ›fungi‹ and other suchlike kinds[70].

Fully as important as the attempt to make sense of a puzzling passage in *De plantis* is a series of miscellaneous passages in *De veget.* Those passages, where taken together, indicate that Albertus recognized that there were natural affinities between various taxa. That recognition, while never systematically developed by Albertus, is revealed in two different ways – by the adoption of a simple linguistic convention and second, by the appropriate selection of plant names to serve as ›exempla‹.

The first of these two taxonomic techniques is so simple that it may not be readily apparent what a deep understanding of plant life it entails. Grammatically the technique consists of nothing more than the juxtaposition of the word ›genus‹ with the name of a plant, usually but not always, in the genitive plural case, for example, »genera juncarum«, »genera boletorum«, »genera prunorum«, »carduorum genera«, »genera leguminis«, »totum genus abietis«, etc. Each of these combinations represents a natural, as opposed to an artificial or arbitrary, grouping of plants, based on the presence of many individual characteristics considered together. Thus, the term ›carduorum genera‹ is a taxonomic name denoting those plants which resemble the ›carduus‹ (Dipsacus spp.) more closely than they resemble any other kind or grouping of plants[71].

The second of the two techniques serves a similar purpose. Among the ›exempla‹ cited by Albertus, there are some in which three or four plant names occur in succession following the introductory ›ut‹ or ›sicut‹. That series of names, in some of the ›exempla‹, denotes plants that modern taxonomists would place in the same family. Thus, in concluding his discussion of the dif-

---

68 »Fungus enim non videtur esse nisi quaedam exhalatio humoris, ex putredine ligni vel alicuius alterius commixti et putridi evaporans, et ad frigus aeris constans et coagulata.« *De veget.* lib. II, tr. 1, c. 2, p. 111 Jess.

69 »Fungus autem debilioris est essentiae omnibus dictis. Et ideo etiam istae species non sunt per eandem rationem recipientes nomen plantae, sed per prius et posterius.« *De veget.* lib. V, tr. 1, c. 2, p. 296 Jess.

70 »Dividentes plantam in subjecta sibi proxime genera, dividimus eam in arbores et arbusta et frutices et olus et herbam et fungum et alia huiusmodi.« *De veget.* lib. II, tr. 1, c. 2, p. 111 Jess.

71 »Aliae autem plantae sine siliqua, in simplici pellicula grani, nuda proferunt grana, in culmis infixa, sicut carduorum genera.« *De veget.* lib. III, tr. 1, c. 1, p. 167 Jess.

ferent kinds of inflorescence, Albertus gives as an exemple ›petrosilinum‹ (Petroselinum crispum [Mill.] Nym.), ›feniculum‹ (Foeniculum vulgare Mill.), ›anisum‹ (Pimpinella anisum L.) and ›siler montanum‹ (Laserpitium siler L.) all of which belong to the ›umbelliferae‹ or carrot family[72]. Again, to illustrate the position of fruit, he cites as an ›exemplum‹, pears, apples, plums, and almonds, all of which are members of the ›Rosaceae‹ or rose family[73]. So far, little has been said directly on the subject of nomenclature and, save for some passing remarks, this was not a subject of specific interest to Albertus. But those remarks en passant indicate that he was aware of the complexities and uncertainties that characterized plant nomenclature until the time of Linnaeus. This is indicated in several different ways in *De veget.*, depending largely on the particular subject under discussion.

The most common of Albertus' remarks concerning nomenclature has to do with synonyms. Because of the Arabic plant names in *De plantis* that Albertus took over and cited in his own ›exempla‹, it was desirable to equate those names with latin plant names. That Albertus did on several occasions, for example, he glossed the strange name ›fingekest‹ by adding, »quod quidam pentafilon vocant«[74] and elsewhere, as an ›exemplum‹, he wrote »ut ramnus qui alaz arabice dicitur«[75]. But Albertus' interest in supplying synonyms where he felt them to be necessary was not confined to Arabic names. In fact, most of the synonymies of *De veget.* are concerned with adding one or more latin synonyms to a latin (or latinized greek) name. Thus, the description of ›altea‹ (Althaea officinalis L.) begins: »Altea graeco nomine dicitur . . . vocatur etiam bismalva . . . vocant autem quidam eandem malvaviscum[76].« Instances of a simpler synonymy are represented by such statements as »marrubium« (Marrubium vulgare L.), »quae alio nomine prassium vocatur[77]« or »uva lupina, quam nos solatrum (Solanum nigrum L.) vocamus[78]«. An equally interesting case is provided by Albertus' explicit references to folk names, that is, the use of plant names as he found it in popular speech, for example: »Ungula caballina (Tussilago farfara L.) est herba, quae vulgo vocatur herba leporis[79].» In describing the leaves of aquatic plants, he provides the following ›exemplum‹: ». . . sicut . . . folia, quasi figuram lentis habentia, quae etiam a vulgo in aliquibus locis lentes paludum vocantur[80].« One final

---

72 *De veget.* lib. III, tr. 1, c. 1, p. 167 Jess.

73 »Secundum diversitatem autem magis specialem quaedam plantarum producunt fructum super folia, sicut piri et mali et cini et pruni et amigdali.« *De veget.* lib. I, tr. 2, c. 4, p. 71. A similar grouping occurs in lib. III, tr. 1, c. 1, p. 166 Jess.

74 *De veget.* lib. I, tr. 2, c. 5, p. 78 Jess.

75 *De veget.* lib. I, tr. 2, c. 6, p. 83 Jess.

76 *De veget.* lib. VI, tr. 2, c. 2, p. 483 Jess.

77 *De veget.* lib. VI, tr. 2, c. 12, p. 539 Jess.

78 *De veget.* lib. VI, tr. 1, c. 35, p. 463 Jess.

79 *De veget.* lib. VI, tr. 2, c. 19, p. 578 Jess.

80 *De veget.* lib. IV, tr. 2, c. 3, p. 243 Jess.

instance may be noted because of the historical importance of the synonymy in the evolution of ›Hochdeutsch‹. The description of a plant, which Meyer correctly identified as ›Achillea millefolium‹-L., begins thus: »Carvi est herba, quae alio nomine achillea vocatur[81].« The name ›carvi‹, in this passage, does not denote ›Kümmel‹ (Carum carvi L.), almost universally known as ›carvi‹ in the Middle Ages[82], but rather is cognate to ahd. garba which appears in nhd. as ›Schafgarbe‹, the modern name of ›Achillea millefolium‹[83].

No account of medieval botanical nomenclature would be complete without mentioning the subject of etymology. Almost all medieval herbals contains one or more chapters in which the description of a particular plant also includes a discussion of the origin and derivation of its name. Many of those etymologies are fanciful and can be traced back to Isidore of Seville's *Etymologiae sive Origines*, lib. XVII. Without always mentioning Isidore by name, Albertus made use of these etymologies because he, like his contemporaries, believed that a knowledge of the meaning of a plant name was useful for purposes of understanding its appearance, properties or use. Thus, ›semperviva‹ (Vinca minor L.) is described in such a way as to not only include a synonym but to explain on phenological grounds the reason for its name[84].

V. Applied Botany

As a result of dramatic changes in the methods and goals of science in the early seventeenth century, and with the introduction, shortly thereafter, of physical apparatus for the purpose of studying details of plant life not readily visible to the unaided senses, *De veget.* was passed by. Like many medieval treatises, it became obsolete because it represented a philosophic approach to the study of nature that was no longer acceptable to practicing scientists. Under these circumstances, the corpus of philosophic botany, codified by Albertus, was pushed aside in favor of those studies that represented a more empirical

[81] *De veget.* lib. VI, tr. 2, c. 4, p. 503 Jess.

[82] Cf. Hans Wölfel (Ed.) *Das Arzneidrogenbuch Circa Instans in einer Fassung des XIII. Jahrhunderts* . . . Diss. Berlin 1939, 35 s. v. de carui; Henrich Ebel, *Der Herbarius communis des Hermannus de Sancto Portu* . . . Würzburg 1940, 11: Carui. Carwe (Am Rand: Carwe); Hortus sanitatis, ed. cit., cap. 115: ›Carui‹ – ›wiesen kummel‹; Willem Daems (Ed.), *Boec van Medicinen in Dietsche.* Leiden 1967, 123 s. v. Carvi es een sadeken.

[83] Cf. Henrich Marzell, *Wörterbuch der deutschen Pflanzennamen.* Lfg. 1. (Leipzig 1937) Sp. 82 who notes, in addition to ›Garbe‹, related forms such as ›gar(a)wa‹, ›gar(e)wa‹, ›Gor‹, ›garwenkrut‹ usw.

[84] »Semperviva dicitur alio nomine pervinca, quae ideo semperviva dicitur, quia in hieme eandem, quam in aestate habuit, retinet viriditatem.« *De veget.* lib. VI, tr. 2, c. 17, p. 568 Jess. The same synonymy occurs, but without an etymology, in lib. II, tr. 2, c. 3, p. 146 Jess.

approach. In retrospect, it can be seen that, despite Albertus' overt philosophic approach, *De veget.* contains an impressive collection of empirical data, many of which had been neglected by his predecessors. But several centuries were to elapse before botanists were able to integrate the various uses of plants noted, but neither systematically collected nor organized by Albertus. In order, therefore, to provide a balanced account of Albertus' contributions to the study of plants, a selection of those data bearing on the subject of applied botany must be included. It is especially necessary to call attention to this side of Albertus' work, because, unlike his philosophic botany, scholars have neglected this more mundane material in favor of the theoretical and philosophical aspects of his writing.

The term ›Applied Botany‹ (Angewandte Botanik) is useful in so far as it covers a variety of subjects, all of which are associated with the uses of plants and plant products by man. An enumeration of those uses clearly was not Albertus' intention. But in the course of his commentary, and especially in the ›digressiones‹ and Book VI, he noted for purposes of illustration and clarification, many uses of plants whose study later became subjects in their own right. This includes not only the alimentary, economic, horticultural and therapeutic uses of plants but also various customs, superstitions, and folk beliefs about plants. It may be noted, parenthetically, that Albertus occasionally refers to ›rustici‹[85] as his source of information and to the ›opinio vulgaris‹[86], in a manner that suggests that such information was a valuable supplement to his own observations and what he found in his texts.

An independent study of alimentary plants and their properties was no more Albertus' primary concern than an enumeration of the many economic uses of wood. Yet the fact that the question was raised in *De plantis* (lib. II, c. 7) concerning why some and not other plants were edible, required comment. Moreover, the fact that certain plants which were poisonous to humans could be eaten by animals and birds with no ill effects, raised further questions regarding the nature of the properties of plants in general[87].

---

[85] *De veget.* lib. I, tr. 2, c. 10, p. 95; lib. VI, tr. 2, c. 19, p. 579. Cf. also the reference to the »artifices, qui mechanici vocantur« (lib. V., tr. 2, c. 5, p. 334), who may have been a source of information. Botanists of a much latetime also sought out the ›rustici‹ for information otherwise unknown, or at least, not readily available. Cf. JERRY STANNARD, *Classici et rustici in Clusius' Stirpium pannonicarum historia.* Festschrift anläßlich der 400jährigen Wiederkehr der wissenschaftlichen Tätigkeit von Clusius im pannonischen Raum. Eisenstadt 1973, 253–269.

[86] »Qui flos (sc. humuli, i. e. Hopfen) propter siccitatem suam conservatur per longitudinem maximam temporis in virtute sua, ita quod vulgaris opinio est, quod nunquam putrescit.« *De veget.* lib. VI, tr. 2, c. 9, p. 525. It is also possible that Albertus' reference to ›communis opinio‹, in connection with ›Gartenkunst‹, means that he obtained information from gardeners; *De veget.* lib. VII, tr. 2, c. 2, p. 643.

[87] »Similiter autem quaedam animalia comedunt quosdam fructus, et quaedam non possunt comedere eosdem; sicut passer jusquiamum comedit, qui homini nocet valde.«

Albertus' philosophic explanation of why some plants are edible by humans followed traditional humoralistic theory[88]. Nonetheless, the belief that edible plants are those whose humors were perfected by the heat of the sun, was counterbalanced by other references to edible plants in which theory played a minor role. For example, Albertus' discussions of the alimentary uses and nutritive properties of ›caulis‹ (Brassica oleracea L. var. capitata), ›nasturcium‹ (Lepidium sativum L.), and ›spinachia‹ (Spinacia oleracea L.) owe more to contemporary dietary habits than to the older Graeco-Roman literature which reflected not only different food preferences but different food plants as well[89].

Albertus' techniques of incorporating such stray pieces of information in order to make his comments more meaningful to his readers, is illustrated also by numerous references to the economic uses of plants. A collection of those references to the economic and domestic uses of plants is a cross section of the material culture of the thirteenth century and indicates an awareness of the practices of the common people that was unrivalled in the Middle Ages. Typical of Albertus' attention to the application of plants in everyday life is a detailed and accurate description of mistletoe (Viscum album L.) which concludes with the remark that hunters use the paste, obtained from the crushed berries in order to capture birds[90]. He noted in passing that the spinous, dried seed capsules of ›virga pastoris‹ (Dipsacus sativus [L.] Honck.) were used in carding woolen cloths[91] and that fullers report that the meal of ›nigella‹ (Agrostemma githago L.) cleaned and whitened woolens, »like the herb called borith«[92]. The latter refers to ›Saponaria officinalis‹-L., a commonly used soap plant in pre-modern times and mentioned in some texts under its vernacular

De veget. lib. I, tr. 2, c. 7, p. 86. The poisonous qualities of ›jusquiamus‹ (Hyoscyamus niger L.) are again referred to in lib. III, tr. 1, c. 1, p. 165; lib. VI, tr. 2, c. 10, p. 527.

88 *De veget.* lib. IV, tr. 2, c. 4, pp. 246–250 Jess.

89 *De veget.* lib. VI, tr. 2, c. 4, p. 493 (caulis); tr. 2, c. 13, p. 541 (nasturcium); tr. 2, c. 17, p. 563 (spinachia).

90 The description of mistletoe concludes: »in hieme granula alba, et substantia eorum componitur ex malleolis, sicut vitis, et haec planta est rarae substantiae, et humor pelliculae interioris, quae est inter corticem et lignum, est viscosus valde, propter quod utuntur ipso aucupes ad viscum quo capiunt aves.« *De veget.* lib. IV, tr. 2, c. 4, p. 253; cf. also lib. V, tr. 1, c. 3, p. 299.

91 »Ipsa autem spinostias optime pectit lanositatem pannorum laneorum.« *De veget.* lib. VI, tr. 2, c. 19, p. 577 Jess.

92 »Fullones etiam quidam tradunt quod farina eius lavat laneos albissime et mundissime, sicut herba, quae vocatur borith.« *De veget.* lib. VI, tr. 2, c. 13, p. 543 Jess.

name, ›herba fullonum‹[93]. Other miscellaneous uses of plants include the preparation of inks and dyes, and as insecticides[94].

Another series of references concern the many different economic uses of wood, for example, the durability of ›alnus‹ (Alnus glutinosa [L.] Gaertn.) when submerged in water for construction purposes[95], the burning properties of ›fagus‹ (Fagus silvatica L.)[96], the use of different woods for the manufacture of bowls and other utensils[97], while a resinous exudation from the bark of ›fibex‹ (Betula pendula Roth.) was used as a lubricant for the axles of wagons[98]. Perhaps the most interesting reference pertains to ›picea‹ (Pinus silvestris L.), one of the most characteristic by German trees. Owing to its resin, the branches supplied the local inhabitants with torches[99]. The account of ›picea‹ concludes with a description of the technique for the extraction of pitch[100]. Judging from the technical details, Albertus may have witnessed such a procedure.

Albertus' references to gardens and garden practices provides further evidence of his ability to utilize his personal experiences. Unfortunately, none of the passages in *De veget.* specifically refers to a monastic or cloister garden, though it is difficult to believe that while traveling in his official capacity, he did not observe many such gardens[101]. At any rate, from two different passages, we can form some idea of what was grown in those gardens[102]. Elsewhere

---

[93] »Radicula called otherwyse Lanaria, is called in greeke struthion, and of the commune Herbaries it is called Saponaria and Herba fullonum, it groweth in certeine gardines of Germany, but I never sawe it in England.« WILLIAM TURNER, *The Names of Herbes*. London 1543, sig. f. vi.

[94] For inks and dyes, cf. *De veget.* lib. VI, tr. 2, c. 2, p. 478, s. v. alterana; tr. 2, c. 8, p. 520, s. v. gauda; tr. 2, c. 17, p. 560, s. v. sandix; tr. 2, c. 19, p. 578, s. v. venae tinctorum. For insecticides, cf. *De veget.* lib VI, tr. 1, c. 6, p. 351, s. v. alnus; tr. 2, c. 2, p. 477, s. v. absinthium; tr. 2, c. 2, p. 480, s. v. abrotanum; tr. 2, c. 13, p. 543, s. v. nigella.

[95] *De veget.* lib. VI, tr. 1, c. 6, p. 351.

[96] *De veget.* lib. VI, tr. 1, c. 20, p. 390.

[97] *De veget.* lib. V, tr. 2, c. 5, p. 334. The soft, porous wood of coniferous trees (?Pinus sp.) is suited to the production of musical instruments, lib. VI, tr. 1, c. 1, p. 343.

[98] *De veget.* lib. VI, tr. 1, c. 20, p. 391.

[99] »Picea, dicitur quasi piccata, eo quod multam picem emittat . . . Et utuntur eo quidam Germanicorum pro facibus et luminaribus.« *De veget.* lib. VI, tr. 1, c. 30, p. 432.

[100] *De veget.* lib. VI, tr. 1, c. 30, pp. 432–433 Jess.

[101] The setting of the legend *Albertus Magnus u. Kaiser Wilhelm* (Die Brüder Grimm, Deutsche Sagen, Nr. 489) is a monastic garden. Since this legend can be traced back to one generation following Albertus' death (H. C. SCHEEBEN, *Albertus Magnus*. Bonn 1932, 201) it may preserve some evidence pertaining to Albertus' interest in gardens. This hypothesis is not incompatible with *De veget.* and the numerous references therein to gardens and garden practice nor with his travels, among which he consecrated various monastic establishments; cf. PAULUS DE LOË O. P., *De vita et scriptis B. Alberti Magni*. Analecta Bollandiana 20 (1901), 273–316.

[102] *De veget.* lib. VII, tr. 1, c. 14, pp. 636sq. »De plantatione viridariorum«; tr. 2, c. 2, pp. 643sq. »De his, quae per cultum domesticantur in agro hortulano.«

he noted that a garden includes a grassy turf[103], perhaps not unlike that found in the pictorial representations of the so-called ›hortus conclusus‹[104].

We have left for last those references to the therapeutic uses of plants and their medicinal properties. Because the majority of those passages, especially those of Book VI, are derived from his literary sources, it is difficult to determine Albertus' own beliefs and the extent of his personal experiences. But no study of *De veget.* would be complete without calling attention to the numerous passages that contain medical, herbalistic, and pharmaceutical data. Albertus frequently mentions various composita, their use for specific diseases or complaints, the means for their preparation, recipes, amounts of the ingredients, adulteration of drugs and the like. It may also be noted in passing, that it is possibly on the basis of those passages which contain technical details and directions for the preparation of composita, that the pseudonymous, but clearly spurious, *Secreta* later became associated with his name[105].

A more detailed study than is possible here might reveal that some of the medical references and therapeutic counsels were part of the general fund of information available at the time, for example the belief that ›petrosilinum‹ was better as medicine than as food[106] or that the medicinal properties of the common cabbage vary depending upon the color of its leaves[107] or that those properties vary with the habitat[108]. Because of the great reliance on substances of plant origin as medicamenta, local plants often competed with the exotica whose high cost prevented many from using them. It is thus possible that Albertus' remark about the quince (Cydonia oblonga Mill.) was applicable to many plants: »coctanus autem sive citonius arbor est apud nos quidem nota, sed virtus eius non omnino cognoscitur a multis[109].«

When we turn from the straightforward, practical application of plants and their products to the beliefs held about those plants and their application, we enter into another world that is neither always easy to understand nor to ac-

---

103 »In medio autem caespites nihil sit arborum . . .« *De veget.* lib. VII, tr. 1, c. 14, p. 638.

104 Cf. H. Faulwetter, *Mittelalterliche Gartenstudien.* Gartenkunst 12 (1910), 192–194.

105 Among the more serious studies on a controversial, and bibliographically complex subject, cf. Meersseman, op. cit., 143–148; Eusebio Gómez, *San Alberto Magno y sus obras en la Universidad de Oxford,* Divus Thomas (Piacenza) 35 (1932), 633–643 (bes. 635–636); Christoph Ferckel, *Die Secreta mulierum u. ihr Verfasser.* Sudhoffs Archiv 38 (1954), 267–274 and Lynn Thorndike, *Further Considerations of the Experimenta, Speculum Astronomiae and De Secretis mulierum ascribed to Albertus Magnus.* Speculum 30 (1955), 413–443.

106 *De veget.* lib. VI, tr. 2, c. 15, p. 551.

107 *De veget.* lib. VI, tr. 2, c. 4, p. 494.

108 With reference to ›menta‹ (Mentha spp.), Albertus states: »Est autem calida et sicca, et propter locum suae generationis est in ea humiditas superflua, et subtilioris est substantiae inter omnia olera, quae comeduntur.« *De veget.* lib. VI, tr. 2, c. 12, p. 538.

109 *De veget.* lib. VI, tr. 1, c. 16, p. 381.

cept. The fact that Albertus recorded a variety of popular beliefs and superstitions does not, of course, imply that he accepted them. It is rather the case that he regarded such material as worthy of inclusion in *De veget.* Since some of the fables mentioned by him were widely known and some of the beliefs equally widespread, they added another dimension to the understanding of particular plants and their properties. It is probably for this reason that Albertus mentioned the superstitious practices of those who studied the magical arts. For similar reasons, he recorded some of the practices of ›incantatores‹ and ›nigromantici‹, several of which involved the use of common plants.

The heterogeneous nature of the folk beliefs, referred to in *De veget.* precludes anything more than a brief survey. But bearing in mind Albertus' overall purpose, it can be seen that some of the beliefs were mentioned because they illustrated a certain property or characteristic of the plant in question. Thus, the belief in the generation of fleas by the fruit of ›ficus‹ (Ficus carica L.) is simply the result of noting the swarm of small insects around the juicy and moist flesh of a fully ripened fig[110]. In a similar fashion, Albertus was probably responding to a local folk belief when he compared the slightly unpleasant odor of the foliage of ›buxus‹ (Buxus sempervirens L.) to human odor[111]. Finally a desire to produce horticultural novelties appears to underlie the belief that a garden rose, when planted in soil to which blood has been added, will produce a flower of similar color[112]. The point of the fables and legends to which Albertus occasionally alludes are not always easy to understand. However, in the case of »salvia silvestvis« (Teucrium scorodonia L.), »alio nomine dicitur ambrosia deorum«, a fable is referred to for the purpose of explaining the unusual synonym[113]. The short account of ›nepita‹ (Nepeta cataria L.) is revealing, both for what it states and for what is left unsaid. The relevant portion consists of one sentence: »Fabulose etiam traditur quod de ea (sc. nepita) impraegnantur cattae[114].« Then, as now, cats delight in rolling in a clump of catnip, and equally often, an unannounced litter will be associated with such behaviour.

Not all of the material, however, that Albertus incorporated was of such an innocent, folkloristic nature. A darker side, so to speak, is represented by some twenty five references to the uses of plants by magicians, necromancers, enchanters and the like. Their practices and some of their beliefs were widely known and despite efforts of the authorities to eradicate them, many survived, in some instances, into recent times. For example, the custom of wearing amulets or carrying various apotropaic devices made from some portion of a

---

110 *De veget.* lib. VI, tr. 1, c. 19, p. 386.
111 *De veget.* lib. VI, tr. 1, c. 9, p. 359.
112 *De veget.* lib. VI, tr. 1, c. 32, p. 448.
113 *De veget.* lib. VI, tr. 2, c. 17, p. 569.
114 *De veget.* lib. VI, tr. 2, c. 13, p. 544.

plant or tree, is several times mentioned by Albertus[115]. Another custom which he mentioned concerned ›barba Jovis‹ (Sempervivum tectorum L.). The belief that when planted on the roof top it will protect the occupants from thunder is typical ›Abwehrzauber‹[116]. To the same class of practices is the belief that the future can be predicted by examining the galls produced on an oak tree[117].

A particularly interesting group of passages refers to the uses of plants by the ›incantatores‹, noted earlier. Thus, a staff made of the wood of ›mespilus‹ (Mespilus germanica L.) was said by the enchanters to be of special value as fighting staves[118].

Another reference to the supposed magical properties of wood is no less easy to understand. At any rate, Albertus reported that branches of ›ramnus‹ (Rubus spp.) served to incite dissension, quarrels and hatred when carried[119]. A final example, which derives from Avicenna, concerns ›lingua arietis‹ (Plantago spp.), now widely distributed throughout western Europe as a weed. Its root, suspended around the neck of a young boy, was believed to be effective in preventing scrofulous sores or nodules[120]. This is an interesting passage because underlying the reported use is a belief in the so-called Doctrine of Signatures which, even today, provides the rationale for the folk-medical employment of certain herbs[121].

---

115 Cf. *De veget.* lib. VI, tr. 1, c. 30, p. 435 (pira); tr. 2, c. 2, p. 483 (artemesia). The use of ›Eibe‹ (Taxus baccata L.) as an amulet was an old custom (cf. TH. v. GRIENBERGER, *Drei westgermanische Runeninschriften.* Zeitschrift f. deutsche Philol. 41 [1909], 419–437) and was probably included among the practices condemned in the anonymus *Indiculus Superstitionum et Paganiarum* (MGH, Capitularia Regum Franc. t. I, ed. A. Boretius). Hannoverae 1883, 223 nr. 10: *De filacteriis et ligaturis.*

116 *De veget.* lib. VI, tr. 2, c. 3, p. 484.

117 *De veget.* lib. VI, tr. 1, c. 31, p. 441. For the history of this idea cf. the late medieval »Phrophezeiung aus dem Gallapfel« in GERHARD EIS (Ed.), *Wahrsagetexte des Spätmittelalters.* Berlin 1956, 53, 69.

118 »Lignum (sc. mespili) autem eius quaerunt incantatores, ut faciant ex eo baculos pugilum, et dicunt ad hoc specialiter valere.« *De veget.* lib. VI, tr. 1, c. 25, pp. 405–406.

119 »Incantator autem dicit, hoc genus ligni portatum schismata et rixas et odio suscitare.« *De veget.* lib. VI, tr. 1, c. 31, p. 444.

120 »Incantator etiam dicit, quod radix eius collo pueri appensa impedit scrophulas.« *De veget.* lib. VI, tr. 2, c. 11, p. 529.

121 Cf. HEINRICH MARZELL, *Das Wegerich-Amulett in der antiken und in der deutschen Volksmedizin.* Sudhoffs Archiv 37 (1953), 307–311.

Acknowledgment. I am grateful to the following for their help in obtaining the material necessary for the completion of this study: Mrs. Marianne Lynch Griffin, Reference Department, University of Kansas Libraries; Fr. Gerald J. Mesmer O.S.B., Librarian, Benedictine College, Atchison, Kansas; Dr. Paul Mai, Bibliothekar, Bischöfl. Zentralbibliothek, Regensburg; and Dr. Johannes Koehler, Regensburg.

XIV

Conclusion

Short of reading *De veget.* in its entirity, one cannot form an adequate idea of Albertus' contributions and the above pages must be regarded as nothing more than a sample. At the risk of too great a degree of simplification, the two most notable characteristics of *De veget.* are the careful attention to detail and second the enormous range of relevant data. That holds true for all of the different and varied aspects of ›scientia plantarum‹ about which Albertus wrote. Whether he was dealing with structure and function, habit and habitat, or nomenclature and taxonomy, he was able to gather together the best material available, organize it in the manner required by a commentary, and incorporate the many details at his disposal. Many of the latter, as we have noted, were the result of his own observations and personal experiences and of the kind not again recorded until several centuries had passed. It is idle to speculate what the course of plant studies might have been had *De veget.* received the attention it deserved. Even though seven hundred years now separate us from *De veget.* one must insist that by any criteria, Albertus left for us a monumental record of his attempt to understand the world of plants.

XV

Identification of the Plants Described by Albertus Magnus, *De vegetabilibus*, lib. VI

The contributions of Albertus Magnus (1193-1280) to philosophy and theology are well known and need not be rehearsed here. [1] Less well-known, however, are his contributions to science, the encyclopedic range of which, prevents any simple generalization. [2] Despite an enormous literature on Albertus and his influence, many problems remain in determining his sources and in the classification of details which are no longer the concern of science. The present study is an attempt to face one such set of related problems, viz. the identification of the plants described by Albertus in Book VI of his *De vegetabilibus libri VII* and the establishment of some of his sources, both direct and indirect. [3] These are interrelated issues because the same plant is often discussed under different names by various authors, many of whom borrowed freely from one another and most of whom ultimately used the same authorities.

It was Albertus' object to comment on each of the Aristotelian texts, scientific as well as philosophic. [4] To that end, he prepared a commentary, unfortunately not firmly dated, on the *Liber de plantis,* then attributed to Aristotle. [5] Today this small text is generally attributed to Nicolaus of Damascus (64 B.C.-post A.D. 4), a peripatetic polymath who presumably had access to Aristotelian writings no longer extant. [6]

The first five books of *De vegetabilibus* treat of the form, nature, and parts of plants in general from a philosophical, theoretical point of view typical of scholasticism. In Book VI, Albertus turned from general philosophic questions to specific plants. This book, save for the introductory and concluding sections, is a series of short descriptions of the plants mentioned (but not described) in the preceding five books. The purpose of Book VI is to provide the reader with sufficient evidence to identify or visualize the plant about which the preceding philosophical analyses turn. [7] These descriptions make up the bulk of Book VI. Taken together, they may be regarded, in form and content alike, as a herbal. In fact, later writers sometimes cited Albertian passages as if they derived from a herbal. [8] Although it was not Albertus' intention to write a herbal, and certainly not his intention to prepare a text for practical or therapeutic purposes, much of the information he used, especially the names and properties, derived from herbals and the herbalistic chapters of leechbooks. Both the general format of Book VI, for example, its alphabetical arrangement, and the individual sections, normally each one devoted to a single plant, clearly demonstrate Albertus' herbalistic sources. [9] Some of his sources are mentioned by name while others must be inferred on the basis of textual parallels.

Because of Albertus' care in describing plants, particularly those indigenous to or naturalized in Western Europe for which he sometimes incorporated empirical data obtained through personal observation, and because of the conjunction of names and synonyms, in Albertus and in his sources, the great majority of the plants described by him can be identified with a high degree of certainty. The descriptions of some of the exotica of Near Eastern origin, however, are not always so easy to identify. [10] Like the majority of medieval writers on medicine, botany, and pharmacy, Albertus had little personal knowledge of the parent plant of such exotica. [11] He was thus forced to rely on his sources, some of whom also had no first-hand experience, and to make inferences based on experience with the dried spices of commerce, for example, cloves, nutmeg, pepper, etc.

The identifications proposed by Albertus' editors and by others, were for their time, excellent. [12] However, in the intervening century, taxonomic and nomenclatural revisions have made obsolete many of the names proposed in the editors' identifications. [13] Moreover, they did not have access to some of the texts now available, each of which has preserved some data not found elsewhere, for example, Rufinus' *De virtutibus herbarum* or the *Herbarius communis* of Hermannus von Heiligenhafen. (see below for complete citation.)

The following identifications are based upon the name used by Albertus in the descriptions of Book VI. When a synonym is used elsewhere by Albertus, that name will be noted in alphabetical sequence but discussed under the lemma-name. The location of Albertus' description is indicated by the first set of numbers, immediately following the common English name of the plant in question. The three numbers to the left of the diagonal line indicate the page of Book VI in Meyer and Jessen's edition; the numbers to the right indicate the section, again following Meyer and Jessen. Following the relevant data from *De vegetabilibus,* other authorities are noted who treat of the same plant. Unless otherwise specified, these passages employ the same name as Albertus' lemma-name; in those cases where a different name is employed, that name immediately follows the reference. Orthographical variants have not been noted except in cases where there is some textual problem or nomenclatural confusion. The authorities used for study are:

AVLC Avicenna, *Liber canonis*. Venice: P. de Paganinis, 1507. (Cited by folio and section).

CAGS Constantinus Africanus, «De gradibus quos vocant simplicium liber» (in his) *Opera medica.* Basel: Henricus Petri, 1536 pp. 342-387. (Cited by page; individual sections are not numbered).

EHCH Heinrich Ebel, ed. *Der Herbarius communis des Hermannus de Sancto Portu und das Arzneibüchlein des Claus von Metry*. Würzburg: Konrad Triltsch, 1940 (=Texte und Untersuchungen zur Geschichte der Naturwissenschaften, 1). (Cited by page).

HLHH Henrik Harpestraeng, *Liber herbarum*. ed. Poul Hauberg. Copenhagen: Hafnia, 1936. (Cited by page and section).

IJDU Isaac Judaeus, *Thesaurus sanitatis de victus salubris ratione alimentorum facultatibus ciborumque varietate ac delectu sive De dietis universalibus et particularibus libri II.* Antwerp: Gaspard Beller, 1607. (Cited by page).

KVMP Konrad von Megenberg, *Das Buch der Natur*. ed. Franz Pfeiffer (Stuttgart, 1861). Facsimile reprint, Hildesheim: Georg Olms, 1962. (Cited by page and section).

MFVH Macer Floridus, *De viribus herbarum*. ed. Ludwig Choulant. Leipzig: Leopold Voss, 1832. (Cited by page and chapter).

RFMS «Flos medicinae scholae Salerni» (in) *Collectio Salernitana*, ed. Salvatore de Renzi. Naples: Sebezio, 1859. Vol. V pp. 1-101. (Cited by page and section).

RUFH Rufinus, *De virtutibus herbarum*. ed. Lynn Thorndike as *The Herbal of Rufinus*. Chicago: University of Chicago Press, 1949. (Cited by page and section; NB not all sections are numbered; in those cases, I cite by page alone).

SINJ *Sinonima* (in) Nicolaus Salernitanus, *Antidotarium Nicolai*. Venice: Nicolaus Jenson, 1471. fols. [51r] - [68v]. (Cited by folio and column).

TCNR Thomas Cantimpratensis, *Liber de natura rerum*. ed. H. Boese. Berlin: de Gruyter, 1973. (Cited by page and chapter).

WARZ Mattaeus Platearius, *Circa instans* (in) Hans Wölfel, ed. *Das Arzneidrogenbuch Circa instans in einer Fassung des XIII. Jahrhunderts aus der Universitätsbibliothek Erlangen*. Diss. Berlin, 1939. (Cited by page; the sections are unnumbered).

Nota: A great many of Albertus' plant names ultimately derive from Pliny, *Historia naturalis* via Isidore's *Etymologiae* and from the «Dioscorides Alphabeticus» (represented by the Colle edition of 1478 of Dioscorides, *De materia medica*.) As far as can be determined, however, Albertus did not directly use either Pliny or Dioscorides; hence I have not recorded the earlier greco-roman forms of the medieval names used by Albertus and the authorities noted above.

NOTES

1. For representative studies, see Martin Grabmann, «Der Einfluss Alberts des Grossen auf das mittelalterliche Geistesleben,» *Ztschr. f. kathol. Theol.* 52 (1928), 153-182, 313-356; G. Meersseman, *Introductio in Opera Omnia B. Alberti Magni* (Bruges: Beyaert, 1931); Franz Pelster, *Kritische Studien zum Leben und zu den Schriften Alberts des Grossen* (Freiburg/Br.: Herder, 1920).

2. See, in general, Hermann Stadler, «Albertus Magnus als selbständiger Naturforscher,» *Forschungen zur Geschichte Bayerns* 14(1906), 95-114; Franz Strunz, «Les sciences naturelles dans l'oeuvre d'Albert le Grand,» *Annales de l'Inst. Guébhard-Séverine* 8 (1932), 245-263.

3. All references to *De vegetabilibus* are to Alberti Magni, *De vegetabilibus libri VII*. edd. Ernst Meyer and Karl Jessen (Berlin: Reimer, 1867).

4. K. Zell, «Albert (sic.) Magnus als Erklärer des Aristoteles,» *Der Katholik* 49/1. Hälfte 1869), 166-178; G. Manser, «Alberts Stellung zur Autorität seiner Vorgänger,» *Divus Thomas* (Freiburg/Br.) 2 (1915), 75-85; Rud. Kaiser, «Zur Frage der eigenen Anschauung Alberts d. Gr. in seinen Philosophischen Kommentaren,» *Freiburger Ztschr. f. Philos. u. Theol.* (IV) 9 (1962), 53-62.

5. Cf. M. Bouyges, «Sur le De Plantis d'Aristote-Nicholas à propos d'un manuscrit arabe de Constantinople,» *Mélanges de l'Université Saint-Joseph, Beyrouth* 9 (1924), 71-89; Lotte Labowsky, «Aristoteles De Plantis and Bessarion,» *Mediaeval and Renaissance Studies* 5 (1961), 132-154; Bertrand Hemmerdinger, «Le De plantis, de Nicolas de Damas à Planude,» *Philologus* 111 (1967), 56-65; H. J. Drossaart Lulofs, «Aristoteles περὶ φυτῶν,» *Journal of Hellenic Studies* 77 (1957), 75-80.

6. Nicolai Damasceni, *De Plantis libri duo Aristoteli vulgo adscripti. Ex Isaaci ben Honain versione arabica latine vertit Alfredus*. ed. E. Meyer (Leipzig: Voss, 1841).

7. Cf. Jerry Stannard «Die Botanik des hl. Albert.» (To appear in *Albertus Magnus-Festschrift*. Köln, 1980.)

8. Jerry Stannard, «Albertus Magnus and Medieval Herbalism» (in) *Albertus Magnus and the Sciences: Commemorative Essays.* ed. James A. Weisheipl (Toronto: Pontifical Institute of Mediaeval Studies) (in press.)

9. Cf. Jerry Stannard, «Medieval Herbals and their Development,» *Clio Medica* 9 (1974), 23-33.

10. Cf. John M. Riddle, «The Introduction and Use of Eastern Drugs in the Early Middle Ages,» *Sudhoffs Archiv* 49 (1965), 185-198 and Jerry Stannard, «Eastern Plants and Plant Products in Medieval Germany,» *Actes du XIIIe Congrès International d'Histoire des Sciences, Moscou, 18-24 août, 1971.* 4e section t. 5 (1974), 220-225.

11. Jerry Stannard, «Greco-Roman Materia Medica in Medieval Germany.» *Bull. Hist. Med.* 46 (1972), 455-468.

12. The identification of most, but not all, of the plant names used by Albertus appear in the *De vegetabilibus*, pp. 684-751. These identifications supersede those in two earlier studies by Ernst Meyer, «Albertus Magnus. Ein Beitrag zur Geschichte der Botanik im dreizehnten Jahrhundert,» *Linnaea* 10 (1836), 641-741; *id.* «Albertus Magnus. Zweiter Beitrag zur erneuerten Kenntniss seiner botanischen Leistungen,» *Linnaea* 11 (1837), 545-595.

13. Much valuable information, including the identification of selected plants, will be found in: Heinrich Balss, *Albertus Magnus als Biologe* (Stuttgart: Wissenschaftliche Verlagsgesellschaft, 1947), pp. 116-170; Stefan Fellner, *Albertus Magnus als Botaniker* (Vienna: Hölder, 1881); Rudolf von Fischer-Benzon, *Altdeutsche Gartenflora* (Kiel: Lipsius and Tischer, 1894); T. A. Sprague, «Plant Morphology in Albertus Magnus,» *Kew Bulletin* 9 (1933), 431-440; *ib.* «Botanical Terms in Albertus Magnus,» *id.* 9 (1933), 440-459; Josef Wimmer, *Deutsches Pflanzenleben nach Albertus Magnus (1193-1280)* (Halle/S.: Buchhandlung des Waisenhauses, 1908). In no case, ought one to trust the identifications proposed by Hermann Fischer, *Mittelalterliche Pflanzenkunde* (Munich: Verlag der Münchner Drucke, 1929), pp. 257-326.

ABIES (1). *Abies alba* Mill.; silver fir.
342/5. The first of four coniferous trees, included by Albertus under *abies*, but now placed under three different genera. The medieval nomenclature of conifers, puzzling and inconsistent, originates with Isidore, *Etymologiae*. ed. W. M. Lindsay (Oxford, 1911); cf. XVII 7, 31 *pinus* and XVII 7, 32 *abies*. KVMP 314/5 distinguishes three kinds cf *abies* of which *abies alba* = Albertus' *abies* (1).

ABIES (2). *Picea abies* (L.) Karst.; Norway spruce.
342/5. Cf. KVMP 314/5 *abies citrina;* TCNR 317/7.

ABIES (3), see PICEA.

ABIES (4), see PINUS.

ABROTANUM. *Artemisia abrotanum* L.; southernwood.
480/280. Cf. AVLC 95/69; CAGS 363; EHCH 3; HLHH 96/34; MFVH 29/2; RFMS 24/1; RUFH 3/2.

ABSINTHIUM. *Artemisia absinthium* L.; wormwood.
477/274. Cf. AVLC 88/1; CAGS 344; EHCH 2; HLHH 90/30; KVMP 380/1; MFVH 30/3; RFMS 24/2; RUFH 2/1; TCNR 343/2; WARZ 12.

ACACIA. *Prunus spinosa* L.; sloe, blackthorn.
405/131...acaciae, quae est pomum spinae nigrae. Often *acacia* denotes, the juice expressed from the fruit, e.g. WARZ 10 acatia ... succus prunellorum immaturorum. Cf. RUFH 5/4; 257 «prunelli»; SINJ 51ra.

ACHAIAE, see GUMMI ARABICUM.

ACHILLEA, see CARVI.

ACORUS, see CALAMUS AROMATICUS.

AFERESIUS, see MIRRA.

AGNUS CASTUS. *Vitex agnus castus* L.; chaste tree.
349/23. Cf. AVLC 91/44; CAGS 378; KVMP 311/1; RUFH 13/20; TCNR 316/5; WARZ 7.

AGRIMONIA. *Agrimonia eupatoria* L.; agrimony.
482/283. Cf. EHCH 5; HLHH 126/51; MFVH 125/2; RFMS 24/5; sometimes known as *eupatorium*, e.g. CAGS 346; RUFH 129/33.

AGYCIUM, see ZUCARUM.

ALEXANDRINA. *Aconitum vulparia* Rchb.; wolf's bane.
510/333...a quibusdam vocatur herba luporum. Cf. SINJ 52rb aconitum...herba leporalis. Rufinus' account of *aconitum* (RUFH 8/8) confuses *Aconitum napellus* L. (cf. infra Napellus) with *A. vulparia* by listing as synonyms herba lepora vel luparia which belong to the present species, not *A. napellus.*

ALHUSAR, see ZUCARUM.

ALKANNA, see CIPRUS.

ALLIUM. *Allium sativum* L.; garlic.
482/284. Cf. AVLC 96/75; EHCH 3; HLHH 60/8; IJDU 473; KVMP 384/5; MFVH 34/5; RFMS 21/15; RUFH 13/21; WARZ 18; described by CAGS 382 as *allium domesticum.*

ALLIUM SERPENTIS. *Allium victoriale* L.; long-root onion.
482/284. This appears to be the earliest reference to *A. victoriale.* Later, under various names, especially *Allermannsharnisch* and *Siegwurz,* the plant frequently occurs in German medical texts, e.g. Hans von Gersdorff, *Feldtbuch der Wundtartzney,* Strassburg: Joh. Schott, 1517), fol. 95va *victoria. sygwurtz.* For the nomenclature and folk beliefs, see Herbert Menhofer, «Die Pflanzennamen Siegwurz, Siegenkummel und Siegminze,» *Mitteilungen für Namenkunde* 1 (1957), 11-14.

ALNUS. *Alnus glutinosa* (L.) Gaertn.; black alder.
350/25. Despite the wide distribution of the alder in Northern Europe and its economic importance due to its durability in water, I have found only one other description in medieval herbals, KVMP 314/6. References to the medicinal uses of alder bark in receptaria are, however, not uncommon, e.g. Franz Pfeiffer, «Zwei deutsche Arzneibücher aus dem 12. und 13. Jahrhundert,» *Stzb. Akad. Wiss. Wien, Phil-hist. Kl.* 42 (1863), 123/23 *erlun rinde* or Gersdorff, *ed. cit.* fol. 89rb alnus. ein orlin boum.

ALOE. Aloe spp.; aloe.
345/13. The most commonly used species was *A. succotrina* Lam. but the dried juice from the fleshy leaves of other species possibly was involved. Some medieval texts, in fact, distinguished by name, several taxa, e.g. CAGS 354. Cf. AVLC 95/66; KVMP 354/1; MFVH 121/77; RFMS 24/6; RUFH 15/23; TCNR 332/2; WARZ 3.

ALOES LIGNUM, see LIGNUM ALOES.

ALSEMENDURI, see LIGNUM ALOES.

ALTEA. *Althaea officinalis* L.; marsh mallow.
483/285 ... vocatur etiam bismalva ... vocant autem quidam eandem malvaviscum. Cf. AVLC 96/76; EHCH 4; KVMP 385/10; MFVH 43/9; RFMS 24/7; RUFH 18/27; WARZ 72 «malva domestica».

ALTERANA. *Alkanna tinctoria* (L.) Tausch.; alkanet.
478/276. Identified by eds. under the former name, *Anchusa tinctoria* L., but the identification is not conclusive. The lemma word appears no place else except KVMP 384/6 alterana haizt verbkraut; the remainder of Konrad's account differs from Albertus' and throws no light on the identification. Albertus' account begins with an etymology, but since *alterana* does not appear in Isidore, other sources must be considered. *Alaternus* appears once in Pliny, *Historia naturalis* 16. 45. 108, and again in Columella, *Rei rusticae VII* 6, 1; both are identified as *Rhamnus alaternus* L. by Jacques André, *Lexique des Termes de Botanique en Latin.* Paris: Klincksieck,

1956, p. 22. Because variants, e.g. *alternus*, appear in the MSS, and because such a term as *alaternus* is easily derived by folk etymology (ab alterando, as Albertus' states), it is possible that *alterana* denotes *Rhamnus* sp. Moreover, many species of *Rhamnus* produce dyes, and Albertus specifically states cum aceto mixta in rubeum tingit colorem.

AMBROSIA DEORUM, see SALVIA SILVESTRIS.

AMIGDALUS. *Prunus dulcis* (Mill.) Webb; almond.
346/16. Albertus, like most herbalists, distinguished between sweet and bitter almonds. Today these are regarded as varieties: *P. dulcis* (Mill.) Webb var. *dulcis* (F. C. Ludw.) Focke and ... var. *amara* (DC.) Buchheim. Cf. AVLC 93/58; CAGS 365; EHCH 6; KVMP 315/7; RFMS 23/11; RUFH 21/31; TCNR 317/8; WARZ 13 «amigdale amare».

AMMONIACUM. *Dorema ammoniacum* Don; ammoniacum.
422/164. Properly the gum-resin from a Near Eastern umbellifer, not the plant itself. Mentioned only in passing by Albertus. With the possible exception of AVLC 89/8, medieval writers knew nothing of the plant beyond the reports which ultimately derive from Dioscorides, *De materia medica* III 84 Wellm. ἀμμωνιακόν. Cf. CAGS 375; RFMS 25/14; RUFH 23/27; WARZ 11.

AMOMUM. *Aframomum melegueta* (Rosc.) Schum.; grains of paradise.
352/29. Cf. AVLC 92/52; CAGS 376; KVMP 356/3; RUFH 22/35; TCNR 333/4.

ANETUM. *Anethum graveolens* L.; dill.
481/282. Cf. AVLC 95/72; CAGS 363; EHCH 4; IJDU 479; KVMP 381/2; MFVH 44/10; RFMS 24/9; RUFH 26/44; TCNR 343/3; WARZ 17.

ANISUM. *Pimpinella anisum* L.; anise.
476/272. anisum alio nomine feniculum romanum vocatur. Cf. AVLC 88/1; CAGS 376; EHCH 5; KVMP 384/7; RFMS 25/10; RUFH 26/43; WARZ 12.

APIUM AQUATICUM. *Ranunculus* spp.
480/281. One of four plants placed by Albertus under the general name *apium*. Among the many species of *Ranunculus*, the two most probable species denoted by the lemma word are *R. flammula* L., lesser spearwort and *R. aquatilis* L., water crowfoot. In distinguishing four taxa, Albertus follows AVLC 93/56 and, in turn, KVMP 382/3 follows Albertus. Except for KVMP 382/3 apium ... wehset in wazzer, other writers describe the plant (or closely related species) under *flammula*: RUFH 137/14; SINJ 58ra; WARZ 53.

APIUM DOMESTICUM. *Apium graveolens* L.; celery.
480/281. Usually, *apium* without an adjective, denotes celery, though IJDU 454 preserves the distinction by writing *apium domesticum*. Cf. AVLC 93/56; CAGS 379; EHCH 4; MFVH 42/8; RFMS 21/11; RUFH 28/49; TCNR 343/4; WARZ 8; KVMP 382/3 apium ... ist haimisch.

APIUM MONTANUM. *Athamanta macedonica* (L.); Spr.
480/281. There seems to be no English common name for this Mediterranean species. Albertus almost certainly followed AVLC 93/56 and was, in turn, followed by KVMP 382/3 apium ... auf pergen.

APIUM SILVESTRE. *Smyrnium olusatrum* L.; alexanders.
480/281; 566/441. semurion est apium silvestre, et de ipso prosecuti sumus. This common plant was known under many names; in addition to Albertus' name (also, AVLC 93/56; IJDU 454), cf. RUFH 303/46 smimium ... semen olisandri ... apium silvestre; SINJ 63rb olisacrum; KVMP 382/3 apium ... in walden.

AQUILEA. *Aquilegia vulgaris* L.; columbine.
521/352. Mentioned only in passing. Although a common plant and much admired for its flower, it is more often referred to in receptaria than described; in one instance, a reasonable woodcut illustration depicting *A. vulgaris* is followed by the

text referring to quite a different plant, *Hortus Sanitatis germanice.* Mainz: Peter Schöffer, 1485. Cap. 162 Egilops - ackley. Columbine is described in late medieval herbals, cf. R. M. Garrett, «Middle English Rimed Medical Treatise,» *Anglia* 34 (1911), p. 182, vv. 726-729 «Off columbyne» and Gösta Brodin, ed., *Agnus Castus: A Middle English Herbal* (Upsala: Lundeqvist, 1950), p. 145 «Columbina». For medicinal uses, cf. Gersdorff, *ed. cit.*, fol. 89v and M. S. Ogden, ed. *Liber de diversis medicinis* (London: Early English Text Society, 1938), p. 42 bibat aquilayam cum aqua aut semen aquilaye.

ARANGUS. *Citrus aurantium* L.; bigarde, Seville orange.
363/54. The introduction of citrus fruits intro Europe is a complex matter. Shortly after Albertus' death, fresh oranges are noted as commercial products by Francesco Balducci Pegolotti, *La pratica della mercatura,* ed. Allan Evans (Cambridge, Mass.: Medieval Academy of America, 1936), p. 293 col. a «aranci freschi». Somewhat later, oranges (arausen) and lemons (lymons) were described by an anonymous pilgrim travelling in the Holy Lands, R. Röhricht und H. Meisner, «Ein niederrheinscher Bericht über den Orient,» *Ztschr. f. deutsche Philologie* 19 (1886), 82. At about the same time, Konrad mentions oranges (aranser) in passing but in such a way as to suggest that he was unclear between oranges and other fruits (KVMP 318/11). In the late Middle Ages, oranges became more common and were frequently noted in Italian medical texts, e.g. Carlo Castellani, ed. *Secreti Medicinali di Magistro Guasparino da Vienexia* (Milano: Stedar, 1959), pp. 36, 165, etc. and Francesco Zambrini, ed. *Il libro della cucina del sec. XIV* (Bologna: Gaetano Romagnoli, 1863), p. 63 «aranci».

ARBOR DELICIARUM, see ARBOR PARADISI.

ARBOR MIRABILIS. *Ricinus communis* L.; castor bean.
347/20. Cf. KVMP 313/4; RUFH 232 «palma christi».

ARBOR PARADISI. *Musa X paradisiaca* L.; plantain.
347/19; 348/21. arbor deliciarum sive pulcherrima; 348/22 arbor deliciarum sive paradisi. Cf. AVLC 136/495 «musa»; KVMP 312/3.

ARBOR PESTIFER, see MIRRA.

ARISTOLOGIA LONGA. *Aristolochia longa* L.; long birthwort.
479/277. Frequently this plant and round birthwort (*A. rotunda* L.) are combined under the general name *aristologia* or *aristolochia.* Cf. AVLC 92/50; CAGS 356; EHCH 3; MFVH 85/41; RFMS 25/13; RUFH 40/76; WARZ 14; KVMP 383/4 «aristologia lang».

ARISTOLOGIA ... SICUT PALMITES VITIS. *Aristolochia clematitis* L.; birthworth 479/277. Cf. MFVH 85/41 aristolochia ... tertia clematis; KVMP 383/4 aristologia ... sam die weinrebeste.

ARISTOLOGIA ROTUNDA. *Aristolochia rotunda* L.; round birthworth.
479/277. For references, see ARISTOLOGIA LONGA; KVMP 383/4 «aristologia sinbel.»

ARNOGLOSSA, see LINGUA ARIETIS (MAIOR and MINOR).

ARTEMESIA. *Artemisia vulgaris* L.; mugwort.
483/286. Cf. EHCH 3; HLHH 88/29; KVMP 385/18; MFVH 28/1; RFMS 25/15; RUFH 42/77; WARZ 14.

ARUNDO MAGNA. *Arundo donax* L.; giant reed.
377/79. Despite the widespread distribution of arundines, there is no certain description of them in other medieval herbals. Because of some textual parallels, AVLC 95/68 may have been one of Albertus' sources, but Albertus also added data based on his own observations.

ARUNDO PARVA. *Phragmites australis* (Cav.) Trin.; reed.
377/79. Probably referred to by Gersdorff, *ed. cit.* fol. 89va as arundo. ein ror.

ASA FOETIDA. *Ferula assa foetida* L.; asafetida.
479/279. A gum resin obtained from incising the root of the parent plant. Widely

used in the Middle Ages and later because of strong, almost nauseous odor of the dried gummous mass. Cf. AVLC 93/53; RUFH 45/82; WARZ 5. In the *Hortus sanitatis,* ed. cit. cap. 41 asafetida is termed *dufels dreck.*

ASA ODORIFERA. *Styrax benzoin* Dryand.; gum benjamin.
479/279. A balsamic resin obtained from incising the trunk of the parent tree. Distinguished from ASA FOETIDA because of its balsamic, fragrant odor. Cf. AVLC 93/53 and Gersdorff, *ed. cit.* fol. 89va assa dulcis, heisst cristian.

ATRIPLEX. *Atriplex hortensis* L.; orach.
483/287 ... crisolocanna alio nomine vocatur. Cf. EHCH 4; IJDU 462; KVMP 385/9; MFVH 67/78; RFMS 25/16; RUFH 48/86; SINJ 51ra; AVLC 105/153 «chrysolocanna.»

AURICULA MURIS. *Veronica officinalis* L.; heath speedwell.
477/275. Cf. AVLC 90/18; KVMP 386/11. Albertus' name frequently occurs in other herbals but denotes a different plant.

AVELLANA, see CORILUS.

AVENA. *Avena sativa* L.; oats.
467/256. Mentioned only in passing, but discussed more fully, from an agricultural standpoint, in lib. VII, 640/127. Cf. RUFH 48; Gersdorff, *ed. cit.* fol. 89rb avena. haberen.

BALAUSTIA, see GRANATUM.

BALSAMUS. *Commiphora opobalsamum* (L.) Engl.; balm of Gilead.
354/33. An odorous, resinous exudation of Near Eastern origin, probably derived from other species of *Commiphora* as well. In 355/36 it is noted that the woody portion is called xylobalsamum; the dried fruit, carpobalsamum; and the expressed juice *(succus),* opobalsamum. Under any of these headings, it is a common ingredient in medieval medicaments. Cf. AVLC 97/85; CAGS 356; EHCH 7; KVMP 358/6; RUFH 50/1; TCNR 334/8; WARZ 19.

BARBA JOVIS. *Sempervivum tectorum* L.; house leek.
484/288. Cf. KVMP 387/14; RUFH 54/6; EHCH 32 «semperviva».

BASILICON ... FOLIA MAIORA. *Ocimum basilicum* L.; basil.
486/293; 544/398. Referred to as *basilicon maius*; 546/403 Oximen. Cf. EHCH 24 «oximum»; WARZ 28 «oximum»; RUFH 55 «basilicon»; RUFH 224/33 «oximum»; KVMP 387/13 «basilicon»; CAGS 349 ocimi duo sunt genera...

BASILICON ... FOLIA PARVA. *Ocimum minimum* L.; bush basil.
486/293. Identification by Albertus' editors. *O. minimum* is now considered by some taxonomists as only a cultivated form of *O. basilicum* L., e.g. Rudolf Mansfeld, *Vorläufiges Verzeichnis landwirtschaftlich oder gärtnerisch kultivierter Pflanzenarten* (Berlin: Akademie-Verlag, 1959), p. 380.

BASILICUS. *Dracunculus vulgaris* Schott; dragon arum.
484/290. basilicus [codd. basiliscus] est herba, quae dracontea vel serpentaria dicitur. A plant of truly unusual appearance, whence its synonyms and associated folklore. Cf. EHCH 13 «dracontea»; HLHH 116/46 «serpentina»; MFVH 99/54 «colubrina»; RUFH 110/2 «dracontea», SINJ 53rb «basiliscus»; WARZ 112 «serpentaria». Cf. K. Ohlert, «Das wundertätige Schlangenkraut in Mythen, Sagen und Märchen,» *Grenzboten* 62 (1903), 95-100.

BDELLIUM. *Commiphora africana* Engl.; bdellium.
429/181. bdellium, gumma palmae silvestris. A gum-resin, from one or more species of *Commiphora* (fam. *Burseraceae*), having nothing to do with palms, and similar to myrrh (see MIRRA). Cf. AVLC 99/116; CAGS 359; KVMP 358/5; RUFH 58/16; TCNR 317/10 and 333/6 «bidella».

BEDEGAR. *Rosa rubiginosa* L.; eglantine.
358/42; 444/210 ... folia spinae, quae bedegar vocatur. Cf. CAGS 352; RUFH 57; KVMP 316/8.

BENEDICTA. *Geum urbanum* L.; herb bennet.
578/470. benedicta ... alio nomine gariofilata vocatur. Mentioned only in passing. Cf. HLHH 106/41; RUFH 146/5 «gariofilata»; WARZ 59 «gariofilata».

BETA. *Beta vulgaris* L.; beet.
485/292. beta ... quae a quibusdam blitus vocatur. AVLC 153/644 «sicla»; EHCH 8; IJDU 463 «blitum»; 464 «sicla»; KVMP 387/15 beta oder blitus; RFMS 21/7 «cicla»; RUFH 60/21 «blitus».

BETONICA. *Stachys officinalis* (L.) Trev.; betony.
484/289. Cf. EHCH 8; HLHH 114/45; KVMP 386/12; MFVH 46/11; RFMS 25/17; RUFH 55/12; WARZ 23.

BISMALVA, see ALTEA.

BLITUS, see BETA.

BORAGO. *Borrago officinalis* L.; borage.
485/291. Cf. CAGS 348; IJDU 461; RFMS 20/4; RUFH 61/25; WARZ 22.

BORITH. *Saponaria officinalis* L.; soapwort.
543/396 ... sicut herba, quae vocatur borith. Mentioned only in passing but well known for its detergent properties, whence its name. Cf. RUFH 63/30 buriht, saponaria, herba fullonum idem est; SINJ 53ra burit, herba fullonum, soponaria idem est.

BOTRUS, see VITIS.

BUXUS. *Buxus sempervirens* L.; box
359/45. Cf. KVMP 316/9; TCNR 318/11.

CALAMENTUM. *Calamintha officinalis* Moench; calamint.
497/309. Albertus distinguishes two taxa: montanum sive fluviale and silvestre. It is not possible to further identify these two taxa because the genus *Calamintha* is itself poorly understood by taxonomists and second, Albertus has borrowed portions of his text from AVLC 106/158 who, along with CAGS 376 distinguish three taxa. Cf. EHCH 12; RFMS 25/20; RUFH 65/4; WARZ 30.

CALAMITES, see STORAX.

CALAMUS AROMATICUS. *Acorus calamus* L.; sweet flag.
376/77; 353/32 ... hoc [amomum] convenit cum virtutibus acori. Cf. KVMP 365/11; TCNR 335/13; as «acorus» cf. AVLC 91/46; CAGS 355; EHCH 4; RUFH 7/6; WARZ 11.

CAMOMILLA. *Matricaria chamomilla* L.; chamomille.
486/294. Cf. AVLC 100/122; CAGS 346; EHCH 9; KVMP 388/16; MFVF 51/14; RUFH 71/17.

CAMPANA, see ENULA CAMPANA.

CAMPHORA. *Cinnamomum camphora* (L.) Sieb.; camphor.
491/300; Cf. AVLC 102/134; CAGS 370; KVMP 395/31; RFMS 25/21; RUFH 63/1; TCNR 344/8; WARZ 25.

CANABUS. *Cannabis sativa* L.; hemp.
575/462. Mentioned only in passing. Cf. AVLC 108/174; EHCH 12; HLHH 78/22.

CANNA IN INDIA. *Bambusa arundinacea* (Retz.) Willd.; bamboo.
505/323. Mentioned only in connection with his description of canna mellis (see ZUCARUM). There are a few references to bamboo in Greek writings (e.g. Herodotus 3. 98; Theophrastus, *Historia plantarum* 4. 11. 13) but Albertus' precise source is undetermined.

CANNA MELLIS, see ZUCARUM.

CANUCA, not identifiable.
506/326. The editors conjecture that this term, which is apparently unrecorded elsewhere, denotes a species of *Filago* or *Gnaphalium*. Both genera contain many species of similar appearance, some of which are not inconsistent with Albertus' brief description.

CAPISTRUM, see RAPISTRUM.

CAPPARIS. *Capparis spinosa* L.; caper
371/68. Cf. AVLC 103/142; CAGS 365; IJDU 473; KVMP 365/13; RFMS 26/24; RUFH 67/7; TCNR 336/15; WARZ 29.

CARDAMOMUM. *Elettaria cardamomum* (L.) Maton.; cardamom.
498/310. Cf. AVLC 106/159; CAGS 347; EHCH 11; KVMP 357/4; RUFH 67/6; 75/32; WARZ 28.

CARDUUS, see VIRGA PASTORIS.

CARDUUS BENEDICTUS. *Cnicus benedictus* L.; blessed thistle.
506/326. Mentioned only in passing for its medicinal properties. Cf. HLHH 108/42 «benedicta alba»; RUFH 74/24.

CARECTUM, see GLADIOLUS AQUOSUS.

CARICA, see FICUS.

CARPOBALSAMUM, see BALSAMUS.

CARVI. *Achillea millefolium* L.; yarrow
503/318. carvi est herba, quae alio nomine achillea vocatur. *Carvi*, < OHG *garbe*, now exists as *Schafgarbe*, the modern German name for *A. millefolium*, and is the only instance of a purely germanic vernacular plant name used by Albertus. In medieval Latin, carvi usually denotes *Carum carvi* L., caraway while *millefolium* is the more usual name for yarrow. Cf. EHCH 20; HLHH 102/39; RUFH 191/23, all of whom describe the plant under the name millefolium or a variant thereof.

CASSIA FISTULA. *Cassia fistula* L.; purging cassia.
375/75. Cf. AVLC 109/197; CAGS 346; KVMP 364/10; RUFH 66/5; TCNR 335/12; WARZ 27.

CASSIA LIGNEA. *Cinnamomum aromaticum* Nees; cassia.
373/71. The bark of this tree, formerly known as *C. cassia* Blume, is the cinnamon of commerce, not the true cinnamon (see CINAMOMUM), which is almost exclusively reserved for purposes of perfumery and medicaments. Cf. CAGS 369; KVMP 363/8; RUFH 67/8; TCNR 335/10; WARZ 31.

CASTANEA. *Castanea sativa* Mill.; chestnut.
360/47. Cf. EHCH 43; KVMP 317/10; RUFH 75/28; TCNR 318/15; AVLC 118/286 «glans»; RFMS 23/10 «glans castanea».

CAUDA EQUI. *Equisetum* spp.; horsetail.
505/325. Although a common and widely distributed plant and useful for various domestic and medicinal purposes, it is not often described in herbals. Cf. AVLC 110/205; RUFH 47 asperella duplex est. una est que dicitur cauda equina...; *Hortus sanitatis, ed. cit.* cap. 221 iparis vel cauda equina - ross zagel; Gersdorff, *ed. cit.* fol. 90ra cauda equina. rosszschwanz. For its uses, see B. Schumacher, ed. *Das Luminare Majus von Joannes Jacobus Manlius de Bosco* (Mittenwald: Arthur Nemayer, n.d.), p. 281 «cauda caballina».

CAUDA PORCINA. *Lithospermum officinalis* L.; gromwell.
508/331. Mentioned only in connection with a differential diagnosis of several compositae. A common plant, known under many names. Cf. RUFH 195/28 «milium solis»; WARZ 60 «granum solis»; *Hortus sanitatis*, ed. cit. cap. 140 cauda porcina-wyss steinbreche; Brodin, *op. cit.*, p. 160 «granium.» For the medieval synonymy, see Willem Daems, *Boec van Medicinen in Dietsche* (Leiden: Brill, 1967), pp. 131, 274.

CAULIS. *Brassica oleracea* L.; cabbage.
493/304. The *locus classicus* of a plant used for alimentary and medical purposes, depending upon mode of administration, dosage, and its frequency. Cf. AVLC 104/143; EHCH 13; IJDU 465; KVMP 394/27; MFVH 77/36; RFMS 20/6; RUFH 75/29.

CAULIS AQUAE, see NENUFAR CROCEUS.

CAULIS MARINUS. *Convolvulus soldanella* L.; sea bindweed.
494/304. Mentioned only in connection with an enumeration of the different plants known as CAULIS. A common plant, but rarely described in herbals. Cf. *Hortus sanitatis, ed. cit.* cap. 390 «soldanella». The earliest record of this plant is Dioscorides II, 122 κραμβὴ θαλάσσια, whence the vernacular name.

CAULIS SILVESTRIS. *Brassica oleracea* L. *var. oleracea* auct.
494/304. Because of the gradations between the wild form of *B. oleracea* and the later cultivars, it is difficult to trace this plant in medieval herbals; possibly it is the *brassica erratica* of Marcellus Empiricus, *De medicamentis liber* ed. Max Niedermann (Leipzig: Teubner, 1916), X, 61 (p. 91). For the synonymy, see Mansfeld, *op. cit.*, p. 78.

CEDRUS. *Cedrus libani* A. Rich.; cedar of Lebanon.
361/49. cedrus ... quae vero et proprio nomine cedrus vocatur. In this way, Albertus emphasized that this tree was not to be confused with the citron, sometimes also called *cedrus* or *cedrum* (see CEDRUS ITALORUM). Cf. KVMP 317/11; RUFH 84/44; TCNR 318/12.

CEDRUS [ITALORUM]. *Citrus medica* L.; citron.
362/51. Italici ... secundum usum linguae latinae cedrum vocant arborem, quae facit poma crocea oblonga magna ... pomum autem hujus arboris vocatur pomum citrinum aut pomum cedrinum; 433/188 Ferunt etiam de quadam arbore, quae vocatur pomarius adae ... The first citrus fruit known in Europe, its nomenclature became a controversial matter, largely, as Albertus intimates, because of the resemblance of, and thus confusion between, *cedrus* < κέδρος and *citrus*. (For a useful summary, see André, *op. cit.*, pp. 78 s.v. cedrus and 93 s.v. citrus.) Cf. AVLC 100/120 «citrum»; RUFH 83/40 «cetrona poma»; TCNR 321/27 «medica»; WARZ 97 «pomum citrinum.»

CEDRUS MASCULINA, not identifiable.
363/54. Mentioned only here, and in counterdistinction to ARANGUS. Since Albertus mentions that its fruit has sterile seeds, it may designate a *Citrus* cultivar resulting from natural hybridization.

CELIDONIA. *Chelidonium majus* L.; greater celandine.
421/163. Mentioned only in connection with describing a yellowish-orange color, sicut est succus celidoniae. Cf. CAGS 381; EHCH 10; KVMP 390/19; MFVH 98/52; RFMS 26/29; RUFH 81/39; WARZ 42.

CELSA, see MORUS.

CENTAUREA MAIOR. *Centaurea centaurium* L.; greater centaury.
498/311. Cf. AVLC 107/162; CAGS 362; KVMP 397/33 «centaurea ... daz ander»; MFVH 99/53; RUFH 79; WARZ 30.

CENTAUREA MINOR. *Centaurium minus* Moench.; lesser centaury.
499/311. Cf. AVLC 107/162; CAGS 362; EHCH 10; HLHH 120/47; KVMP 397/33; MFVH 99/53; RFMS 26/27; RUFH 77/36; TCNR 344/10; WARZ 30.

CENTINODIA. *Polygonum aviculare* L.; knotgrass.
504/322. The wide distribution of this insignificant, low growing, weedy plant may be the reason why it was neglected by most herbalists. Cf. RUFH 83 «centumnodia»; Brodin, *op. cit.* 144 «centenodium»; *Hortus sanitatis, ed. cit.* cap. 302 «proserpinata ... wegdred.»

CEPA. *Allium cepa* L.; onion.
487/295. Cf. AVLC 100/123; EHCH 9; HLHH 62/10; IJDU 473; KVMP 388/17; MFVH 75/23; RFMS 21/16; RUFH 86/47.

CEPA MURIS, see SQUILLA.

CEPA SILVESTRIS ?*Allium* sp.
378/80. Mentioned only as an ingredient in a medicament. The name does not appear to occur elsewhere. André, *op. cit.*, p. 80 notes one occurrence of *cepa agrestis; allium silvestre,* which may be the same as Albertus' plant, occurs in CAGS 382; RUFH 14/21.

CERASUS. Prunus cerasus L.; sour cherry.
381/88. Cf. RFMS 22/3; RUFH 88/48; TCNR 318/14.

CICER. *Cicer arietinum* L.; chick pea.
490/299. Albertus distinguishes *cicer domesticum and c. silvestre.* According to Mansfeld, *op. cit.*, p. 193 a wild form is unknown; that, plus the fact that four sub-species and sixty-five varieties have been described, suggests that Albertus is referring to two related cultivated forms of *C. arietinum.* Cf. AVLC 102/132; EHCH 12; KVMP 389/18; RFMS 20/16; RUFH 96/61.

CICOREA. *Cichorium intybus* L.; chickory.
504/321. cicorea, quae et sponsa solis vocatur... florem habens... qui dionisia vocatur. Cf. KVMP 394/28; RFMS 21/13; RUFH 119/6 «eliotropium»; listed as «sponsa solis»: CAGS 353; EHCH 31; WARZ 109.

CICUTA. *Conium maculatum* L.; poison hemlock.
502/317. Cf. CAGS 376; EHCH 9; MFVH 112/65; RUFH 94/58 «cicuta nigra»; WARZ 36.

CICUTA AQUATICA. *Cicuta virosa* L.; cowbane.
502/317. Cf. RUFH 95/58 «cicuta alba»; *Hortus sanitatis, ed. cit.*, cap. 87 «cicuta ... wontzerling» combines this plant with the foregoing.

CIMINUM. *Cuminum cyminum* L.; cumin.
492/303. Albertus, following AVLC 103/140, distinguishes five taxa: *c. nigrum, citrinum, persicum, silvestre,* and *domesticum;* AVLC (in Gerard of Cremona's translation) notes *persicum* and *silvestre,* the remaining three presumably identical to the Arabic names. Since *Cuminum* is a genus of only two species, it is not possible without fuller information on the Arabic terms to identify any of the taxa other than CIMINUM DOMESTICUM and CIMINUM SILVESTRE. Cf. CAGS 374; EHCH 10; ib. 12; IJDU 478; KVMP 396/32; MFVH 116/69; RFMS 27/35; RUFH 88/49; TCNR 344/9; WARZ 35.

CIMINUM DOMESTICUM. *Carum carvi* L.; caraway.
493/303. Except for KVMP 396/32 (which follows Albertus' account closely), caraway is described under *carvi* (which must be distinguished from Albertus' CARVI). Cf. AVLC 103/141; CAGS 372; EHCH 11; IJDU 479; KVMP 396/32 «haimisch kumel»; RFMS 26/25; RUFH 68/12; WARZ 35.

CIMINUM SILVESTRE. *Lagoecia cuminoides* L.
493/303. Except for the reference in AVLC 103/140, on which Albertus' account is based, this is the only known account of an eastern mediterranean native, known both to the Greeks and Romans, e.g. Dioscorides, 3. 60 κύμινον ἄγριον; Pliny 19. 47. 161 «cuminum silvestre»; Scribonius Largus, *Compositiones.* ed. G. Helmreich (Leipzig: Teubner 1887), cap. 119 «cuminum silvaticum».

CINAMOMUM. *Cinnamomum zeylanicum* Bl.; Ceylon cinnamon.
368/64. In sect. 65, Albertus mentions *cinamomum mendosum;* this is a commercial term, designating one of the many grades of cinnamon then (and now) on the world market, not a taxonomic entity. Cf. AVLC 101/128; CAGS 368; EHCH 10; IJDU 480; KVMP 361/7; MFVH 118/73; RFMS 26/31; RUFH 93/57; TCNR 334/9; WARZ 33.

CINUS. *Prunus domestica* L. ssp. *insititia* (L.) Schneid.; damson, bullace 381/88. The taxonomy and nomenclature of the plum is complex and continues to be unstable because of small gradations and near-absence of wild forms. For a summary of the problems involved in distinguishing taxa, reported in earlier times as color-varieties, see, Rud. v. Fischer-Benzon, *Altdeutsche Gartenflora* (Kiel: Lipsius and Tischer, 1894), pp. 152-154 and Johannes Hoops, *Waldbäume und Kulturpflanzen im germanischen Albertum* (Strassburg: Trübner, 1905), p. 299. Probably the damson plum is included by many medieval writers under PRUNUS.

CIPARUS. *Cyperus longus* L.; galingale.
496/308. Cf. AVLC 105/149; CAGS 362; EHCH 10; MFVH 93/48; RUFH 89/51; SINJ 55rb; WARZ 37.

CIPRUS. *Lawsonia inermis* L.; henna
505/324; 356/38. ...oleum de alkanna. Described *sub verbo* alc(h)anna: AVLC 92/51; RFMS 33/95; RUFH 17/26; WARZ 16.

CIRPUS. *Scirpus* spp.; rushes and sedges.
503/320. cirpus iuncus in locis palustribus crescens, sicut et carectum... The genus *Scirpus (Cyperaceae)*, estimated at some 300 species, is widespread in marshy land and served many local, domestic needs, especially in north-western Europe. Cf. KVMP 390/21.

CITONIUS, see COCTANUS.

CITRULUS. *Citrullus lanatus* (Thunb.) Matsum. et Nakai; watermelon.
501/315. Cf. IJDU 447; KVMP 391/22; WARZ 41 (combined description: cucurbita et citruli.)

COCTANUS. *Cydonia oblonga* Mill.; quince.
381/89. coctanus autem sive citonius. Cf. AVLC 106/156; KVMP 319/13; RUFH 92/54; as mala citonia: EHCH 22; RUFH 183/9; WARZ 76.

COLOQUINTIDA. *Citrullus colocynthis* (L.) Schrad.; colocynth.
378/81. Cf. AVLC 102/131; CAGS 367; EHCH 11; KVMP 365/12; RUFH 99/66; TCNR 336/14; WARZ 26.

CONSOLIDA MAIOR. *Symphytum officinale* L.; comfrey.
509/332. Mentioned only in connection with the shape of its leaf as a means of distinguishing between it and ENULA CAMPANA. Cf. RUFH 24/39 anagallis ... herba est quam alii simphitum dicunt; *Hortus sanitatis, ed. cit.* cap. 95 consolida maior; see also, Leo Jules Vandewiele, ed., *De Herbarijs: Middelnederlandse Handschriften uit de XIVe Eeuw* (Brussel: Palais der Academien, 1965), p. 239 *sub voce,* Consolida maior.

CORIANDRUM. *Coriandrum sativum* L.; coriander.
494/306. Cf. AVLC 104/144; EHCH 9; IJDU 476; KVMP 395/30; MFVH 68/29; RFMS 27/32; RUFH 102/68; TCNR 344/7; WARZ 42.

CORILUS. *Corylus avellana* L.; hazel, filbert.
415/150. corilus ... quae est arbor nota, habet nuces, quae avellanae dicuntur. Described under «avellanae»: AVLC 91/43; EHCH 6; RUFH 49/89; under «nux» or «nuces»: KVMP 334/32; MFVH 130/7; TCNR 322/32.

CORONA REGIS. *Hypericum perforatum* L.; St. John's wort.
503/319 ... alio nomine perforata vocatur, graece autem dicitur ypericon. Cf. AVLC 123/365 «ipericon»; KVMP 391/24; RUFH 152/4 «herba sancti johannis»; ib. 346/3 «ypericum».

COTULA FOETIDA. *Anthemis cotula* L.; stinking mayweed.
486/294. Mentioned only in order to contrast with CAMOMILLA, which superficially resemble one another; the odor of the foliage is, however, unmistakably different. Cf. RUFH 99; RFMS 33/95: foetet amarisca (*A. cotula*); redolet similis camomilla.

CRASSULA, see ORPINUM.

CRISOLOCANNA, see ATRIPLEX.

CROCUS. *Carthamus tinctorius* L.; safflower.
488/297. Cf. AVLC 107/163 «cartamus»; WARZ 36 «crocus»; RUFH 105/80 «crocus orientalis»; *Hortus sanitatis, ed. cit.* cap. 133 «cartamus ... wilder saffron».

CROCUS HORTENSIS. *Crocus sativus* L.; saffron.
488/297. Usually known simply as *crocus,* e.g. EHCH 11; IJDU 481; KVMP 392/25; RFMS 27/33; TCNR 343/5; WARZ 36; AVLC 106/157 «crocus hortulanus»; CAGS 363 «semen croci hortensis»; RUFH 105/80 «crocus ortensis.»

CUBEBAE. *Piper cubeba* L. fil.; cubebs.
380/86. Cf. AVLC 103/138; CAGS 349; EHCH 11; KVMP 366/14; RFMS 27/34; RUFH 106/83; TCNR 336/16; WARZ 33.

CUCUMER. *Cucumis sativus* L.; cucumber.
500/314. Cf. AVLC 108/180; IJDU 446; RUFH 107/85.

CUCUMER ASININUS. *Ecballium elaterium* (L.) A. Rich.; squirting cucumber.
502/316. Cf. AVLC 108/181; RUFH 117/4 «elacterium»; *ib.* 265/4 «qucumer salvaticus»; WARZ 50 «elacterium.»

CUCURBITA. *Cucurbita pepo* L.; pumpkin.
499/312. Cf. AVLC 108/179; IJDU 447; KVMP 393/26; RUFH 107/84; TCNR 344/6; WARZ 41.

CYPRESSUS. *Cupressus sempervirens* L.; Italian cypress.
364/56. Cf. AVLC 105/150; CAGS 350; KVMP 318/12; RUFH 93/56; TCNR 318/13; WARZ 33.

CYPRESSUS SILVESTRIS, see JUNIPERUS.

DACTILUS, see PALMA.

DAMASCENA, see PRUNUS.

DAUCUS. *Daucus carota* L.; carrot.
506/328. daucus est creticus et daucus asininus. Albertus' description provides no basis for distinguishing the two taxa as different species and, in the absence of synonyms, the two taxa may be regarded as two forms or races of the common carrot, known for its variability in response to different edaphic factors. Whether *daucus creticus* and *daucus asininus* designated different species in Albertus' sources, remains to be determined; for the medieval synonymy see Daems, *op. cit.* p. 269 «dijstel». Cf. AVLC 111/218; CAGS 369; EHCH 14; IJDU 469 «baucia domestica»; RFMS 20/2 «baucia»; *ib.* 76/10 «daucus»; RUFH 109/1; SINJ 56rb «daucus creticus»; WARZ 45; *Hortus sanitatis*, ed. cit. capp. 62 «baucia ... moren»; 147 «daucus ... wilde moren.»

DAXUS, see TAXUS.

DIONISIA, see CICOREA.

DIPTANNUS. *Dictamnus albus* L.; burning bush.
506/327. Cf. CAGS 380; EHCH 14; KVMP 398/34; RUFH 114/7; TCNR 345/11; WARZ 46.

DRACONTEA, see BASILICUS.

DRAGANTUM. *Astragalus gummifer* Labill.; tragacanth.
384/94. A true gum, derived from one or more species of *Astragalus*, a genus of *ca.* 1600 species, many of which are native to the Near East. Cf. AVLC 111/124; CAGS 360; KVMP 366/15 «diadragagantum»; RUFH 111/3; TCNR 336/17; WARZ 44.

DUMUS ?*Ononis* spp.
383/93. Tentatively identified by the editors as *Ononis spinosa* L. and *O. repens* L., through earlier E. Meyer «Albertus Magnus. Zweiter Beitrag zur erneuerten Kenntniss seiner botanischen Leistungen,» *Linnaea* 11 (1837), 567 identified «dumus» as *Crataegus oxyacantha* L. The latter, now known as *C. laevigata* (Poir.) DC., or hawthorn, is noted by Albertus under SPINA FAGINA. If DUMUS is to be regarded as *Ononis* spp., cf. RUFH 270 «resta bovis» and Brodin, *op. cit.* p. 201 «reta bouis».

EBENUS. *Diospyros ebenum* Koen.; ebony.
384/95. Cf. AVLC 112/230; KVMP 321/14; RUFH 115; TCNR 319/16.

EBULUS. *Sambucus ebulus* L.; dwarf elder.
474/267; 510/333. Mentioned only for purpose of comparison with two other plants. Cf. EHCH 15; MFVH 125/3; RUFH 115/1; WARZ 52.

EDERA. *Hedera helix* L.; ivy.
385/97. Cf. KVMP 321/15; RUFH 115/2; *Hortus sanitatis* cap. 163 «edera arborea.»

ELLEBORUS ALBUS. *Veratrum album* L.; white hellebore.
511/335. This, and the following plant, are often grouped together in one chapter under the general name of «elleborus» but are distinguished in the text by separate names. Cf. AVLC 113/244; CAGS 371; EHCH 14; KVMP 399/36; MFVH 101/56; RFMS 28/44; RUFH 116/3; WARZ 51.

ELLEBORUS NIGER. *Helleborus niger* L.; black hellebore.
510/333. In addition to references to ELLEBORUS ALBUS cf. AVLC 113/243; CAGS 371; MFVH 104/57.

ENDIVIA HORTULANA. *Cichorium endivia* L.; endive.
508/331. Cf. AVLC 112/233; CAGS 353; IJDU 452 «endivia domestica»; EHCH 14; RUFH 290 «scariola»; WARZ 47.

ENDIVIA SILVESTRIS. *Taraxacum officinale* Web.; dandelion.
508/331, endivia silvestris ... sunt duo modi ... quorum unus habet lac, quem vocant rostrum porcinum. Cf. AVLC 112/233; IJDU 452; RUFH 273 «rostrum porcinum»; RFMS 33/95.

ENULA CAMPANA. *Inula helenium* L.; elecampane.
509/332; 397/118. Mentioned only in passing as CAMPANA. In medieval herbals, usually listed simply as «enula»; cf. AVLC 113/240; CAGS 367; EHCH 14; HLHH 68/13; MFVH 89/44; RFMS 27/36; WARZ 48; RUFH 117/5 «elempnium»; SINJ 56vb «elemiium.»

EPITHIMUM. *Ajuga chamaepitys* (L.) Schreb.; ground-pine.
507/330. This little labiate was known under many names, though not described by all herbalists. Cf. RUFH 68/11 «camepitheos»; *ib.* 264 «quercula minor»; *Hortus sanitatis* cap. 102 «camepitheos»; Brodin, *op. cit.* p. 165 «iva»; Vandewiele, *op. cit.* cap. 39 «camepitheos.» It must not be confused with ἐπίθυμον (Dioscorides, IV, 177) or «epithymum» (Pliny 26. 35. 56), both of which denote the parasitic dodder, *Cuscuta epithymum* (L.) Murr.

ERSIMUM. *Sisymbrium* spp.
566/439. Undoubtedly an error in transcribing «erysimum» (e.g., Pliny XVIII 22, 96 where it is mentioned in connection with *sisamum*, as Albertus does likewise) or ἐρύσιμον (Dioscorides II, 158.) Cf. CAGS 359 «sisymbrium» and RUFH 301 «sinsibrium», though the latter may refer to a different plant.

ERUCA. *Eruca vesicaria* (L.) Cav. ssp. *sativa* (Mill.) Thell.; rocket.
507/329. eruca ... est duorum modorum, hortensis videlicet et silvestris. The *Circa instans* of Matthaeus Platearius distinguishes two taxa (WARZ 52), domestica and silvestris as does RUFH 126/28. Mansfeld, *op. cit.* pp. 95-96 notes several varieties and subvarieties, but the evidence in medieval herbals is insufficient to identify the domestic and wild forms. Cf. AVLC 112/231; EHCH 14; IJDU 457; KVMP 398/35; MFVH 70/31.

ESCULUS, see MESPILUS.

ESULA. *Euphorbia esula* L.; leafy spurge.
512/336. esula est non herbae species, sed genus. This statement probably means that Albertus grouped several spurges of similar appearance, e.g. *E. esula* L., *E. cyparissias* L. etc. under the general name ESULA, while other spurges are placed under TITUMALUS. Cf. RUFH 127/31; WARZ 51; SINJ 56vb esula similis est minori titimallo. *Hortus sanitatis* cap. 158 «esula ... wolftz milch» also served as a general term for several euphorbiae; for an attempt to distinguish between the several species, see RFMS 33/95.

FABA. *Vicia faba* L.; broad bean.
512/337. Cf. AVLC 114/245; EHCH 15; HLHH 78/21; KVMP 402/39; MFVH 128/6; RFMS 20/16; *ib.* 27/37; RUFH 130/1.

FABA EGYPTIA. *Nelumbo nucifera* Gaertn.; East Indian lotus.
512/337. faba [sc. *Vicia faba* L.] est herba ... sed egyptia faba et nabathia sunt magis probate in virtutibus fabarum. Known in antiquity as *faba aegyptia* (Pliny, XVIII 30, 121) or αἰγύπτιος κύαμος (Dioscorides II, 106), the name was later transferred to lupine e.g. RUFH 131 faba egyptiaca, id est lupinus. Albertus does not describe either of the two *fabae* and no other medieval descriptions have been located nor has Albertus' source been identified.

FABA NABATHIA, not identifiable.
512/337. The only reference to this plant is in connection with FABA EGYPTIA. The editors tentatively identify *f. nabathia* as *Vicia* spp. (Index rerum p. 715) but provide no evidence nor references to other occurrences of the name.

FAGUS. *Fagus silvatica* L.; beech.
389/105. Cf. KVMP 323/17; TCNR 319/19.

FASEOLUS. *Phaseolus vulgaris* L.; common or kidney bean.
515/341. Cf. AVLC 115/253; RUFH 131/2.

FENICULUM. *Foeniculum vulgare* Mill.; fennel.
517/346. Cf. AVLC 117/281; CAGS 364 «marathri semen»; EHCH 15; IJDU 460; KVMP 400/37; MFVH 56/17; RFMS 27/38; *ib.* 29/55 «marathrum»; RUFH 131/4; TCNR 345/12; WARZ 55.

FENICULUM ROMANUM, see ANISUM.

FENUGRAECUM. *Trigonella foenum-graecum* L.; fenugreek.
515/342. Cf. AVLC 114/251; RFMS 28/39; RUFH 131/3; WARZ 56.

FIBEX. *Betula pendula* Roth; white birch.
390/107. fibex ... a quibusdam vocatur arbor, quam antiqui vocant miricam; 410/140 de mirica ... superius in tractatu de fibice dictum est. For the synonymy, cf. Leo Jordan, «Ein mittelniederdeutsches Pflanzenglossar». *Ztschr. f. deutsche Wortforschung* 3 (1902), 354.55 vibex vel pinnosa. berke. Cf. KVMP 331/28 «mirica» and *Hortus sanitatis* cap. 421 «vibex ... byrck.»

FICUS. *Ficus carica* L.; fig.
385/98; 386/100. ficus ... immatura ... carica vocatur. Cf. AVLC 117/283; EHCH16; KVMP 322/16; RFMS 23/7; RUFH 136/13; TCNR 319/18; SINJ 54ra carice. ficus sicce. At the end of his account of FICUS PHARAONIS Albertus mentions *silvestres ficus* (389/104.) These are the wild figs, formerly called «caprifigs», necessary for the pollination of the fruit of *F. carica.*

FICUS FATUA, see FICUS PHARAONIS.

FICUS PHARAONIS. *Ficus sycomorus* L.; sycomore fig.
387/102. sunt ... quaedam ficus ... quae vocantur ficus pharaonis; 411/144 s.v. MORUS: ... et dulce quidem morum in omnibus habet fere effectum ficus. It is doubtful whether Albertus ever saw a fruit of the sycomore fig tree, which is native to the Near East, including the Levant. As *sycomorus,* it is mentioned several times in the O.T., e.g. *Amos* 7. 14 and once in the N.T., *Luke* 19. 4; Isidore 17. 7. 20 «sycomorus» confuses the sycomore fig with the mulberry (see MORUS) but describes the sycomore fig as ficus aegyptia (XVII 7, 17) following Pliny 13. 14. 56. The source of Albertus' name has not been determined. Cf. AVLC 123/362 «jumeiz»; KVMP 347/47 «sicomorus ... der wild veigenbaum.» The fruit of the sycomore fig has been described as sweet, but inferior to that of *F. carica;* in fact, unless the skin of the immature fruit is punctured or scraped, it «will secrete a quantity of watery juice and will not ripen», H. N. Moldenke, *Plants of The Bible* (New York: Ronald, 1952), p. 108. Hence the name, *ficus fatua,* for the insipid fig-like fruit.

FICUS SILVESTRIS, see FICUS.

FLOS CAMPI, see OCULUS PORCI.

FRAXINUS. *Fraxinus excelsior* L.; ash.
391/108. Cf. EHCH 16; KVMP 324/18; RUFH 138/22; TCNR 319/20. See also LINGUA AVIS.

FRUMENTUM. *Triticum aestivum* L.; wheat.
518/348, 523/357; in 467/256 and in lib. VII *passim,* Albertus uses *triticum.* Cf. AVLC 114/250; EHCH 34 «triticum»; KVMP 402/40; RUFH 139.

FUMUS TERRAE. *Fumaria officinalis* L.; fumitory.
518/347. Cf. AVLC 117/282; CAGS 351; EHCH 15; RUFH 142/24; WARZ 54.

FUNGI, not identifiable.
516/343. Any attempt at identifying which of the many mushrooms and/or bracket fungi Albertus had in mind is sheer guess work. Cf. RUFH 141/23.

FUNGI MELIORES, not identifiable.
516/343. Among the most common of the edible mushrooms is *Psalliota campestris* (L.) Fries, the common mushroom. The same indefiniteness characterizes other medieval accounts, e.g. AVLC 116/275; IJDU 471 «fungi ... non mortiferi»; RFMS 24/4 «agaricus»; RUFH 10/16 «agaricum», WARZ 10 «agaricus». Konrad's account of «fungi oder swammen» is noteworthy because he specifically notes the common morel, *Morchella esculenta* Pers. ... haizent ze dautsch maurochen oder ... morhen.

FUNGUS MUSCARUM. *Amanita muscaria* (L.) Pers.; fly agaric.
517/345. No other medieval descriptions recorded.

FUNIS PAUPERUM, see HUMULUS.

GALANGA. *Alpinia* spp.; galangal.
395/114. Properly the scented rhizome of one or more species of a large, mostly tropical genus of *Zingiberaceae.* The two species most commonly imported into Europe were *A. galanga* (L.) Willd. and *A. officinarum* Hance. Cf. AVLC 120/321; CAGS 372; EHCH 16; HLHH 48/1; KVMP 368/18; MFVH 117/70; RFMS 28/41; RUFH 145/3; TCNR 337/21; WARZ 57.

GALBANUM. *Ferula gummosa* Boiss.; galbanum.
392/110. Properly the gum resin, imported from the Near East. Cf. AVLC 120/319; CAGS 375; KVMP 367/16; RUFH 144/2; TCNR 337/18; WARZ 38.

GALLA, see QUERCUS.

GARIOFILATA, see BENEDICTA.

GARIOFILOS, see UNGULA CABALLINA.

GARIOFILUS. *Syzygium aromaticum* (L.) Merr. et Perr.; cloves.
395/115. Cf. AVLC 120/318; CAGS 357; EHCH 17; HLHH 56/4; KVMP 367/17; MFVH 117/72; RFMS 28/43; RUFH 145/4; TCNR 337/20; WARZ 56.

GAUDA. *Reseda luteola* L.; weld.
520/352. Cf. RUFH 150/15.

GELOVEX, see PINUS.

GENTIANA. *Gentiana* spp.
521/353. Which of the some 400 species is intended, is impossible to say, but doubtlessly *G. lutea* L., yellow gentian, is included. Cf. AVLC 118/288; CAGS 367; EHCH 16; RUFH 147/7; WARZ 57.

GERGUERS, see MILIUM: PANICUM.

GLADIOLUS. *Iris germanica* L.; bearded iris.
522/355. Cf. RUFH 157/6 «iris» and Gersdorff, fol. 42r «blow gilgen.»

GLADIOLUS AQUOSUS. *Iris pseudacorus* L.; yellow flag.
522/355. et quoddam [sc. gladiolus] crescit in locis aquosis; *ibidem,* ... locus, in quo crescit multitudo gladioli aquosi, vocatur carectum, et aliquando per metonymiam locus pro locato ponitur, et gladiolus aquosus carectum vocatur. (cf. also 503/320). Cf. EHCH 17 «iris»; KVMP 403/42 «gladiolus»; RUFH 74 carectum. est herba que nascitur in aqua et scindit manus.

GLANS, see QUERCUS.

GRAMEN. *Carex* spp.; sedges.
524/358. It is impossible to specify which of the 1500-2000 species of *Carex* are in question. In lib. VII, Albertus refers to «gramina lata aspera», but what this includes remains unclear; possibly other genera of the *Cyperaceae* are involved here and in Albertus' references to CIPARUS, CIRPUS, and SEGES. Cf. AVLC 120/313; RUFH 12/18 «agrostis.» The latter term includes *Cynodon dactylon* (L.) Pers. or bermuda grass, which may have been included by Albertus.

GRANATUM. *Punica granatum* L.; pomegranate.
397/118. granatum ... vocatur malum punicum ... Habet florem rubeum sicut rosa, et ille vocatur balaustia. Of frequent occurrence in medieval herbals, it may be cited (i) *sub voce* «balaustia»: AVLC 99/113; CAGS 365; RUFH 52/2; WARZ 21; (ii) «granatum»: AVLC 120/320; RFMS 23/9 or (iii) «mala granata»: EHCH 22; KVMP 329/25; RUFH 182/8; TCNR 321/29; WARZ 77.

GRANUM SPONSI, see NENUFAR CROCEUS.

GUMMI ARABICUM. *Acacia senegal* (L.) Willd.; gum arabic.
393/112. gummi [sc. arabicum], quod vocatur achaiae [cod. *acaie*] apud Corinthum. In Pliny XXIV 67, 109 *acacia* (e Dioscoride I, 101 ἀκακία ἐν Αἰγύπτῳ ... τὸ κόμμι) means both the tree and the gum produced there from; thus except for *achaiae?* > acatitia(e) > ἀκακίαι, Albertus' description leaves little doubt that he is describing gum arabic, the best known of the exotic gums in the Middle Ages. Cf. AVLC 120/317 «gummi»; KVMP 369/19; RUFH 150/14 «gummi»; TCNR 338/22; WARZ 59.

HERBA LEPORIS, see UNGULA CABALLINA.

HERBA LUPORUM, see ALEXANDRINA.

HERBA PEDICULARIS, see STAFISAGRIA.

HERBA SALOMONIS, see SALOMONICA.

HERMODACTILUS. *Colchicum autumnale* L.; autumn crocus.
524/359. Cf. AVLC 122/354; CAGS 379; RUFH 153/5; WARZ 61.

HUMULUS. *Humulus lupulus* L.; hop vine.
525/361; 576/465. quaedam autem est species eius humuli [sc. volubilis], quae vocatur funis pauperum. Cf. KVMP 404/43; RUFH 176 lupuli, id est, volubilis; *ib.* 141 funis pauperis ... lupulus et livertisius idem sunt; *Hortus sanitatis* cap. 215 «humulus ... hoppen». The chapter begins: Humulus sive volubilis magna latine; AVLC 160/735 mentions but does not describe *volubilis magna.*

IABRO, see MANDRAGORA FEMINA.

IAMENUM, see OLUS IAMENUM.

IUSQUIAMUS. *Hyoscyamus niger* L.; henbane.
526/362. Cf. AVLC 123/360; CAGS 384 «hyoscyamus»; EHCH 18; KVMP 404/44; MFVH 108/61; RUFH 159/9; *ib.* 346/1 «yusquianus»; TCNR 345/13; WARZ 61.

JUJUBER. *Zizyphus jujuba* Mill.; jujube.
364/55. Mentioned for purpose of comparison with ARANGUS. Cf. AVLC 124/371; EHCH 18; RUFH 161.

JUNCUS, see CIRPUS.

JUNIPERUS. *Juniperus communis* L.: juniper.
398/121. juniperus ... etiam cypressus silvestris vocatur. Cf. AVLC 124/370 iuniperus ... est cupressus silvestris; HLHH 130/53 «grana juniperi»; KVMP 325/20; MFVH 131/10; RFMS 28/46; RUFH 92 «cupressus silvestris»; *ib.* 160/10; TCNR 320/22; WARZ 63.

KEBULI, see MIRABOLANI.

LABRUSCA. *Vitis vinifera* L. ssp. *sylvestris* (Gmel.) Berger; lambrusca 467/255 (at the conclusion of his description of vitis) ... de uva agresti quae labrusca vocatur. Cf. RUFH 34 ampelos melena ... id est, vitis nigra eademque labrusca. The distinction between *labrusca* and Albertus' VITIS SCLAVA is problematic.

LACTUCA. *Lactuca sativa* L.; lettuce.
527/364; 528/364 called *lactuca domestica* (in counterdistinction to LACTUCA SILVESTRIS); 528/365 included in *lactucae verae* (in counterdistinction to LACTUCA ASINI.) Cf. AVLC 131/452; EHCH 18; HLHH 126/50; IJDU 449 «lactuca domestica»; KVMP 405/46; MFVH 60/20; RFMS 21/8; RUFH 162/2; WARZ 68.

LACTUCA ASINI, not identifiable.
528/365. On the strength of Albertus' description and Rufinus' statement: lactuca asinina est cardo ... (RUFH 164), Albertus' plant may be a thistle of the genus *Carduus* or *Cirsium*. The editors identify *l. asini* as *Anchusa tinctoria* L., the obsolete name for *Alkanna tinctoria* (L.) Tausch, which we accept as designating Albertus' ALTERANA. Cf. AVLC 131/453.

LACTUCA SILVESTRIS. *Lactuca serriola* L.; prickly lettuce.
528/364. Mentioned only in connection with LACTUCA. Cf. IJDU 449; RUFH 163.

LAPPA. *Arctium lappa* L.; great burdock.
533/376. Albertus' description may also cover the smaller species, *A. minus* Benth.; the two species are often combined under one heading in medieval herbals. Cf. RUFH 55/8 «bardana»; Brodin, *op. cit.* p. 173 «lappa-burrus or clote»; Garrett, *op. cit.* p. 182 «scharpe burre»; *Hortus sanitatis* cap. 226 «lappacium ... groisz klett.»

LAPPATIUM. *Rumex patientia* L.; patience dock.
534/377. Other species of the genus, often used as culinary herbs, may be included. Cf. EHCH 18 «lapacium»; MFVH 111/63 «paratella ... herba lapathi»; RFMS 28/48 «lapathum acutum»; RUFH 164/3 «lapatium (who describes three taxa); WARZ 67 «lapacium».

LAUDANUM. *Cistus ladanifer* L.; gum cistus.
394/113. Laudanum or ladanum < λάδανον (Diosc. I 97 s.v. κίσθος) is a resinous exudation obtained from several different species of *Cistus*. Cf. AVLC 127/432; CAGS 347; RUFH 165/6; WARZ 64.

LAURUS. *Laurus nobilis* L.; laurel.
400/123. Cf. AVLC 119/303 «granum lauri»; *ib.* 131/455; CAGS 370; EHCH 19; HLHH 128/52 «lauribacca»; KVMP 327/21; RUFH 166/7; TCNR 320/23; WARZ 69.

LENS. *Lens culinaris* Medik.; lentil.
531/372. Cf. AVLC 130/449; RUFH 169/20 «lenticula»; WARZ 70.

LENS SILVESTRIS. *Medicago sativa* L. ssp. *falcata* (L.) Arcang.; yellow alfalfa.
531/372. Cf. AVLC 130/449.

LENTISCUS. *Pistacia lentiscus* L.; mastic.
400/125; 403/130 «mastix». The highly prized resin *(mastix)* is often described in medieval herbals under its own name or under the name of the parent tree *(lentiscus.)* Cf. AVLC 131/454; *ib.* 133/465; CAGS 354; IJDU 481; KVMP 328/24; RUFH 169/19; *ib.* 180/6; TCNR 321/24; WARZ 69; *ib.* 73.

LIGNUM ALOES. *Aquilaria agallocha* Roxb.; aloes wood.
344/11 «aloes»; 467/257 «xyloaloes est lignumaloes»; 345/12 incensum (*sc.* aloe) autem est odoriferum valde, ita ut etiam altaribus pro thymiamate imponatur; in 468/257 occurs a list of eight Arabic names but they designate commercial sources or, as Albertus states, sunt a diversis locis Indiae, a quibus veniunt (sc. multa alia xyloaloes) denominata. Cf. AVLC 161/744; CAGS 355; KVMP 355/2 «aloes»; RUFH 16/24; TCNR 332/3; WARZ 4.

LILIUM. *Lilium candidum* L.; madonna lily.
530/370. Cf. AVLC 130/447; EHCH 18; KVMP 406/47; MFVH 61/22; RFMS 28/50; RUFH 170/23; WARZ 66.

LILIUM COELESTE, see YREOS.

LINARIA. *Linaria vulgaris* Mill.; toadflax.
512/336. Mentioned only for the purpose of comparison with ESULA: et habet folia sicut linaria, sed differt in hoc, quod esula lactescit, et non facit linaria. This passage

indicates that Albertus either had access to Salernitan writings or used an anthology. The Salernitan passage differs in word order but states essentially the same fact: esula lactescit, linaria lac dare nescit, RFMS 33/95. Cf. RUFH 173 who has a slightly different comparison.

LINGUA ARIETIS MAIOR. *Plantago maior* L.; greater plantain.
529/368. Albertus distinguishes between the greater and the lesser plantain (LINGUA ARIETIS MINOR), as do most medieval herbalists; moreover, often they use the general term *arnoglossa* to cover both the larger and smaller forms, though sometimes *arnoglossa* is a synonym for the larger form alone. Albertus' use of *arnoglossa* (521/353) seems to be used as a synonym for *lingua arietis maior*. Cf. AVLC 128-439; CAGS 355 «arnoglossa»; HLHH 104/40 «plantago»; RFMS 30/67 «plantago»; RUFH 38/75 «arnoglossa»; MFVH 36/3.

LINGUA ARIETIS MINOR. *Plantago lanceolata* L.; ribwort plantain.
529/368. In addition to the references cited above for *arnoglossa*, cf. AVLC 128/439; EHCH 25 «plantago lanceolata.»

LINGUA AVIS ?*Fraxinus excelsior* L.; ash.
528/366 herba est folia habens longa. Albertus' short account resembles Avicenna's (AVLC 128/437) except that the latter mentions cortices ius, a phrase that normally refers to trees, not herbs. Precisely what plant or tree, Avicenna was describing is difficult to determine, but in Western Europe *lingua avis* denoted the seed of the ash tree, as EHCH 19 demonstrates: lingua avis ... esschensaed. The name *bird's tongue* was used by William Turner (1515-1568) in the same sense, James Britten and Robert Holland, *A Dictionary of English Plant-Names* (London: Trübner, 1886), p. 44. Cf. CAGS 351; RUFH 172/27; WARZ 67.

LINGUA BOVIS. *Anchusa officinalis* L.; alkanet.
529/367. Cf. AVLC 128/438; RUFH 174/30; cited as *buglossa*: EHCH 8; MFVH 74/34; RFMS 25/19.

LINGUA CERVINA, see SCOLOPENDRIA.

LINUM. *Linum usitatissimum* L.; flax.
563/435. semen lini. Elsewhere in *De vegetabilibus*, when Albertus is referring to the plant itself, not merely the seed, he uses *linum* (e.g. 94/191; 295/13 etc.) Cf. AVLC 149/606; KVMP 422/80; RUFH 172/26.

LIQUIRICIA. *Glycyrrhiza glabra* L.; liquorice.
401/126. Cf. AVLC 130/448; CAGS 347; EHCH 19; RFMS 29/51; RUFH 170/24; WARZ 65.

LIVISTICUM. *Levisticum officinale* Koch; lovage.
517/346; 519/349; 551/414 etc. Mentioned only in connection with describing other plants or as an ingredient in composita. Cf. EHCH 19 «libisticum»; RFMS 28/49; RUFH 169/21; WARZ 70.

LOLIUM, see ZIZANIA.

LUF. *Colocasia esculenta* (L.) Schott var. *antiquorum* (Schott) Hubb. et Rehd.; taro. 532/374. Following AVLC 128/442, Albertus distinguishes two forms, «crispa» and «plana»; of the latter he states (532/374): plana autem propter similitudinem serpentariae vocatur LUF SERPENTIS. It is unlikely that Albertus ever saw the edible tuber (taro) or the parent plant; his description follows AVLC 128/442, even though Albertus begins his description luf, ut dicit Dioscorides (cf. Diosc. II, 167 ἄρον τὸ καλούμενον παρὰ Σύροις λοῦφα.) Cf. IJDU 469; RUFH 295 «serpentaria».

LUF SERPENTIS, see LUF.

LUPINUS. *Lupinus* spp.; lupine.
533/375. Which of the several species of lupine Albertus is referring to is impossible to say, although probably it included the common *L. albus* L. and *L. luteus* L. For the same reason, Albertus' *lupinus silvestris* (533/375) cannot be more precisely identified. Cf. AVLC 131/451; CAGS 352; EHCH 19; RFMS 29/52; RUFH 175/33; WARZ 69.

MACIS, see MUSCATA.

MAIORANA. *Majorana hortensis* Moench; marjoram.
537/384. Cf. AVLC 134/477; CAGS 371 «sansucus»; EHCH 21; RUFH 176/1; WARZ 82.

MALUM PUNICUM, see GRANATUM.

MALUS. *Malus pumila* Mill.; apple.
401/127. Cf. AVLC 145/569 «poma»; EHCH 22 «mala marciana»; RFMS 22/2 «pyra et poma»; as «mala marciana»: KVMP 329/26; RUFH 184/10; WARZ 77.

MALVA. *Malva rotundifolia* L.; dwarf mallow.
534/378. Cf. CAGS 364; HLHH 122/48; IJDU 464; KVMP 407/49; MFVH 109/62; RFMS 29/54; RUFH 177/2; WARZ 72 «malva silvestris.»

MALVAVISCUM, see ALTEA.

MANDRAGORA FEMINA. *Mandragora autumnalis* Bertol.; mandrake.
535/379 mandragora est herba, cuius radix iabro vocatur; 535/379 est autem in mandragora masculus et femina. The original reason for distinguishing between male and female forms is unclear and there is no consistency in medieval descriptions regarding that distinction. Albertus' description, however, clearly points to *M. autumnalis.* Cf. AVLC 124/368 «iabrol»; KVMP 406/48; RUFH 178/3; TCNR 345/14; WARZ 75; *Hortus sanitatis* cap. 257 «mandragora ... alrun man.»

MANDRAGORA MASCULUS. *Mandragora vernalis* Bertol.; mandrake.
535/379. In addition to the references for MANDRAGORA FEMINA cf. *Hortus sanitatis* cap. 258 «mandragora ... alrun frauwe.»

MANNA, not identifiable.
471/262 (at the end of Albertus' description of ZUCARUM): est autem quiddam simile zucaro, quod vocatur etiam zucarum, et cadit in terra, quae vocatur alhusar de aere, et est species mannatis et roris, et est sicut frusta salis; in his discussion of LADANUM (394/113) he notes non esse aliquid plantae vel arboris, sed potius esse de genere roris, sicut mel et manna et id, quod vocatur tereniabin. There is a considerable literature on manna owing partly to the fact that as a biblical subject it has attracted many, if not always critical notices, and partly to the fact that a number of different plants, plus some insects, produce a saccharine exudation. Each of these various substances has served medical and alimentary needs, each has been known locally as manna, each varies in color due to impurities, and many of them have entered commerce in an adulterated form (see, Alfred Ebert, «Beiträge zur Kenntnis einiger seltener Mannasorten und verwandter Körper,» *Ztschr. d. allgem. Österr. Apotheker-Vereines* 46 [1908], 427-530.) While the origin of several kinds of manna mentioned in Greco-Roman texts can be reasonably established, it is not clear which of many possible sources Albertus used. His account is similar but not identical to those of the *Circa instans* (WARZ 72) and Avicenna (AVLC 136/493). That one of his sources ultimately was of Arabic provenience is proven by the name *tereniabin* (see Pierre Guigues, «Les noms arabes dans Sérapion, Liber de simplici medicina,» *Journal Asiatique* 6 [1905], 58/360 «men» and 84/497 «tereniabin»). It is probable, therefore, that Albertus is referring to at least two different substances; knowing the origin of neither, he confused the exudate of the common manna ash, *Fraxinus ornus* L., which he easily may have known when in Italy or southern France (see Paul Dorveaux, «La récolte de la manne à Cinisi (Sicile)» *Bulletin des Sciences Pharmacologiques* 16 [1914], 107-111) with *tereniabin,* whatever may have been the origin of the latter (see, Max Meyerhof, «The earliest mention of a manniparous insect,» *Isis* 37 (1947), 32-36. Cf. CAGS 347; RUFH 179/4; *Hortus sanitatis* cap. 267 «manna ... hymmeldauwe.»

MARMACORA ?*Melissa officinalis* L.; balm.
536/382. The tentative identification is that of the editors. In lib. I, 94/191 *turego* (codd. *citrago, citrigo* etc.) is also identified as *M. officinalis* (adn. ad loc.); *citrago et varr.* are attested forms (André, *op. cit.* 92) but their connection with *marmacora*

is not at all evident. Albertus' source is clearly AVLC 133/469 «marmacora», but no other examples have been located. The closest parallel is *marolmabor* in *Hortus sanitatis* cap. 250 «melissa ... muter krut» which clearly denotes *M. officinalis*. For references to *M. officinalis* as *melissa* cf. RUFH 189/17; SINJ 61rb; WARZ 83.

MARMOREA, not identifiable.
537/383. The editors note, without comment (p. 537 adn.) that *marmorea* was once identified as *Anagyris foetida* L., but since that plant is poisonous, the identification seems inconsistent with Albertus' short description in which it said that the juice *(succus)* is drunk. No other occurrence of the lemma word has been noted.

MARRUBIUM ALBUM. *Marrubium vulgare* L.; horehound.
539/389. Usually cited simply as *marrubium*: EHCH 20; HLHH 94/33; KVMP 408/51; MFVH 87/42; TCNR 346/16; WARZ 78; as *prassium*: CAGS 361; RFMS 30/69; RUFH 231/6.

MARRUBIUM NIGRUM. *Ballota nigra* L.; black horehound.
539/389. Cf. Brodin, *op. cit.* p. 182 «marubium rubium ... rede horroun.»

MASTIX, see LENTISCUS.

MELLILOTUM. *Melilotus officinalis* (L.) Pall.; ribbed melilot.
537/385. Cf. AVLC 131/456; CAGS 349; RUFH 185/12; WARZ 72.

MELONES. *Cucumis melo* L.; muskmelon.
501/314 melones, quae alio nomine pepones vocantur. Cf. IJDU 442; RUFH 191.

MENTA. *Mentha* spp.; mint.
538/386. Probably Albertus included *M. arvensis* L. and *M. sativa* L. under the general term. Cf. AVLC 136/499; CAGS 359; EHCH 20; HLHH 124/49; IJDU 461; KVMP 408/50; MFVH 93/47; RFMS 29/56; TCNR 346/15.

MENTA IN AQUIS NASCITUR. *Mentha aquatica* L.; water mint.
538/386. Cf. EHCH 32 «sinsimbrium»; RUFH 53/5 «balsamita»; WARZ 112 «sisimbrium.»

MENTASTRUM. *Mentha silvestris* L.; horse mint.
538/387. Cf. TCNR 346/15; WARZ 73 «menta.»

MESPILUS. *Mespilus germanica* L.; medlar.
405/133. mespilus est arbor nota, quae alio nomine esculus corrupte vocatur. Cf. EHCH 22; KVMP 333/30; RFMS 23/8; RUFH 205/11 «mespilla»; TCNR 319/17 «esculus».

MEU. *Meum athamanticum* Jacq.; wild dill.
476/272 ... herba quae meu vocatur, a quibusdam autem dicitur radix ursi. Cf. RUFH 187/15; SINJ 61rb; WARZ 76.

MILIUM. *Panicum miliaceum* L.; millet.
523/327. gerguers est genus milii, quod quidem tres species habere dicitur, sed apud nos duae sunt notae: illa videlicet, quae vulgariter milium vocatur ... Cf. EHCH 22; HLHH 74/19; KVMP 403/41 «gegrues ... milium»; RUFH 194/27.

MIRABOLANI.
410/141 mirabolani sunt multiplices. Albertus' classification, somewhat simplified, is as follows:

1. mirabolani citrini immaturi: *Terminalia citrina* Roxb.
2. nigri indi: unripe fruits of *Terminalia chebula* Retz.
3. illi, qui ultimi in maturitate, sunt pinguiores: *Phyllanthus emblica* L.
4. alii, qui dicuntur kebuli: *Terminalia chebuli* Retz.
5. alii, qui vocantur synii: *Terminalia bellerica* Roxb.

Cf. AVLC 98/95 «beiliricus»; *ib.* 132/461 «myrabolani»; CAGS 345; RFMS 29/59; RUFH 193/25; WARZ 80.

MIRICA, see FIBEX.

MIRRA. *Commiphora myrra* (Nees) Engl.; myrrh.
406/135; 407/136. (in reference to its adulteration) ... commiscetur quodam lacte cuiusdam arboris pestiferae, quae vocatur aferesius. Albertus has paraphrased his source (AVLC 134/479) and, in the process, has misunderstood the process of adulterating myrrh, a gum-resin, with a water-soluble gum. When this is mixed with water, a whitish emulsion results (cf. AVLC *loc. cit.*) ... hoc quidem lacticinium (the adulterated mixture before the excess moisture is evaporated) nominatur alferisius (sic!) Cf. CAGS 359; EHCH 21; KVMP 369/20; RFMS 29/58; RUFH 199/36; TCNR 338/23; WARZ 82.

MIRTILLUM. *Vaccinium myrtillus* L.; bilberry.
553/416. Mentioned only in connection with the description of the fruit of PASTUS COLUMBARUM ... habet granum sicut mirtillum. The only other medieval reference noted is Gersdorff, who refers to the berries' therapeutic usefulness twice: fol. 92vb and 95rb «schwartzheydelber.»

MIRTUS. *Myrica gale* L.; bog myrtle.
408/138. Cf. EHCH 21 «myrrus»; KVMP 331/29; RUFH 194/26 «mirice ... galle».

MORUS. *Morus nigra* L.; black mulberry; *M. alba* L.; white mulberry.
411/143 morus ... habet fructus, latine vocaretur rubus ... In late latin, *celsa* became a synonym of *sycaminum* (see Konrad Hofmann und T. M. Auracher, edd. «Der Longobardische Dioskorides des Marcellus Virgilius,» *Romanische Forschungen* 10 [1899], p. 102; lib. I. cap. 138 «De sicamina ... id est celsa») but was subpect to some confusion; hence Albertus' synonymy of *ficus fatua* and *celsa* (388/103). Cf. AVLC 136/502; EHCH 22 «mora celsi»; KVMP 330/27; RFMS 22/5; RUFH 196/32; SINJ 55rb «celsa. mora domestica»; TCNR 321/28; WARZ 83.

MORUS ASSENICA, not identifiable.
373/72. Avicenna ... sententia talis est, quod cassia sit cortex radicis mori assenicae. Avicenna's text actually reads *mori alceni* (AVLC 106/155). No other record of this name has been found.

MORUS DULCIS, see FICUS PHARAONIS.

MUSCATA. *Myristica fragrans* Houtt.; nutmeg.
412/146; 413/146 flos eius a quibusdam macis esse dicitur, fructus autem est nux muscata. It was not established until the seventeenth century that mace and nutmeg were derived from the same tree. As a consequence, in most medieval and renaissance herbals, there are two entries, one for mace, and one for nutmeg. Cf. AVLC 132/459; 137/506; CAGS 355; EHCH 21, 23; HLHH 54/3; KVMP 371/21, 22; RFMS 29/57; RUFH 179/5; 207/14; TCNR 338/24-26; WARZ 82, 85.

NAPELLUS. *Aconitum napellus* L.; monkshood.
540/391 napellus est napo marinus. Cf. AVLC 137/503; KVMP 411/56; RUFH 204/4.

NAPELLUS MOYSI, not identifiable.
541/392. Albert borrows the name and some of the medicinal uses from AVLC 137/504. No other citation located.

NAPO. *Brassica napus* L.; colza, rape.
540/390. Cf. RUFH 204/6 and Vandewiele, *op. cit.* 440/168 «rapen.»

NAPO MARINUS, see NAPELLUS.

NARCISCUS. *Narcissus* spp.; narcissus.
542/394. Cf. AVLC 137/514; RUFH 202/2; SINJ 62va.

NARDUS. *Nardostachys jatamansi* (Wall.) DC.; spikenard.
417/154. Cf. KVMP 409/52; RUFH 202/3; TCNR 346/17; SINJ 62va «nardo stachium.» In his description of PISTACIA, Albertus likens the odor of the wood to odorem nardi pisticae sive spicatae (433/188); the latter, often termed *spica nardi* is identical to Albertus' NARDUS. Cf. AVLC 153/648 «spica»; CAGS 348 «spica nardi»; MFVH 119/75 «spica»; WARZ 105 «spica ... spica nardi».

NARDUS PISTICA, sive SPICATA, see NARDUS.

NASTURCIUM. *Lepidium sativum* L.; garden cress.
541/393. Cf. AVLC 137/513; EHCH 23; KVMP 410/53. Often in medieval herbals «nasturcium» denotes water cress (*Nasturtium officinale* R.Br.), e.g. RUFH 201/1 but the *Hortus sanitatis* cap. 278 clearly distinguishes between *nasturcium ... kresse* and *nasturcium aquaticum* or water cress.

NENUFAR ALBUS. *Nymphaea alba* L.; white water-lily.
542/395. Cf. CAGS 361; RFMS 29/61; WARZ 84; KVMP 410/54 «nenufar ... weiz.»

NENUFAR CROCEUS. *Nuphar lutea* (L.) Sm.; yellow water-lily.
542/395 nenufar croceus ... ideo quia aquaticus est, vocatur a quibusdam caulis aquae, et granum eius vocatur granum sponsi. Cf. AVLC 138/519 nenufar ... caulem aque ... granum sponsi; EHCH 23; KVMP 410/54 «nenufar ... gel»; MFVH 127/5 «alga palustris sive nymphaea»; RUFH 74 «caulus aque»; *ib.* 205/9 «nenufar».

NEPITA. *Nepeta cataria* L.; catnip.
543/397. Cf. MFVH 52/15; RUFH 204/8.

NIGELLA. *Agrostemma githago* L.; corn cockle.
543/396. Cf. AVLC 138/527; CAGS 375; KVMP 411/55; RFMS 29/62; RUFH 206/13; WARZ 86.

NUX. *Juglans regia* L.; walnut.
414/147. Cf. AVLC 137/505; CAGS 358 «nucleus»; EHCH 43; KVMP 333/31; RFMS 22/1; RUFH 207/17.

NUX HENDEN, not identifiable.
416/153 ... non est nux, sed terra quaedam granosa ad modum ciceris. This is Albertus' only reference; it is a considerably shortened version of AVLC 137/507. The substance described by Avicenna cannot be identified with certainty nor can the «nux henden - ieuz henden» of Serapion, *Practica Ioh. Serapionis* (Lyon: Jac. Myt, 1525), fol. 133vb, the only other record noted. Guigues, *op. cit.* 541/287 identified Serapion's ieuz henden with the mangosteen *(Garcinia mangostana L.)* but it appears strange to describe a commonly eaten fruit as «adeps terre» or «petra terre» (Serapion, *loc cit.*)

NUX INDICA. *Cocos nucifera* L.; coconut.
416/152. Cf. AVLC 137/509; RUFH 207/15; SINJ 62rb; WARZ 85.

NUX MUSCATA, see MUSCATA.

OCULUS PORCI. *Tragopogon porrifolius* L.; salsify.
546/404 oculus porci est flos qui flos campi vocatur. Albertus' description leaves no doubt about the identification, although no other instances of the lemma name have been noted and only one record of *T. porrifolius:* Tullia Gasparrini Leporace et al., edd. *Un inedito erbario farmaceutico medioevale,* (Firenze: Olschki, 1952), p. 46, «herba salsifica.» Albertus' synonym, «flos campi,» is frequently applied to other plants, e.g. Brodin, *op. cit.* pp. 150, 173, 174, etc.

OLEA, see OLIVA.

OLEANDER. *Nerium oleander* L.; oleander.
422/165. Cf. AVLC 139/532; KVMP 328/22 «lorander»; TCNR 321/25 «lorander.»

OLEASTER. *Olea europaea* L. ssp. *sylvestris* (Mill.) Rouy; wild olive.
423/167; 421/162 at the conclusion of the article on OLIVA, Albertus mentions, in passing, gumma olivarum silvestrium. Cf. KVMP 334/33; TCNR 323/25.

OLIBANUM, see THUS.

OLIVA. *Olea europaea* L.; olive.
417/155; 419/158 dicunt quidam, arborem vocari oleam, et fructum olivam, liquorem oleum, et os, quod in ipsa est, vocari amurcam. Cf. AVLC 140/534; EHCH 25; KVMP 335/34; RUFH 209/2; TCNR 323/34.

OLIVA ETHIOPIAE, see STORAX.

OLIVA SILVESTRIS, see OLEASTER.

OLUS IAMENUM ?*Amaranthus lividus* L.; wild blite.
553/417 pes locustae est herba similis oleri iameni. The identification is that of the editors, and while it may be correct, no corroborative evidence has been located, beyond the fact that the young leaves of many amaranths are eaten locally as a pot herb. Albertus' name derives from AVLC 121/330 «olus aliameni» (i.e. «the vegetable of the Yemeni.»

ONICHA.
424/168. Albertus is substantially correct in his paraphrase of AVLC 97/82 «De blacte bizantinie»: onicha ... est de genere blactae byzantiae ... et non est aliquid plantae, sed testa ostreae ad modum unguis humani polita, et optimi odoris, praecipue quando pulverizata proiicitur in ignem. That the opercula of various marine gastropods are so employed in rituals, see, E. Gerardin, «A propos des opercules de murex dits 'ongles odorants,'» *Bulletin des Sciences Pharmacologiques* 31 (1924), 50-52 and G. Petit, «Parfums et remèdes tirés d'opercules de gastéropodes marins,» *Parfumerie Moderne* (1923), pp. 57-62. It is interesting to note that Albertus' account of «onycha» in his *De mineralibus* is quite different, see *Book of Minerals.* trans. Dorothy Wyckoff (Oxford: Clarendon Press, 1967), p. 109. Cf. KVMP 372/23; SINJ 63rb; TCNR 339/27.

OPOBALSAMUM, see BALSAMUS.

OPOPONACUM. *Opopanax chironium* Koch; opopanax.
421/163. Opopanax is an oleo-gum resin derived from a plant closely related to the common parsnip and native to the mediterranean basin. Cf. AVLC 139/531; CAGS 373; RUFH 219/26; WARZ 87.

ORDEUM. *Hordeum vulgare* L.; barley.
544/399. Cf. AVLC 141/539; EHCH 24; KVMP 413/59; RUFH 216/24; WARZ 89.

ORIGANUM. *Origanum vulgare* L.; oregano.
544/398. Cf. AVLC 141/536; CAGS 372; EHCH 24; HLHH 98/35; MFVH 81/38; RUFH 215/23; WARZ 88.

OROBUM. *Vicia sativa* L.; common vetch.
545/401 orobum est herba, quae a quibusdam vocatur vicia avium; 579/472 vicia est herba sicut legumen ... Cf. EHCH 24; KVMP 424/84 «vicia»; RUFH 218/25.

ORPINUM. *Sedum telephium* L.; orpine.
546/402 orpinum est herba, quae communiter crassula vocatur. Cf. KVMP 412/57; TCNR 347/18; RUFH 106 «crassula maior.»

OZIMEN, see BASILICON ... FOLIA MAIORA.

PALMA. *Phoenix dactylifera* L.; date palm.
425/169; 425/170 ... nucleus palmae, qui os dactilorum suorum vocatur; 426/173 ... nucleus palmae ... vocatur dactilus. Cf. AVLC 143/552; EHCH 43 «dactili»; KVMP 337/35; RUFH 113/5 «dactili»; TCNR 323/36.

PANICUM. *Setaria italica* (L.) P. Beauv.; Italian millet.
523/357 GERGUERS est genus milii ... apud nos duae sunt notae, illa, quae vulgariter millium vocatur; et illa, quae vocatur panicum. Cf. EHCH 28; RUFH 230/4; KVMP 403/41 «gegrues ... panicum.»

PAPAVER CAMPESTRE. *Papaver rhoeas* L.; corn poppy.
553/419. Cf. AVLC 145/570; Gersdorff fol. 93vb papaver rubeum ... rot kornrossen.

PAPAVER HORTENSE. *Papaver somniferum* L.; opium poppy.
553/419 papaver est duarum specierum, hortense videlicet et campestre. Et [sc. p. hortense] est album et nigrum, sed melius et sanius est album; 528/364 (in reference to lactuca silvestris) in virtute est papaveris nigri. The distinction between the white and black poppy, based on the color of the seed coat, has no taxonomic validity today. Because of the medicinal and culinary uses of the products of the poppy, frequently it is listed twice in medieval herbals. As «opium» cf. AVLC 139/529; CAGS 384; RUFH 214/21; WARZ 87; as «papaver» cf. AVLC 145/570 «papaver

hortulana»; CAGS 346; *ib.* 386 «semen nigri papaveris»; EHCH 26; HLHH 66/12; KVMP 414/61; MFVH 71/32; RFMS 30/63; RUFH 227/2; TCNR 347/20; WARZ 92.

PAPAVER MARINUM CORNUTUM. *Glaucium corniculatum* (L.). Rud.; red-horned poppy.
554/420. No other certain medieval description recorded, though Hermann Fischer, *Mittelalterliche Pflanzenkunde* (München: Verlag der Münchner Drucke, 1929), p. 270 identifies an illustration of «glaucium» by Benedetto Rinio, prepared *ca.* 1410, as *G. corniculatum*. It is equally probable that the plate refers to *G. flavum* Crantz, see Otto Penzig, *Contribuzioni alla Storia della Botanica* (Milano: Hoepli, 1905), p. 127 «papaver cornutum vel ceratitis.»

PAPAVER NIGRUM, see PAPAVER HORTENSE.

PASTINACA. *Pastinaca sativa* L.; parsnip.
362/51 mentioned only in passing; in lib. VII, 646/147 agricultural practices are noted. There is some evidence to suggest that the authors of medieval herbals did not always distinguish between the carrot and the parsnip; the roots of the wild forms, the ancestors of our modern cultivars are very similar. Cf. EHCH 26; IJDU 469 «baucia sylvestris»; MFVH 80/37; RFMS 21/9; RUFH 52/3 «bautia»; WARZ 22 «baucia.»

PASTUS CAMELORUM, see PASTUS COLUMBARUM.

PASTUS COLUMBARUM, not identifiable.
553/416 pastus columbarum vel camelorum dicitur herba, cuius semen columbae quaerunt, et cameli herbam; the passage continues with a description based on AVLC 145/565. Whatever plant or plants Avicenna was referring to (cf. the editors' note p. 553) was already a puzzle in the late Middle Ages as is indicated by Simon of Genoa, *Synonyma medicinae* (Venice: Anima Mia, 13 november 1486), fol. [72rb]: pastus columbarum et camelorum apud Avi[cennam]; quid sit ignoro; arbitrabatur tamen Judeus esse salviam quod non credo. No other medieval descriptions located.

PENTAFILON. *Potentilla reptans* L.; cinquefoil.
555/421 pentafilon graeco nomine est herba, quae latine vocatur quinquefolia ... Cf. AVLC 143/554; RUFH 235/11.

PEPONES, see MELONES.

PEREDIXION, a fabulous tree.
437/198 narrant de quadam arbore quae graece peredixion vocatur ... (Albertus continues by repeating a fable concerning the tree and the birds which eat its fruits, and concludes) sed hoc non est probatum satis per certum experimentum, sicut cetera, quae hic scribuntur; sed in scripturis veterum invenitur. The ultimate source for this fabulous tree, is the *Physiologus*, cap. 34 περὶ δένδρου περιδεξίου, ed. Friedrich Lauchert, *Geschichte der Physiologus* (Strassburg: Trübner, 1889), pp. 264-265. Albertus' account is followed by KVMP 374/25; TCNR 340/29.

PERFORATA, see CORONA REGIS.

PERSICA. *Prunus persica* (L.) Batsch; peach.
437/199. Cf. AVLC 145/571; KVMP 342/41; RFMS 22/6; RUFH 237/15.

PERVINCA, see SEMPERVIVA.

PES CORNICIS. *Plantago coronopus* L.; crowfoot plantain.
447/215. Mentioned only in passing. No other medieval description noted, although the common medieval folknames «pes corvi» and «pes corvinus» may sometimes designate *P. coronopus.*

PES CORVI. *Ranunculus* spp.
553/418. Cf. AVLC 145/567; RUFH 236 «pes corvinus»; Brodin *op. cit.* p. 199 «pes corvi silvatici ... white crowfoot.»

PES LOCUSTAE, not identifiable.
553/417 pes locustae est herba similis oleri iameni; 529/367 mentioned in a comparative description of lingua bovis. Albertus' source is AVLC 145/566. No other

medieval references located but two arabic synonyms are mentioned by Moritz Steinschneider, «Gafiki's Verzeichniss einfacher Heilmittel,» *Virchows Archiv f path. Anat.* 77 (1879), p. 536/56 «rage», «algerat.» The identification as *Campanula rapunculus* L., proposed by H. Fischer, *op. cit.* p. 263, is a mere guess, unsupported by any evidence.

PETROSILINUM. *Petrosilinum crispum* (Mill.) Nym.; parsley.
531/413. Cf. AVLC 144/560; EHCH 27; HLHH 86/28; IJDU 456; KVMP 413/60; RUFH 233/7; TCNR 347/19; WARZ 99.

PICEA. *Pinus silvestris* L.; scotch pine.
432/187 picea dicitur quasi piccata,...; 342/5 mentioned as «abies tertia». One of the economically most important European conifers, it is seldom described in herbals, though its products are. Cf. KVMP 314/5 «abies resinosa»; RUFH 245/23 «pix»; *ib.* 103/74 «colofonia»; TCNR 324/38 «pinus»; WARZ 98 «pix»; *Hortus sanitatis* cap. 134 «colofonia ... grex beche».

PINEA, see PINUS.

PINUS. *Pinus pinea* L.; stone pine.
432/186; 342/5 mentioned as «abies quarta»; 523/356 gelovex non est herba sed potius nux, quae est in pomo pini; 416/153 ... nux pineae etc. Cf. AVLC 119/308 «granum pini»; KVMP 338/37; RFMS 30/65 «pinea»; RUFH 243/20 «pinee»; WARZ 93.

PIPER. *Piper longum* L.; long pepper and *P. nigrum* L.; black pepper.
435/195. Cf. AVLC 144/558; CAGS 385 «piperis semen»; EHCH 27; HLHH 58/7; IJDU 480; KVMP 372/24; MFVH 114/66; RFMS 30/66; RUFH 238/16; TCNR 339/28 «piperis arbor»; WARZ 91.

PIPER AQUAE. *Polygonum hydropiper* L.; water pepper.
551/412; 583/480 described under the name «zinziber caninum»; in lib. IV, 250/80 described as «persicaria». Cf. AVLC 144/559; RUFH 237/14 «persicaria»; *Hortus sanitatis* cap. 428 «ydropiper ... wasser pfeffer»; Brodin p. 196 «persiccaria ... arsmert».

PIRETUM. *Achillea ptarmica* L.; sneezewort.
550/411. The identification is that of the editors. Cf. AVLC 144/556. No other medieval description has been noted for *A. ptarmica* and the common medieval plant name «piretrum» («piretron», «pyretrum» etc.) denotes an entirely different plant.

PIRUS. *Pyrus communis* L.; pear.
433/190. Cf. AVLC 143/550; KVMP 340/39; RFMS 22/2 «pyra et poma»; RUFH 245; WARZ 97.

PISA. *Pisum sativum* L.; pea.
513/337 and 531/372 (in both cases mentioned only in connection with other legumes); in lib. VII, 641/131 agricultural practices are described for «pisum». Cf. HLHH 76/20; RFMS 20/16; *Hortus sanitatis* cap. 319 «pisa heyssen erbeyss.» Despite the infrequent references to peas in herbals, they were a staple food stuff and hence are often mentioned in medieval cook books, for example, Anita Feyl, «Das Kochbuch des Eberhard von Landshut (erste Hälfte des 15. Jhs.),» *Ostbairische Grenzmarken* 5 (1961), 355.

PISTACIA. *Pistacia vera* L.; pistachio.
433/188. Cf. RUFH 244/21; WARZ 98.

PLATANUS. *Acer pseudoplatanus* L.; plane tree maple.
430/183. Cf. KVMP 338/36; TCNR 324/37.

POLIPODIUM. *Polypodium vulgare* L.; oak fern.
549/408. Cf. AVLC 142/544; CAGS 368; EHCH 28; RUFH 248/27; SINJ 63vb; WARZ 95.

POLIUM. *Teucrium polium* L.; hulwort.
548/407. Cf. AVLC 142/546; CAGS 357; EHCH 26; RUFH 249/28; WARZ 98.

POMARIUS ADAE, see CEDRUS [ITALORUM].

POMUM CEDRINUM, see CEDRUS [ITALORUM].

POMUM CITRINUM, see CEDRUS [ITALORUM].

POPULUS. *Populus alba* L.; white poplar; *P. canescens* L.; grey poplar.
431/185. Cf. EHCH 28; KVMP 339/38; RUFH 138/20 «folia populi»; TCNR 325/39.

PORRINUM. *Allium* sp.
482/284 (the third of the three *species* of *allium* described by Albertus): id autem, quod vocatur porrinum, est compositae virtutis inter allium et porrum. There are many species of the large genus *Allium* (*ca.* 450 species) which might fit such a description. Cf. AVLC 96/75 «alleum»; CAGS 386 «allium sylvestre.»

PORRUM. *Allium porrum* L.; leek.
549/409. Cf. AVLC 143/549; EHCH 26; HLHH 64/11; IJDU 458; KVMP 415/63; MFVH 49/13; RFMS 22/17; RUFH 253/35.

PORRUM SILVESTRE ?*Allium ursinum* L.; ramsons.
550/410. The tentative identification is that of the editors. Albertus' description might equally well apply to several alliaceous plants, for example, *A. fistulosum* L., the latter, perhaps the plant described as *Unlauch* by Hildegard of Bingen, *Psysica*, I cap. 81 (PL 197:1163).

PORTULACA. *Portulaca oleracea* L.; purslane.
548/406. Cf. AVLC 142/543; EHCH 26; IJDU 462; KVMP 416/64; MFVH 59/19; RFMS 30/67; RUFH 251/32; WARZ 96.

PRASSIUM, see MARRUBIUM.

PRUNA ARMENA. *Prunus armeniaca* L.; apricot.
439/201. Cf. AVLC 141/541 «pruna» (the chapter continues by stating armena sunt meliora omnibus (sc. prunis.); cf. Albertus' armena tamen omnibus sunt meliora. However, Albertus also derived some of his data secundum Galenum (439/202); cf. Galen «De simplicium medicamentorum temperamentis» lib. VII c. 18 (Galeni, *Opera omnia*, ed. C. G. Kühn 12 [Leipzig: Cnobloch, 1826], p. 76.) Galen's account derives from Dioscorides I 115.5 μηλέα ἀρμενιακά. Whether Albertus ever saw a fresh apricot is problematic, but he could have learned about it also from Pliny XV 11, 41 «pruna armeniaca» and Isidore XVII 7, 7 «malum armeniacum», both of whom he cites elsewhere.

PRUNUS. *Prunus domestica* L.; plum.
438/201; toward the conclusion of his description of *prunus*, Albertus refers to *damascena* (sc. *pruna*.) The latter, formerly identified as *P. oeconomica* Borkh., are now regarded as a subspecies of *P. domestica* (Mansfeld, *op. cit.* 137). Cf. AVLC 141/541; KVMP 341/40; RFMS 22/4; RUFH 257/40; WARZ 94.

PSILLIUM. *Plantago afra* L.; flea bane seed.
547/405. Cf. AVLC 142/542; CAGS 363; KVMP 416/66; RUFH 246/25; TCNR 347/23; WARZ 94.

PULEGIUM. *Mentha pulegium* L.; pennyroyal.
555/422. Cf. HLHH 98/36; KVMP 416/65; MFVH 54/16; RFMS 30/70; RUFH 256/38; WARZ 97.

PYONIA FEMINA. *Paeonia officinalis* L.; garden peony.
551/414. The so-called male and female forms are often listed under the general heading of «paeonia», «pionia» etc. Cf. AVLC 144/562; CAGS 358; EHCH 28; KVMP 414/62; MFVH 95/49; RFMS 30/64; RUFH 240/18; TCNR 347/21; WARZ 92.

PYONIA MAS. *Paeonia mascula* (L.) Mill.; coralline peony.
551/414. See the references for PYONIA FEMINA.

QUERCUS. *Quercus robur* L.; English oak.
440/203; 440/204 fructus eius glans vocatur; 443/209 galla (the entire section is devoted to the galls or oak apples, produced by chemicals liberated in the process of ovipositing by various hymenopterus insects, but treated as plant products during the Middle Ages; hence in some herbals there are separate chapters devoted to *galla* as well as to glan(des), the acorn.) Cf. CAGS 355; EHCH 17 «galla»; *ib.* 17 «glandes»; KVMP 343/42; RFMS 28/42 «galla»; RUFH 143/1 «galla»; *ib.* 149/12 «glandes»; *ib.* 265/3 «quercus»; TCNR 325/40; WARZ 60 «galla».

QUINQUEFOLIA, see PENTAFILON.

RADIX. *Raphanus sativus* L.; radish.
556/423. A common garden plant, cited under many different names in medieval herbals. Cf. EHCH 29 «radix rafani»; HLHH 70/15 «raphanus minor»; IJDU 467; KVMP 418/68 «raphanus ... radicem»; RUFH 266/1 «rafanum greci»; WARZ 101.

RADIX URSI, see MEU.

RAMNUS. *Rubus fruticosus* L.; bramble and *R. caesius* L.; dewberry.
443/210; 411/143 (in his description of the fruit of the mulberry or MORUS) he states that the mulberry tree ... habet fructus; latine autem vocaretur rubus congrue, sicut et rubum vocamus vepres, quae mora dulcia habent. Cf. AVLC 146/579 «rubus»; CAGS 349; EHCH 8 «batus»; *ib.* 21 «mora spinea»; MFVH 130/8 «batus sive rubus»; RFMS 31/75 «rubus»; RUFH 277/15 «rubus»; WARZ 104 «rubus».

RAPA. *Brassica rapa* L.; turnip.
556/424. Cf. EHCH 29; HLHH 72/16 «rapula»; IJDU 466; KVMP 419/69; RFMS 20/5; RUFH 267/2.

RAPHANUS. *Armoracia rusticana* Gaertn., Mey. et Schreb.; horse radish.
557/425. Cf. AVLC 147/581; HLHH 70/14 «raphanus maior»; WARZ 101; *Hortus sanitatis* cap. 338 «raffanus sive radix ... merretich.»

RAPISTRUM. *Raphanus raphanistrum* L.; charlock.
540/390 napo est radix, quae comeditur, ... et olus eius est fere sicut capistrum (codd. rapistrum); 556/424 rapa est herba ... et olus eius sicut rapistrum. In the *Index* p. 739 is printed «Rapistrum (s. capistrum) = Raphanus raphanistrum L.» This, plus the textual variations, and the fact that *rapistrum* is an attested plant name (cf. RUFH 268) but not «capistrum,» suggests that Albertus' *rapistrum* is the better reading. Cf. *Hortus sanitatis* cap. 344 «rapistrum ... hederich» and Gersdorff, fol. 94rb. «rapistrum ... haderich.» For MHD. *hederich* (et varr.) cf. Georg Pritzel u. Carl Jessen, *Die deutschen Volksnamen der Pflanzen* (Hannover: Ph. Cohen, 1882), p. 327.

REUBARBARUM. *Rheum* spp.; rhubarb.
444/211. From the time of Dioscorides' description, III, 2 ῥα. οἱ δὲ ῥῆον καλοῦσι to Henri Baillon, «Sur l'organisation des rheum et sur la rhubarbe officinale,» *Comptes-rendus, Association française pour l'avancement des sciences* (Bordeaux, 1872) (Paris, 1873), pp. 514-529, there was much speculation concerning the source of the true rhubarb. Since the dried, sliced roots passed through many middlemen, and consequently, were often adulterated, (a fact to which Albertus alludes, *loc. cit.*), the parent plant (*R. officinale* Baill.) was a matter of conjecture. Cf. AVLC 147/585; CAGS 354 «rheum»; RFMS 30/73 «rheubarbarus»; RUFH 268/6; WARZ 102.

RIZUM. *Oryza sativa* L.; rice.
558/427. Cf. AVLC 146/574; EHCH 30 «rysi»; KVMP 419/70; RUFH 270/9 «risum»; *ib.* 215/22 «oriza».

ROSA. *Rosa* spp.; rose.
445/212. The genus has been a favorite subject of monographers but there is little agreement concerning the taxonomy of roses mentioned by medieval writers, partly due to the numerous hybrid roses, partly due to the symbolic associations of the flower. At any rate, Albertus' reference to «rosas albas» suggests, at least, *R. alba* L. Cf. AVLC 146/575; CAGS 344; EHCH 29; HLHH 80/23; KVMP 344/44 «rosarius»; MFVH 60/21; RFMS 30/74; RUFH 270/10; TCNR 325/42; WARZ 99.

ROSA CAMPESTRIS. *Rosa arvensis* Huds.; field rose.
445/213. For reasons stated above under ROSA, it is not always possible to identify the roses mentioned in medieval writings. Cf. RUFH 273/11 «rosa canina» (Rufinus' description may include species not included in Albertus' account.)

ROSA FOETIDA ?*Rosa foetida* Herrm.; fox rose, fire rose.
447/216. Mentioned only in passing at the conclusion of the long section on roses. The present writer has collected from old gardens and farmyards, a yellow, malodorous rose which was provisionally identified as *R. lutea Mill.*, now better termed *R. foetida* Herrm. Whether this is the same as Albertus' *rose foetida* is problematic, although no other European rose has the rank odor. No other medieval citation located.

ROSTRUM PORCINUM, see ENDIVIA SILVESTRIS.

RUBEA TINCTORUM. *Rubia tinctorum* L.; madder.
560/429. Cf. AVLC 147/580; as «rubea» cf. CAGS 351; RUFH 277/16; WARZ 102. See below VENAE TINCTORUM.

RUBUS (1), see MORUS.

RUBUS (2), see RAMNUS.

RUTA. *Ruta graveolens* L.; rue.
559/428. Cf. AVLC 146/578; EHCH 28; HLHH 84/26; KVMP 417/67; MFVH 39/7; RFMS 31/76; RUFH 274/14; TCNR 348/24; WARZ 103; as «ruta domestica» cf. CAGS 386; IJDU 477.

RUTA SILVESTRIS. *Peganum harmala* L.; wild rue.
559/428. A Near Eastern plant which Albertus could not have seen; his account derives from AVLC 146/578 but Albertus did not recognize that the same plant is also described as «harmel» in AVLC 122/340. For the nomenclatural problems, see Jerry Stannard «The Plant Called Moly,» *Osiris* 14 (1962), 254-307. Cf. CAGS 386; IJDU 477; as «piganum» cf. RUFH 245; SINJ 63va; *ib.* 64ra.

SALIX. *Salix alba* L.; white willow.
449/218. Cf. AVLC 156/686; CAGS 358 «salices»; EHCH 32; KVMP 347/46; MFVH 133/11; RFMS 31/77; RUFH 138/21 «folia salicis»; TCNR 326/45.

SALOMONICA, not identifiable.
567/444 salomonica est herba Salomonis. Except for AVLC 147/583, Albertus' source, no other citations have been located.

SALVIA DOMESTICA. *Salvia officinalis* L.; garden sage.
569/450. Almost always cited simply as «salvia»; cf. EHCH 30; HLHH 82/25; KVMP 421/76; MFVH 64/24; RFMS 31/78; RUFH 284/10; TCNR 349/28; WARZ 112.

SALVIA SILVESTRIS. *Teucrium scorodonia* L.; wood or rock sage.
569/450 silvestris salvia alio nomine dicitur ambrosia deorum. Cf. RUFH 22 ambroxana: facit folia viridia ... et folia et stipes ut salvia domestica; *Hortus sanitatis* cap. 22 «ambrosia vel ambrosiana ... oder wilde selbe»; Willy Braekman, *Medische en Technische Middelnederlandse Recepten* (Gent: Koninklijke Academie voor Nederlandse Taal- en Letterkunde, 1975), pp. 236/761 recipe wilde salye die men heet ambrosia ... In some medieval herbals (e.g. EHCH 30; cf. SINJ 56vb) «salvia silvestris» denotes *Agrimonia eupatoria* L.; Albertus knew the latter plant as AGRIMONIA.

SAMBACUS. *Jasminum sambac* (L.) Ait.; Arabian jasmine.
396/115. In many medieval herbals and materia medica texts, there is a deepseated confusion between *sambac(us)* < pers. *Zanbaq* and lat. *sa(m)bucus*. The former was native to India and well known to Arabic writers (but not always distinguished from *J. officinale* L.); in the West, however, it was only a bookname. To compound the problem, Constantine has two entries for *sambucus* (CAGS 361; 362) of which the second refers to *J. sambac* while AVLC 151/619 «sambucus» (without an Arabic synonym) seems, in part, to refer to *J. sambac*. Cf. SINJ 68va «zambacea.»

SAMBUCUS. *Sambucus nigra* L.; elder.
450/220. Were it not for the nomenclatural confusion with *sambacus*. the well known and widely distributed elder would not have been confused in the literature, e.g. EHCH 32 «sambucus» which combines both plants in one description. For the etymologies of *sambacus* and *sambucus* see, Helmut Genaust, *Etymologisches Wörterbuch der botanischen Pflanzennamen* (Basel: Birkhäuser, 1976), p. 327. Cf. CAGS 361; KVMP 348/48; MFVH 125/3; RFMS 31/80.

SANDALUS ALBUS. *Santalum album* L.; white sandalwood.
451/222. Cf. AVLC 154/658; CAGS 369; KVMP 375/28; RFMS 31/79; RUFH 287/16.

SANDALUS RUFUS. *Pterocarpus santalinus* L. fil.; red sandalwood.
451/222. Distinguished from preceeding (except in RFMS 31/79) by the addition of the appropriate adjective denoting the color of the heartwood. Cf. AVLC 154/658; CAGS 369; KVMP 375/28; RUFH 287/16; TCNR 340/32.

SANDIX. *Isatis tinctoria* L.; woad.
560/430. The source of an esteemed blue dye; Cf. KVMP 419/71; MFVH 101/55 «gaisdo».

SATIRIA. *Orchis mascula* L.; long purples.
571/454. Possibly Albertus' *satiria* (< σατύριον) includes other species of the ORCHIDACEAE (see, TESTICULUS CANIS and TESTICULUS VULPIS), as was the case with other medieval autors. Cf. CAGS 379; EHCH 32; RUFH 281/3; WARZ 109.

SATUREGIA. *Satureja hortensis* L.; savory.
569/449. Cf. EHCH 30; KVMP 420/73; MFVH 63/23; RUFH 283/8; WARZ 113.

SAXIFRAGA. *Pimpinella saxifraga* L.; burnet saxigrage.
570/452. Cf. EHCH 33, HLHH 112/44 «pipinella»; KVMP 420/75; RUFH 278/1; TCNR 349/27; WARZ 111.

SCAMONEA. *Convolvulus scammonia* L.; scammony.
564/437; scamonea est succus volubilis [Volubilis (1)] cuiusdam, 577/465 ... volubilis, cuius succus est scamonea. Scammony is properly the resin, extracted from *C. scammonia* and shipped to European markets as dried exudations. Cf. AVLC 152/636; CAGS 369 «diagridium»; RUFH 288/20; SINJ 56r «diagridium id est scamonea»; WARZ 42 «diagridium.»

SCOLOPENDRIA. *Phyllitis scolopendrium* (L.) Newm.; hart's tongue.
565/438: scolopendria ... quae dicitur lingua cervina. Cf. AVLC 153/638; EHCH 33; RUFH 306/53.

SECACUL. *Malabaila secacul* (Mill.) Boiss.
568/445: secacul...est species yringi. Cf. AVLC 155/672; RUFH 296/32; SINJ 65v.

SEGES. *Carex* Spp.; sedges.
496/308 ciparus est herba, quae in foliis est similis porro vel segetibus viridibus. Cf. CIPARUS and GRAMEN.

SEMPERVIVA. *Vinca minor* L.; periwinkle.
568/447 semperviva dicitur alio nomine pervinca. Cf. KVMP 422/78; RUFH 255 «provenca»; WARZ 106.

SEMURION, see APIUM SILVESTRE.

SENABRUD, not identifiable.
434/193 narrat Avicenna de quodam genere piri ... quod senabrud vocatur. Cf. AVLC 143/550 ... species [*sc.* pirorum] que dicitur scenabrud.

SENE. *Cassia senna* L., senna.
474/267 mentioned only as an example of a cholagogue. The dried leaves of *C. senna* (and probably *C. angustifolia* Vahl.) were frequently used in materia medica. Cf. AVLC 106/155 «cassia»; KVMP 363/9 «cassiana»; RUFH 297/34; WARZ 113.

SERPENTARIA, see BASILICUS.

SERPILLUM. *Thymus serpyllum* L.; creeping thyme.
508/330, 572/457. Mentioned only for purpose of comparison. Cf. EHCH 31; MFVH 82/39; RUFH 296/33; WARZ 113.

SETHYN. *Acacia sp.;* acacia.
451/223 sethyn arbor est, et de speciebus est albae spinae. KVMP 346/45 «sethim»; TCNR 326/43. An anonymous, early fifteenth-century pilgrim states: hoult sichim [ex hebr. *sethim*] dat is gelych eychen houltz, ind is swartz as pech...Reinh. Röhricht u. Heinr. Meisner, «Ein niederrheinischer Bericht über den Orient,» *Ztschr. f. dte. Philol.* 19 (1886), 1-86 (ref. p. 81.)

SIGIA, see STACTE.

SILER MONTANUM. *Laserpitium siler* L.; sermontana.
569/448. Cf. AVLC 152/634 «siseleos»; CAGS 364 «seseleon»; EHCH 32 «siseleos vel silermontanum»; KVMP 419/72; RFMS 31/83; RUFH 165/5 «lasar»; *ib.* 298/38 «siseleos»; WARZ 107.

SILIGO. *Secale cereale* L.; rye.
545/399: siligo est species ordei; agricultural practices concerning rye are described in *lib.* VII, 640/128. Although rye was a dietary staple (for example, KVMP 413/59 *s.v.* «Von der gersten»), it is rarely treated singly in herbals, cf. *Hortus sanitatis,* cap. 362 «siligo - rucken korn».

SINAPIS HORTULANA. *Sinapis alba* L.; white mustard.
568/446: sinapis est olus notum...et est silvestris et hortulana. Cf. AVLC 156/683; CAGS 386; EHCH 31; HLHH 60/9; IJDU 481; KVMP 422/79, MFVH 75/35; RFMS 32/88; RUFH 299/42; WARZ 108.

SINAPIS SILVESTRIS. *Sinapis arvensis* L., charlock
568/446 (see SINAPIS HORTULANA.) A common plant, much used as a famine food, but rarely referred to for medicinal purposes, until later; for example, L. J. Vandewiele, ed., *De herbarijs* (Brüssel, 1965) 297/66.

SOLATRUM. *Solanum nigrum* L.; black nightshade.
567/442: solatrum...a quibusdam uva lupi vocatur; 463/248: uva lupina, quam nos solatrum vocamus. Cf. AVLC 154/655; CAGS 365 «strignum»; EHCH 30; MFVH 107/60 «maurella...strignum», RFMS 32/84; RUFH 198 «morella»; *ib.* 305/50 «strignum»; WARZ 106 «strignum...solatrum.»

SOLSEQUIUM, see SPONSA SOLIS.

SORBUS. *Sorbus domestica* L.; service tree.
451/224. Cf. *Hortus sanitatis* cap. 385 «sorbe - sprebern».

SPARAGUS. *Asparagus officinalis* L.; asparagus.
452/225. Cf. AVLC 150/611; IJDU 468 «asparagus»; RFMS 32/85; RUFH 47/84 «asparagus»; *ib.* 291/24; WARZ 110.

SPARGULA. *Spergula arvensis* L.; spurrey.
570/453. Cf. RUFH 292/25.

SPELTA. *Triticum spelta* L., spelt.
520/351. Cf. *Hortus sanitatis* cap. 363. A common staple and sometimes used for medical purposes, for example, Gersdorff, *ed. cit.* 94va «speltz oder dünckel»; RUFH, *Praefatio;* vita continetur ut in herba frumenti, sicalis, spelta et ceteris.

SPINA, see BEDEGAR.

SPINA ALBA, see SETHYN.

SPINA FAGINA. *Crataegus laevigata* (Poir.) DC; hawthorn.
406/134: spina magna...et vocatur vulgariter spina fagina. Cf. EHCH 29 «ramnus... haghedorn».

SPINA MAGNA, see SPINA FAGINA.

SPINA NIGRA, see ACACIA.

SPINACHIA. *Spinacia aleracea* L.; spinach.
563/434. Cf. AVLC 149/604; EHCH 33; IJDU 463; RFMS 21/10; RUFH 299/40.

SPONSA SOLIS (1). *Calendula officinalis* L.; calendula.
570/451: sponsa solis sive solsequium. Cf. HLHH 102/38 «solsequium»; TCNR 348/26.

SPONSA SOLIS (2), see CICOREA.

SQUILLA. *Urginea maritima* (L.) Bak.; squill.
561/431: squilla...quae vocatur cepa muris. Cf. AVLC 148/597; *ib.* 154/655; CAGS 362; KVMP 421/77; RFMS 32/87; RUFH 298/39 «silla»; *ib.* 303/47. For further references, see Jerry Stannard, «Squill in Ancient and Medieval Materia Medica,» *Bull. New York Acad. Med.* 50 (1974), 684-713.

SQUINANTUM. *Cymbopogon schoenanthus* (L.) Spreng.; camel grass.
561/432: squinantum...est duorum generum, campestre et aquaticum. Cf. AVLC 148/598; CAGS 372; RUFH 304/48; WARZ 108.

SQUINANTUM AQUATICUM. *Juncus* spp.; rushes.
561/432 (see SQUINANTUM). To which of the more than 300 species of *Juncus* Albertus refers to, is impossible to say. They were widely used for domestic purposes but rarely are mentioned otherwise; Cf. Gersdorff, 91vb «juncus - ein byntz».

SQUINANTUM CAMPESTRE, not identifiable.
See above, SQUINANTUM.

STACTE. *Liquidambar orientalis* Mil.; liquid storax.
452/226: stacten autem gumma esse dicitur eliquata de mirra; 453/227: storax...hoc genus vocatur a Constantino sigia. There is much confusion in the early literature concerning the geographical origin and the parent plant of this substance, but a few points seem positive. The commercial product, *storax liquida,* derives from *L. orientalis* and is not to be confused with *storax calamites* (or *s. solidus*) (see STORAX). Secondly, the odoriferous gum-resins were frequently adulterated; this applies especially to myrrh, hence the belief that stacte is a form of myrrh (cf. A. Lucas, «Notes on myrrh and stacte,» *Journal of Egyptiam Archaeology* 23 (1937), 27-33.) Finally, the unusual names sigia, sagia, ygia, etc. are said to derive from ζυγία (F. Flückiger and D. Hanbury, *Pharmacographia.* London, 1874, p. 242). In classical Greek, ζύγία denotes *Acer Campestre* L. or maple and there is a superficial resemblance between the leaves of the maple and *L. orientalis* (cf. H. Youngken, *Textbook of Pharmacognosy.* Philadelphia: Blakiston, 1948, fig. 194.) I have not been able to verify the occurrence of ζύγία in th list of simples recorded by Paulus Aegineta or Aetius Amidensis, despite Flückiger and Hanbury's assertions *(loc. cit.)*. A cognate form occurs in a puzzling gloss in a sixteenth century Byzantine botanical glossary: σтακτὴ-ζύγαία τὸ μαστιχίν (A. Delatte, *Anecdota Atheniensia* [Liège, 1939], p. 426. 15) while Simon of Genoa, *Synonyma Medicinae* (Venice: Anima Mia, 13 nov. 1486), fol. 84vb., preserves some knowledge of the Greek; sigia: storax liquida. Cf. CAGS 351: styrax duobus modis est, liquidum videlicet et dicitur sigia, est et calamenthum; *Hortus sanitatis* cap. 378 storax ein gummi—...grece sugia; KVMP 374/26; RUFH 307/56 storax liquida, id est, sagia; *ib.* ygia, id est, storas liquida; SINJ 66r storax liquida, id est, sagia.

STAFISAGRIA. *Delphinium staphisagria* L.; stavisacre 564/436:
564/436: stafisagria alio nomine dicitur uva passa montana...interficit pediculos... propter quod etiam herba pedicularis vocatur. Cf. AVLC 151/622; CAGS 371; EHCH 32; KVMP 420/74; RUFH 279; TCNR 348/25.

STORAX. *Styrax officinalis* L.; storax.
453/227: storax similiter est gumma quaedam ut tradunt veteres...posteriorum autem quidam magis expertorum tradiderunt storacem gummam esse olivae ethiopiae. In the older literature, storax was often termed *storax solidus* or *s. calamites* in order to distinguish it from storax liquidus (see above, «stacte».) Cf. AVLC 148/600; CAGS 351; KVMP 375/27; RUFH 307/56; SINJ 54ra «conficia rubea»; TCNR 340/31.

STYCADOS. *Lavandula stoechas* L.; French lavender.
562/433. Cf. AVLC 149/601; CAGS 350 «stichas»; WARZ 109.

SYNIUS, not identifiable.
410/141; 411/142. Mentioned only in connection with mirabolani, of which it is said to be a form. Albert's source is AVLC 132/461 «De myrabolanis». No other occurrence of the term has been noted.

SYSAMUM. *Sesamum indicum* L.; sasame.
565/439. Cf. AVLC 154/650; RUFH 303 «sisama» (not RUFH 302 «sisamum» which refers to a different plant): SINJ 66va.

TAPSIA. *Thapsia garganica* L.; deadly carrot.
572/456. Cf. CAGS 387; KVMP 423/81; RUFH 311/11; WARZ 115.

TAXUS. *Ilex aquifolium* L.; English holly.
455/229: taxus...a quibusdam enim dicitur taxus, ab aliis antem daxus. Although holly is a common, widespread tree, much used for decorative purposes, and played a large role in folk belief, it is seldom described in herbals. Cf. KVMP 325/19 «ilex»; TCNR 320/21 «ilex».

TEREBINTUS. *Pinus cembra* L.; Swiss stone pine.
455/230; 456/231 (referring to the resin) terebintina vocatur lacrima eius. Cf. KVMP 349/51; TCNR 327/47.

TERENIABIN, see MANNA.

TESTICULUS CANIS. *Orchis militaris* L.; military orchid.
573/459: testiculus canis est alia species satiriae, et est in foliis et stipite sicut testiculus vulpis. Cf. AVLC 158/718.

TESTICULUS VULPIS. *Orchis* sp.; orchid.
572/458: testiculus vulpis est quaedam species satiriae. Cf. AVLC 158/717; RUFH 86 «cepe canis, satyrion, testiculi vulpis...»

THAMARISCUS. *Taxus baccata* L.; yew.
454/228. Cf. AVLC 156/693; KVMP 349/52 «taxus oder daxus»; TCNR 327/48 «taxus».

THIMON. *Thymus vulgaris* L.; thyme.
572/457: thimon sive thimus est herba. Cf. AVLC 158/14 «thymus»; CAGS 368; RUFH 315/14 «timum»; TCNR 327/46.

THIMUS, see THIMON.

THUS. *Boswellia sacra* Flueck.; frankincense.
457/238; 344/12 (aloes lignum used as incense): incensum est odoriferum valde, ita ut etiam altaribus pro thymiamate imponatur; also *thus* (< θύς) is called *olibanum*, e.g. 362/50; 404/130; 407/136 *et saepe*. Cf. AVLC 141/535 «olibanum»; CAGS 357; EHCH 34; KVMP 376/29; MFVH 120/76; RFMS 32/89; RUFH 212/18 «olibanum»; TCNR 341/33; WARZ 90 «olibanum».

THYMIAMA, see THUS.

TILIA. *Tilia* Spp.; linden.
456/232; sunt apud nos duo genera, 455/229: (Just as taxus is called by some daxus) sicut et dilia dicitur a quibusdam, et ab aliis tilia. It seems probable that the two «genera» noted by Albertus are *T. platyphyllos* cop. (large leaved linden) and *T. cordata* Mill. (small leaved linden) but the description does not permit a positive differential diagnosis. Cf. KVMP 350/53 «tilia oder dilia»; TCNR 327/49.

TITIMALUS. *Euphorbia lathyris* L.; caper spurge.
581/476: mentioned only in passing as herbae titimallorum. Probably titimalus, like esula, refer to several species of spurge of similar appearance. Cf. CAGS 387; RUFH 315/15; WARZ 114.

TREMULA. *Populus tremula* L.; aspen.
389/103; 418/157. In both instances, the bark and the wood are mentioned for the sake of comparison with another tree. Despite being a common tree in Northwest Europe, it is, apparently, not described in any medieval herbal, though it was certainly known, cf. the 15th century gloss: hec tremulus - hespe tre, John Earle, ed. *English Plant Names from the Tenth to the Fifteenth Century* (Oxford, 1880), p. 54.

TRIBULUS. *Rosa canina* L.; dog rose.
358/42: ...aliud genus spinae, quod quidam vocant tribulum. Probably *R. canina* is included under *rosa* by many medieval writers but in at least one case it seems to be recognized as a distinct species and described accordingly, cf. W. Daems, ed. *Boec van Medicinen in Dietsche.* Leiden, 1967, 145/72.

TRIFOLIUM. *Trifolium pratense* L.; red clover.
537/385 mentioned only; a longer discussion dealing with its agricultural uses occurs in *lib.* VII, 633/111. Cf. EHCH 34; RUFH 317/18; Brodin *ed. cit.*, p. 178 thre lewyd gres [with] a red flour.

TRIPODIUM, not identifiable.
499/311: haec (sc. centaurea maior) habet folia tripodi similia. As a plant name, *tripodium* has not been located in any other medieval botanical text.

TRITICUM, see FRUMENTUM.

TUBER. *Tuber melanosporum* Vittad., truffle.
571/455. Cf. AVLC 157/698.

ULMUS. *Ulmus procera* Salisb.; elm.
467/256. Although a common, widely-distributed tree and much used of a variety of domestic purposes, it is rarely described in herbals, but frequently mentioned as an ingredient in compound drugs (e.g. RUFH pp. 137 and 153). Cf. KVMP 353/55; TCNR 329/52.

UNGULA CABALLINA. *Tussilago farfara* L.; coltsfoot.
578/470. ungula caballina est herba, quae vulgo vocatur herba leporis...Et sua radix habet saporem gariofilorum, et in hoc convenit cum ea, quae benedicta et alio nomine gariofilata vocatur; 396/117...quod vocant gariofilos, est herba, quae vocatur ungula equi, et apud vulgum vocatur herba leporis, cuius radix declinat ad saporem gariofilorum. Two different plants are confused because 1. the folk name «herba leporis» was applied to *T. farfara* and *Geum urbanum* and 2. the similarity between «ungula caballina» and «ungula equi»: In short, the names «benedicta», «gariofilata» and «gariofilos» refer to *G. urbanum* (cf. RUFH 146/5), «ungula caballina» and «ungula equi» refer to *T. farfara* while «herba leporis» may refer to either. Cf. RUFH 332; SINJ 67vb.

UNGULA EQUI, see UNGULA CABALLINA.

URTICA COMMUNIS. *Urtica dioica* L.; nettles.
574/460. Albertus correctly distinguishes two taxa of *urticae vivae*: urtica communis quae maior est et urtica graeca, quae minor est. Usually in medieval writings the two taxa are grouped together under «urtica.» Cf. EHCH 35; HLHH 93/31; RUFH 328/9; MFVH 33/4; KVMP 423/82; RFMS 32/90; AVLC 159/725.

URTICA GRAECA. *Urtica urens* L., annual nettles.
574/460. See the references under URTICA COMMUNIS.

URTICA MORTUA. *Lamium purpureum* L.; dead nettle.
574/461. Although this plant resembles the stinging or burning nettles (see URTICA VIVA), it does not produce the stinging sensation, hence its name; but for this very reason it is rarely described in herbals. Cf. RUFH 328/9 «urtica mora» [? u. mortua], Gersdorff, *ed. cit.* 95rb: urtica mortua - doub nessel.

URTICA VIVA. *Urtica* spp.; nettles.
574/460. A class-name, designed to distinguish the true nettles (*Urtica* spp.) from the dead nettles (*Lamium* spp.). Cf. above URTICA COMMUNIS and URTICA MORTUA.

UVA AGRESTIS, see LABRUSCA.

UVA LUPI, see SOLATRUM.

UVA LUPINA, see SOLATRUM.

UVA PASSA MONTANA, see STAFISAGRIA.

VALERIANA. *Valeriana officinalis* L.; valerian.
519/349. Mentioned only in passing. Frequently known as «fu» from Dioscorides' φοῦ (I, 11 Wellm.) Cf. AVLC 115/261; CAGS 348; RUFH 318/1; SINJ 57vb; WARZ 54. (Any of the above references may also include or refer to other species of *Valeriana*, several of which were often confused with the «true nard», cf. NARDUS, and, in general A. Meunissier, «Nards antiques et modernes,» *Parfumerie Moderne* 12 [1922], 255-268.)

VENAE TINCTORUM, not identifiable.
578/469: Venae tinctorum quaedam sunt plantae, quae extenduntur sicut venae, et utuntur eis tinctores pannorum. It is tempting to identify this plant with *Rubia tinctorum* or madder (cf. RUBEA TINCTORUM.) However, the Arabic sources, on which Albertus relied, are not sufficiently decisive. See Moritz Steinschneider, «Gafikis Verzeichniss einfacher Heilmittel,» *Virchows Arch. f. path. Anat.* 77 (1879), 533/18 where several alternative identifications are proposed.

VEPRES, see RAMNUS.

VERBENA (1). *Anagallis arvensis* L.; scarlet pimpernel.
579/471: et est duorum modorum: una enim habet florem parvulum croceum, et altera habet parvulum eiusdem figurae azurinum. The two color-forms, formerly designating separate species, are now regarded as color-variants of one species, see Jerry Stannard, «Botanical Nomenclature in Gersdorffs Feldtbüch der Wundtartzney,» *Science, Medicine and Society in the Renaissance. Essays to Honor Walter Pagel.* ed. Allen Debus 1 (New York: Science History Publications, 1972), pp. 87-103 (ref. p. 102, n. 51). It is not clear why Albertus applied the name *verbena* to two quite distinct species, both of which were well-known; it is possible, however, that the confusion resulted from Albertus' reliance on an oral source, reporting on the magical uses of both *Anagallis* and *Verbena* [cf. Verbena (2)]. Although *Anagallis* was a common plant, it is rarely described in herbals, cf. AVLC 91/33; Brodin *ed. cit.*, p. 166 «ipia maior»; F. Holthausen, «Medizinische Gedichte aus einer Stockholmer Handschrift». *Anglia* 18 (1896) 312/5 «pipnielle». For the magical uses of *Anagallis* which, inter alia, exhibit a personal aquaintance with the plant, ef. Franz Pfeiffer, «Zwei deutsche Arzneibücher aus dem 12. und 13. Jahrhundert,» *Stzb. Akad. Wiss. Wien,* Philos-hist. Cl. 42 (1863), 143, lines 19-22 and Robert Garrett, «Middle English Rimed Medical Treatise». *Anglia* 34 (1911), 169, vv. 207 sqq.

VERBENA (2). *Verbena officinalis* L.; verbena vervain.
484/289: hanc (sc. betonicam) multum quaerunt nigromantici sicut et verbenam, dicentes, eam habere signa divinationis; *lib.* V, 338/118 qui in magicis student... betonica divinationem praebere dicitur et verbena... On the nomenclature and rituals associated with verbena, see Jerry Stannard «Magiferous Plants and Magic in Medieval Medical Botany,» *Maryland Historian* 8 (1977), 33-46. Cf. EHCH 34; HLHH 100/37; KVMP 424/83; MFVH 105/58; RUFH 57/13.

VICIA. *Vicia sepium* L.; bush vetch.
546/401 (in his description of *orobum,* Albertus states) sed casta seminis eius non est adeo longa sicut viciae. No other medieval description reported.

VICIA AVIUM, see OROBUM.

VIOLA. *Viola odorata* L.; violet.
575/464; in lib. V, 338/117 it is called *viola vera* in order to distinguish it from *viola crocea* (*Cheiranthus cheiri* L., which does not appear in *lib.* VI.) Cf. AVLC 159/726; CAGS 344; EHCH 35; HLHH 80/24; KVMP 424/85; MFVH 83/40; RFMS 32/91; RUFH 326/8; TCNR 349/29; WARZ 116.

VIRGA PASTORIS. *Dipsacus sativus* (L.) Honck; fuller's teasel.
577/466: virga pastoris est carduus quidam. Cf. AVLC 160/736; CAGS 374; EHCH 35; RFMS 32/92; RUFH 329/12; WARZ 116.

VIRGA PASTORIS MASCULUS. *Dipsacus sylvestris* Huds.; wild teasel.
577/467. No other medieval description recorded.

VITICELLA. *Bryonia dioica.* Jacq.; bryony.
462/245; mentioned only for purposes of comparison with *vitis alba.* Cf. MFVH 126/4 «Bryonia»; RUFH 331/17; WARZ 116; in addition, cf. Daems, *ed. cit.* 157/92.

VITIS. *Vitis vinifera* L.; grape vine.
458/236; the cluster of grapes are called *botrus* in 485/290. Cf. AVLC 160/734; EHCH 42 «uva»; KVMP 350/54; RUFH 330/13; TCNR 327/51; WARZ 97 «passuli».

VITIS ALBA. *Clematis vitalba* L.; traveller's joy.
462/245. Cf. RUFH 331.

VITIS SCLAVA. *Vitis* spp.
431/183; 461/242; 464/251. Mentioned only in connection with *vitis* and other vines. Probably a species of wild grape or, less likely, a cultivated species growing unattended. See above LABRUSCA.

VITIS SILVESTRIS. *Tamus communis* L.; black bryony.
463/248, mentioned in 378/81. A common plant, similar in appearance to *Bryonia dioeca* (see above, VITICELLA) but taxonimically unrelated. No other medieval description recorded.

VOLUBILIS (1), see SCAMONEA.

VOLUBILIS (2). *Calystegia sepium* (L.) R. Br.; hedge bindweed.
576/465. No other medieval description recorded, but possibly included under *volubilis* by RUFH 332 who distinguishes three taxa.

VOLUBILIS MAGNA, see HUMULUS.

XYLOALOES, see LIGNUM ALOES.

XYLOBALSAMUM, see BALSAMUS.

YPERICON, see CORONA REGIS.

YREOS. *Iris florentina* L.; Florentine iris.
579/473: yreos est herba, quae vocatur lilium coeleste apud latinos. Cf. AVLC 123/359; MFVH 88/43; RUFH 156/5; WARZ 63.

YRINGUM. *Eryngium campestre* L.; field eryngo.
568/445: secacul est herba, quae est species yringi; *lib.* VII, 633/111: Diversa tamen animalia diversum desiderant pastum, sicut asinus yringum. etc. Cf. RUFH 158/7 «iringi» and *Hortus sanitatis, ed. cit.* cap. 429 «Yringus-kruss distel».

YSOPUS. *Hyssopus officinalis* L.; hyssop.
581/477. Cf. AVLC 124/367; CAGS 370 «hyssopus»; EHCH 17 «isopus»; HLHH 84/27; KVMP 405/45; MFVH 90/45; RFMS 28/45; RUFH 158/8; WARZ 62.

ZEDOARIA, see ZIDUARIUM.

ZIDUARIUM. *Curcuma zedoaria* (Berg.) Roxb.; zedoary.
582/478 (described as ZIDUARIUM): 584/482 (described as ZEDOARIA). The semidry rhizomes of *C. zedoaria* resemble those of ginger and, depending on the quality, realized different prices. Cf. AVLC 161/745; CAGS 374; EHCH 36; HLHH 58; KVMP 426/87; MFVH 117/71; RFMS 32/93; RUFH 348/2; TCNR 349/31; WARZ 117.

ZINZIBER. *Zingiber officinale* Rosc.; ginger.
525/360 mentioned only as an ingredient in a medicament. Cf. AVLC 161/746; CAGS 367; EHCH 35; HLHH 52/2; IJDU 480; KVMP 425/86; MFVH 116/68; RFMS 33/94; RUFH 295/30 «sengiber»; *ib.* 347/1 «zinziber»; TCNR 349/30; WARZ 117.

ZINZIBER CANINUM, see PIPER AQUAE.

ZIRUMBER. *Zingiber zerumbet* (L.) Rosc. ex Sm.; zerumbet.
583/481. Cf. AVLC 161/747. «Zerumbet» is described by Serapion, an author Albertus possibly might have known, though he does not cite him by name; Ioh. Serapionis *Practica* (Lyon: Jac Myt, 1525), fol. 162v, cap. cclxxi «De zerumbet.»

ZIZANIA. *Lolium temulentum* L.; darnel; tares.
583/479: Zizania est herba, quae crescit in frumento, et alio nomine vocatur lolium. Cf. AVLC 162/756; CAGS 361; EHCH 19 «lolium»; KVMP 426/189; RUFH 175/32 «lolium.»

ZUCARUM. *Saccharum officinarum* L.; cane sugar.
470/260; 505/323 «Canna mellis.» Cf. AVLC 162/757; EHCH 36; KVMP 426/88; RUFH 72/18 «canamellis»; *ib.* 349/3 «zucarum»; TCNR 349/32; WARZ 118.

XVI

BARTHOLOMAEUS ANGLICUS AND THIRTEENTH CENTURY BOTANICAL NOMENCLATURE

One of the most popular of the late medieval encyclopedias was the *De proprietatibus rerum* of Bartholomaeus Anglicus (fl. 1220-1240), probably written shortly before 1250. Twenty-three incunable editions were printed and it was translated into French, Dutch, Spanish, and English (1). Although his dates and even his nationality have been the object of considerable disagreement, his influence on later writers is generally acknowledged (2). Despite the attention given to him by bibliographers and literary historians, few detailed studies have been devoted to those portions of his encyclopedia dealing with plants (3). The purpose of the present paper, therefore, is to examine one aspect of Bartholomaeus' method in the belief that his use of plant names is a good example of an older and now outmoded approach to botany.

Following a long tradition, Bartholomaeus was interested primarily in words, not in the objects denoted by those words. Both by his uncritical and almost total reliance on previous writers and by his acceptance of an etymological method as proper to an understanding of plants, Bartholomaeus appears old fashioned compared to his slightly younger contemporary, Albertus Magnus. Like the other great encyclopedist of the thirteenth century, Vincentius Bellovacensis, Bartholomaeus was essentially a compiler and added little new or original material. There is no certain evidence that he examined any of the plants discussed in Book XVII, *De plantis et earum proprietatibus.* Rather, those chapters derive principally from Pliny, Dioscorides, Isidore, and the *Circa Instans* of Platearius, though another score of authorities, classical as well as medieval, are occasionally cited. His purpose throughout was to comment on and explain the properties of those objects mentioned in the Scriptures. Such a book, befitting one who was regarded by his contemporaries as a celebrated Franciscan preacher, was designed for the use of young scholars bewildered by the great number of texts available to them (4). Despite the self-imposed limitations upon the contents, Bartholomaeus did not restrict himself to the plants mentioned in the Bible. As in other books of the *De proprietatibus rerum*, secular Latin authors, and occasionally an Arabic writer, are cited as authorities.

Approximately 154 different plants are discussed in alphabetical order in the 197 chapters comprising Book XVII (5). Opinions vary as to the exact number because sometimes two or three "genera" are discussed in one chapter and the same plant is sometimes listed under two different headings (6). Moreover, plant names are introduced in chapters which deal with other plants (7). The remaining chapters deal with words denoting foodstuffs and beverages (e.g., *carica*, *uva passa*, *vinum*), agricultural matters (e.g., *propago*, *quisquilis*, *vinea*), drugs and other preparations (e.g., *mirtum*, *tipsana*, *thimiama* (8)), and organs or parts of plants (e.g., *palmes*, *stipula*, *tirsus*, *vimen*). The chapters themselves tend to follow a pattern by usually including such information as the name and synonyms (Latin and Greek), the etymology of the name heading the chapter, a brief description of the plant or some of its noteworthy characteristics, and its medical use plus a few miscellaneous remarks.

The majority of the plant names are of classical origin, though sometimes this is obscured by a strange orthography, e.g. *rampnus (rhamnus)* and *lappate* (? *lapathum*), while others may be due to the printer's unfamiliarity with Greco-Roman plant names (9). Among the plant names of non-classical origin the following may be mentioned: *aloa* (10), *esula* (11), *fumus terre* (12), *scopa*, *sethim*, and *thima* (13). To be certain, some of the plants denoted by the preceding names, were known in antiquity, but under different names. At least two vernacular names occur, *cauchetrape* (14) and *brith* (15), and one Hebrew term, *sethim* (16). In addition to the non-classical plant names, Bartholomaeus includes several plants which have not been identified with certainty in the Biblical flora, e.g., *anisum* and *nux avellana*. The inclusion of these plants appears to depend on the fact that they were among the most common of European domesticated plants and known to nearly everyone. Others, however, such as *agnus castus*, *aristologia*, and *nux* (i.e., the walnut, *Juglans regia* L.) were mentioned in one of the several Biblical *glossae* consulted by Bartholomaeus. Moreover, since some of these plants were associated with Christian symbolism, their inclusion was doubly justified (17).

Many of the plants of both the classical authors and the Bible were indigenous to areas for which little reliable information then existed (18). The occurrence of such phrases as *nascitur in Syria* or *gignit in India* were borrowed from his sources and should not be taken to mean that Bartholomaeus necessarily had any personal or empirical knowledge of the plant in question. At best, he may have known of the dried products of some exotic trees and plants used as spices, drugs, and incense and locally available in the larger cities. To this group belong *armomum*, *balsamum*, *bedellium*, *cinamomum*, *galbanus*, *gariophili*, *mirra*, *nardus*, *nux muscata*, *storax* and *thus*.

Bartholomaeus' use of synonyms derived for the most part from Pliny, though Isidore and Platearius recorded synonyms for some of their plant names. In virtue of the synonyms, plus the careful citation of his sources, frequently by book and chapter, nearly all of the plants mentioned in *De proprietatibus rerum* can be indentified. As it was Bartholomaeus' purpose to facilitate an understanding of the plants and trees mentioned in the Scriptures, his use of synonyms was understandable. Yet, as always, he was dependent upon his sources and when he departed from them it is difficult to determine the accuracy of his synonyms. In the case of lappa (*Arctium lappa* L.), he noted that it was called *philantrophos* (sic !), following Isidore's φιλάνθρωπος. Then he goes on to add *a multis vocatur virgula caballina*, a form not encountered in any of his sources which I have examined [19]. Another strange synonymy which equated *menta* (*Mentha* sp.) with *colocosia* (sic !) (*Nelumbium speciosum* Willd.) can probably be explained by a hasty reading of an unpunctuated text of Isidore in which *colocasia*, without any further information, immediately preceded *menta agrestis* [20]. Apparently, Bartholomaeus understood the two terms as synonyms, despite the fact that he continued by paraphrasing Isidore's text in which καλαμίνθη and *nepeta* were stated to be synonymous with *menta agrestis*.

Bartholomaeus' preoccupation with words is well illustrated by his adoption of an etymological method as an integral part of a plant's description. Following his favorite authority, Bartholomaeus usually copied Isidore's etymologies, though on occasion he expanded them. For example, while Isidore was content to say that *dracontea* (*Arum Dracunculus* L.)... *draconis imitetur*, Bartholomaeus wrote *draguncia... immitetur vipere et draconis* [21]. Today a knowledge of the origin of a plant name may appear to be irrelevant to the botanist, but in the Middle Ages the derivation of words was taken seriously as one means of obtaining knowledge of natural objects. It is instructive to note, that even in those cases where a knowledge of the etymology of a plant name might explain some significant factor of a plant's behavior, Bartholomaeus chose to follow his authorities without testing the information. Thus, in his chapter on *elitropium* (*Heliotropium* sp., perhaps *H. europaeum* L. or *H. villosum* Desf.), there is no evidence that he appreciated Pliny's explicit statement of why *heliotropium* received its name [22].

NOTES

(1) A. C. Klebs, *Incunabula Scientifica et Medica*. Hildesheim, 1963. Numbers 149-154.

(2) Cf. L. Delisle, « Traités divers sur les propriétés des choses », *Histoire Littéraire de la France*. Vol. 30 (Paris 1888), pp. 353-365.

(3) Cf. E. S. Burgess, "History of Pre-Clusian Botany in its Relation to Aster", *Memoirs of the Torrey Botanical Club*, 10 (1902), pp. 284-292.

(4) Lib. XIX fin. All references are to the *editio princeps*, Basel [anon. pr. *ca.* 1470] (Klebs 149.1). In the absence of signatures or foliation, I have cited this edition by the chapter headings in the original orthography and have supplied the number thereto. Occasionally I have consulted the later edition of Strassburg [Georg Husner] 1491 (Klebs 149.11) which differs in orthography but has numbered chapters.

(5) E. H. F. MEYER, *Geschichte der Botanik.* Vol. 4 (Königsberg, 1857), p. 89 states that 144 plants are discussed by Bartholomaeus but does not provide their names.

(6) E.g., *daphirus* (cap. 48) and *laurus* (cap. 89) = *Laurus nobilis* L.; *balaustia* (cap. 21) and *malogranata* (cap. 101) = *Punica Granatum* L.

(7) *Mirobalani*, the fruits of *Moringa arabica* Pers are discussed *s.v. Palma* (cap. 116); *nux muscata*, nutmeg (*Myristica fragrans* Houtt.) is discussed *s.v. Nux avellana* (cap. 109).

(8) *Mirtum est unguentum de mirra* (cap. 103); *tipsana* (cap. 169) = πτισάνη, crushed or peeled barley; *thimiama* (cap. 172) = θυμίαμα, generic term for incense of unspecified origin.

(9) *Daphirus* (cap. 48) = *daphnis* (ed. 1491); *armomum* (cap. 8) = *ammomum* (ed. 1491); *rutha* (cap. 141) = *ruta* (ed. 1491).

(10) *Aloa* or *aloa arbor* (cap. 5) is probably *Aquilaria agallocha* Roxb. and may be identical with the ἀγάλοχον of DIOSCORIDES I 22 Wellm. or with the 'hlwt of the Old Testament.

(11) *Esula* (cap. 56) is probably *Euphorbia* sp. Cf. PLATEARIUS *Circa Instans* (ed. Lyons 1537) fol. 95v : *una species [scil. titimallorum]... dicitur esula.*

(12) *Fumus terre* (cap. 69) is *Fumaria officinalis* L., known in antiquity (DIOSC. IV 109 *s.v.* καπνός; PLINY XXV 99, 156 *s.v. capnos, fumus*).

(13) *Thima* (cap. 165) = *thyna* (ed. 1491) is *Tetraclinis articulata* (Vahl) Masters though from the reference to Solomon's palace it is clear that Bartholomaeus thought it was the mysterious *almug* or *algum* of the Old Testament.

(14) *Saliunca est herba... a saliendo dicta, eo quod facit saltare se calcantes, et ideo dicitur gallice cauchetrape...* (cap. 145).

(15) *Brith* (*s.v. Scopa* cap. 159) is possibly a printer's error for *birth.* From the fact that it is spelled *birck* in the 1491 ed., it may identified with the birch, *Betula alba* L. Cf. ALBERTUS MAGNUS, *De vegetabilibus* ed. Jessen et Meyer (Berlin, 1867), p. 390 n.

(16) *Sethim* (cap. 150) apparently designates the *shittim*, of the Old Testament, *Acacia* sp., perhaps *A. seyal Delile* or *A. raddiana* Savi.

(17) As a sample of Christian symbolism, Bartholomaeus notes that in the walnut *generaliter forma crucis interius est impressa* (cap. 108 s.v. Nux). A trace of an ancient pagan superstition is preserved in Bartholomaeus' recommendation that the seeds of willow carried in the pocket will cause sterility (cap. 143 *s.v. Salix*). Later, the willow was adapted to Christian belief as a symbol of chastity. Cf. Hugo RAHNER, « Die Weide als Symbol der Keuschheit in der Antike und im Christentum », *Zeitschrift für katholische Theologie* 56 (1933), pp. 231-253.

(18) Cf. Jerry STANNARD, "Discorides and Renaissance Materia Medica", *Analecta Medico-Historica* 1 (1966), 1-21.

(19) *Lappa* (cap. 95); for φιλάνθρωπος (= *philantropos* ed. 1491), cf. ISIDORE, *Etymologiae*, XVII 9, 66-67.

(20) *Menta* (cap. 106). Cf. ISIDORE XVII 9, 82.

(21) *Draguncia* (cap. 50). Cf. ISIDORE XVII 9, 35.

(22) *Elitropium* (cap. 54). Cf. PLINY, *Historia Naturalis* XVIII 67, 252; XXII 29, 57.

XVII

Vegetable Gums and Resins in Medieval Recipe Literature

There are many sources for the nomenclature, origin and uses of vegetable gums and resins in the Middle Ages. Among the more obvious are lexica, glossaries and encyclopedias, herbals and leechbooks, plus various documents pertaining more directly to pharmacy. Together these texts provide a wide range of information, not all of which is accurate, however, concerning the botanical origin and geographical provenance of the gums as well as their Greek, Latin and vernacular synonyms and miscellaneous information regarding such practical pharmaceutical matters as price, detection of adulteration, directions for storage, admissible substitutions, etc.

Another source, but one which has not been as closely examined as the aforementioned, is the large but amorphous body of recipes which is widely scattered throughout different literary genres. Among the innumerable medieval recipes, those found in the plague tracts are especially noteworthy because of the frequent references to gums and resins of exotic origin. These substances were commonly employed as ingredients in different classes of plague remedies. In the form of fumigants or in the preparation of pomanders, a wide range of gums was used. Externally, in the form of unguents they were applied to the buboes. But the most popular and widespread use of gums occured in connection with the preparation of theriaca and mithridatium. The latter two composita contained, in addition to many herbs and other simplicia, approximately ten gummous substances each, though there are, naturally, many variations.[1]

But the plague tracts are not our only sources for recipes. They occur throughout a wide range of *Fachliteratur,* for example in herbals and leechbooks, surgical texts, regimina and consilia, while naturally antidotaria and receptaria are veritable storehouses of relevant information.[2]

With such sources at our disposal, let us begin by considering some of the more frequently mentioned gums, resins and related substances. Of the approximately 30 gummous substances named in medieval recipe literature, the most common are galbanum, gummi arabicum, mastix, myrrh, olibanum, terbentina and tragacantha. There exist, of course, references to many other gums. Most of them were of exotic origin, for example, asa foetida, bdellium, opobalsamum, opopanax, sagapenum, sanguis draconis, sarcocolla, storax etc. There were, moreover, a few gummous substances derived from indigenous species, for example, prunus (including cerasus), hedera, and pix or resina pini. Occasionally one also finds references to such unusual substances as gumma senapii, gummi martagon, and gumma salicis.

During the Middle Ages there was little possibility for Europeans to travel in the gum producing regions of the Near East and East Africa. Despite a few partial descriptions of gummiferous species in travel accounts[3], there was almost no reliable information concerning the botanical origin of most of the exotic gums. Legends, myths, and exaggerations were not easily distinguished from the truth[4]. It was, in fact, not until the nineteenth century that the taxonomy of the gummiferous species was put on a scientific basis, and even in the present century, new data have caused some revisions[5].

As a consequence of the ignorance surrounding the origin of exotic gums, medieval writers often copied or paraphrased the accounts of gums found in Greco-Roman authors, particularly DIOSCORIDES, PLINY, and GALEN[6]. Not in all cases, however, can the precise classical source be identified. In some instances the claims made for the gums seem to be medieval fabrications[7].

Although little was known of their origin, dried gums were not unknown in Western Europe and some of the exotica, for example, mastix, myrrh, and olibanum appear to have been reasonably common[8]. It is, however, difficult to establish with certainty how common were the various gums. Despite our general knowledge of the trade routes and major import centers, little is known of local distribution systems, beyond the accounts of the great trade fairs[9]. Moreover, the frequent inclusion of, say, opobalsamum in medieval recipes is no guarantee that it was always available locally or that the writer (or compiler) of the recipe had any first-hand knowledge of it such that he could distinguish it from other gummous masses. It is always possible that the writer was following an older recipe collection which he copied, or modified, without regard to the availability, cost or nature of the ingredients[10].

There is, however, some evidence concerning the availability of gums and resins in Western Europe. This derives from pharmaceutical, medical and economic documents which supplement our understanding of medieval recipes. For example, it is explicitly stated, by several writers, that ammoniacum, mastix, myrrh, tragacantha and other gums can be found or purchased *in apoteca*[11]. Colophonia, moreover, is reported to be available, not only from the apothecary but from the shoemaker as well[12].

Further evidence concerning the availability of gums is found in apothecary inventories and price lists. Thus, the inventory of the contents of the shop of GUILLAUME LEFORT, apothecary of Dijon (ob. 1439) contains about 15 gums, resins and related substances plus theriaca, mithridatium, and numerous unguents, some of which, like ung. marciaton or ung. agrippa, contained many gums[13]. The items in LEFORT'S inventory are also assessed, as are those mentioned in the inventory of JOHN HEXHAM, a London apothecary who was hanged in 1415 for coining false money[14].

Other kinds of economic documents provide information concerning the importation of exotica. Thus, FRANCESCO PEGOLOTTI notes that the Florentine merchants in the mid-fourteenth century handled, as part of their commerce, such substances as asa foetida, gum arabic, frankincense, 2 varieties of tragacanth, 3 varieties of mastix etc.[15] PETER KARBOW, a Hansa merchant in Venice, wrote to a friend in Bruges in 1409 describing among other goods, including two varieties of incense[16].

Finally, the various ordinances provide a somewhat idealized picture of what substances should be stocked by all apothecaries within the political boundaries for which they were official. These ordinances include a variety of both exotic and indigenous gums. Thus the Frankfurter Liste (ca. 1450) mentions 22 gums while the slightly later Nördlinger Register (ca. 1480) contains an even larger number[17]. It is interesting to note that in the latter text, bdellium, galbanum, and sagapenum are listed under the heading of laxatives rather than gums, thus indicating an evolution in pharmaceutical practice foreshadowing the early pharmacopoeias. The latter, of course, also list many gums and resins among the simplicia[18].

Because of high costs and an uncertain supply of gums locally, there was a great temptation for falsification and adulteration. While explicit references to adulterated gums are not common[19], many passages indirectly attest to such practices[20]. For example, references to clean, pure or selected gums imply that of the several commercial grades available, often distinguished by color alone, only those of the highest quality were to be used[21]. Another form of evidence concerning fraudulent preparations is the fact that several tests are mentioned for the detection of suspicious gummous substances[22].

Just as the pleasant and pungent aroma of the gums determined many of their medicinal uses[23], so their physical properties determined their pharmaceutical and economic uses[24]. Accordingly, gums occupy a central position in such *preparata* as unguents, plasters, cerates, and powders designed for topical application. Recipes for collyria, particularly, contained many gums[25]. Among the more common external uses of gums, either singly or in combination with other simples, were wounds, local infections, fractures, contusions and assorted blemishes, incrustations, etc.

But gums were also employed internally, for example as ingredients in electuaries and pills and in acetous, aqueous, cerevisial and vinous potions. In the form of inhalations, incense and other aromata were highly regarded as plague remedies[26]. Moreover, a wide assortment of gums and resins were employed in embalming. Thus, for the burial of PHILIP THE VI in 1350, 11 gums, bitumen, wax, and scented aromata were required[27].

Closely related to the recipes of a medicinal nature are those designed for cosmetic and alimentary purposes. Many of them, containing different gums and resins, occur in late medieval receptaria and books of secrets. A good example is the *Ornatus mulierum,* a thirteenth-century anglo-norman text[28]. Among the beauty preparations containing gums are facial cremes, depilatories, hair restorers, and dyes. Even medical texts occasionally contain such recipes, for example, the use of mastix as a breath sweetner[29] or tragacantha as a depilatory[30]. Gums are not common in alimentary recipes but their use as thickening agents or sweetners was perhaps more common than the published texts indicate[31].

There is also a group of miscellaneous recipes dealing with the preparation of various substances useful in household or domestic economy in which gums play an essential role. Thus, the gum from the plum tree occurs in a recipe for making a black dye[32]. In a similar fashion, the gum of the cherry tree occurs in a recipe for the preparation of a red dye[33]. Gummi arabicum, likewise, is used in the preparation of ink[34], by illuminators[35] and in the preparation of a blue color[36]. Other miscellaneous uses include the use of the resinous exudation of aloes for protecting clothes against *vermes*[37], presumably moths, and laudanum the exudate of several species of *Cistus* as an ingredient in soap, a use which, incidentally still continues, for laudanum is one of the best fixatives derived from the vegetable kingdom[38]. An unusual use of gums occurs in the so-called *mittelalterliches Hausbuch, circa* 1480. In a recipe for making "glass" pearls, both gummi arabicum and mastix are employed[39]. In order to make beads, *pro pater noster,* a paste is prepared containing gummi arabicum, tragacantha, and *cerasa,* presumably the gum of the cherry tree[40].

In conclusion, it may be said that, as a consequence of the common ritual or ceremonial uses of incense, some gums were occasionally credited with paramedical virtues[41] which physicians and others were not slow to exploit[42].

References

1) The basis for the preparation of *theriaca Andromachi* and *mithridatium Damocratis* is Galen, *De antidotis* (vol. XIV, 32—45, 115—119 ed. Kühn). As an example of variation, cf. WILLY BRAEKMAN, Medische en Technische Middelnederlandse Recepten. Gent, 1975 p. 187 nr. 576.
2) JERRY STANNARD, Botanical Data and Late Medieval *Rezeptliteratur* (in press).
3) As an example, cf. R. RÖHRICHT und K. MEISNER, Ein niederrheinischer Bericht über den Orient. Ztschr. f. dte. Philologie 19 (1886), 81—86.
4) As an example of the kind of legends which circulated widely, cf. *Hortus sanitatis germanice* Mainz: PETER SCHÖFFER 1485 cap. 119 (Camphora) and 170 (Euforbium — eyn gummi also genant).
5) Cf. ALEXANDER TSCHIRCH, Was ist Sarcocolla? Schweiz. Apoth. Ztg. 58 (1920), 113—116.
6) Cf. JERRY STANNARD, Greco-Roman Materia Medica in Medieval Germany. Bull. Hist. Med. *46* (1972), 455—468.
7) For example, the illustrations of various gummiferous species in the *Hortus sanitatis.* In a similar fashion, cf. the recipe "Wiltu machen sangwinen draconum" which incorporates, *inter alia,* the blood from the right arm of a man aged 24 years, G. JENNY, Alte Recepte und Hausmittel. Alemannia *19* (1891), p. 35.
8) Both *thus* and *laudanum* appear in a pauper's recipe for a pomander, KARL SUDHOFF, Pestschriften aus den ersten 150 Jahren nach der Epidemie des "schwarzen Todes" 1348. Sudhoffs Archiv *5* (1911), p. 386 lines 44—46.
9) Cf. FÉLIX BOURQUELOT, Etudes sur les Foires de Champagne. Mém. Acad. des Inscrip. et Belles Lettres (ii) t. 5 pte. 1 (1865), 1—335.
10) This may explain, for example, the otherwise strange phrase in an anglosaxon recipe for haemorrhoids (?): nim murran [i. e. myrrh] tha wyrt. *Lacnunga* cp. XIX edd. J. H. G. GRATTAN AND CHARLES SINGER, *Anglo-Saxon Magic and Medicine.* London 1952 p. 102. A similar phrase occurs in BRAEKMAN, *op. cit.* p. 249 nr. 847: Neem woertelen van ... mirren ...
11) For ammoniacum and mastix, see AGI LINDGREN, *Ein Stockholmer Mittelniederdeutsches Arzneibuch.* Stockholm, 1967 p. 116 nr. 136: armoniacum vnde den mastix in der apteken koepet ... Tragacantha, as the principal ingredient in an electuary (dyatragagantum) was commonly available, cf. JEAN DE GARLANDE, *Dictionarius* 44 (ed. AUG. SCHELER, Jahrb. f. roman. u. engl. Literatur *6* (1865), p. 152: Appotecarii haec habent in appoteca ... diadragagantum; cf. HEINR. WERNECK, Kräuterbuch des Joh. Hartlieb. Ostbairische Grenzmarken *2* (1958), p. 79 nr. 5. The gum itself (τραγάκανθα, cf. Diosc. III, 20 Wellm.) was also available, cf. SVEN NORRBOM, *Das Gothaer mittelniederdeutsche Arzneibuch und seine Sippe.* Hamburg 1921 p. 191 nr. XLVI: dragantum, dat vindestu in der apotheken. Many gums are mentioned as available in the apothecary shop by GERSDORFF, *Feldtbuch der Wundtartzney* Strassburg 1517 cf. JERRY STANNARD, Hans v. Gersdorff and some Anonymous Strassburg Apothecaries. *Pharmacy in History 13* (1971), 55 sq.
12) Tak ... gume of the sour plumtre [and] white pik þat þir spicers calles pik album ... MARGARET OGDEN, ed., *Liber de diversis medicinis.* London 1938 p. 54; LINDGREN, *op. cit.* p. 140 nr. 240: recipe j punt pekes also de

schomaker hebben ... Perhaps the same substance is denoted by CENNINO D'ANDREA CENNINI, *Il Libro dell' Arte*, ed. D. V. THOMPSON, New Haven 1932, I, 35: togli dagli speziali sei oncie di ragia di pino.

13) PAUL DORVEAUX, *Inventaires d'anciennes pharmacies dijonnaises.* Dijon 1892.

14) G. E. TREASE and J. H. HODSON, The Inventory of John Hexham, A Fifteenth-Century Apothecary. Medical History 9 (1965), 76—81.

15) FRANCESCO BALDUCCI PEGOLOTTI, *La pratica della mercatura.* ed. ALLAN EVANS. Cambridge, Mass., 1936, pp. 293 sq.

16) WILH. STIEDA, *Hansisch-Venetianische Handelsbeziehungen im 15. Jahrhundert.* Rostock, 1894, pp. 121—124. For contemporary prices for some of the gums, cf. F. A. FLÜCKIGER, Dokumente zur Geschichte der Pharmacie. Arch. d. Pharm. *207* (1875), 422—437 et passim.

17) F. A. FLÜCKIGER, Die Frankfurter Liste. Arch. d. Pharm. *201* (1872), 433—464, 508—526; ibid., Das Nördlinger Register. Arch. d. Pharm. *211* (1877), 97—115. A series of such documents, entitled, "Die Preisverzeichnisse des Grazer Codex 311" is in the process of being edited by WOLFGANG-HAGEN HEIN in *Pharmazeutische Zeitung* 118 (1973), 1146—1148; 1510—1512; 119 (1974), 500—502.

18) Cf. the list of gums and gummous substances, (32 in all), entitled "Delle gomme vsvali" in *Nuovo Receptario* (Firenze 1498; facs. repr. Firenze 1968) fol. $C^{iv\ recto}$. By contrast, only 6 gums are listed in the "Index Simplicium" of VALERIUS CORDUS, *Dispensatorium* (Nürnberg 1546; facs. repr. Mittenwald 1934) sig. $a5^{v}$-6^{r}. In the text proper, however, most of usual gums are included.

19) The *Hortus sanitatis germanice* notes that asa fetida (cap. 41), bdellium (cap. 77) and galbanum (cap. 202) are subject to adulteration. Storax is said to have been adulterated with shavings of iris and honey, ERWIN MÜLLER, *Der Traktat Liber iste.* Würzburg 1942 p. 6. ALBERTUS MAGNUS mentions some of the adulterants of balsamum, *De vegetabilibus*, edd. E. MEYER et C. JESSEN. Berlin 1867 p. 356. The actual recipe for the alteration of galbanum is included in a mid-fourteenth century antidotarium, CLOVIS BRUNEL, Recettes pharmaceutiques d'Avignon en ancien provençal. Romania 87 (1966), p. 520 nr. 47.

20) For example, the stipulation "nemt vrayen wieroeck" in a recipe for an all-purpose ointment, WILLY BRAEKMAN, *Middelnederlandse Geneeskundige Recepten.* Gent 1970. p. 152 nr. 169.

21) In surgical dressings and wound plasters, *lauter hartz* is often stipulated by PETER OF ULM, *Die Cirurgia* ed. GUNDOLF KEIL. Diss. Heidelberg 1960 Recipes nr. 5, 24, 91 etc. Likewise, terbentina is described as *lavada, monda* and *mondissima,* CARLO CASTELLANI, *Secreti Medicinali di Magistro Guasparino da Vienexia.* Milano 1959, Recipes nr. 47, 55, 46. For myrrh, cf. such phrases as *mirra pura* (E. BONDURAND, Fragment des recettes médicales en langue d'oc. Romania 12 (1883), p. 102 nr. 3), and *mirra electa,* GERHARD EIS, Altdeutsche Rezepte von spätmittelalterlichen Verfassern aus Handschriften und Frühdrucken. Med. Mschr. 11 (1957), p. 253 nr. 11c.

22) A simple organoleptic test for the determination of *balsamum bonum* is described by SALADINI DE ASCULO, *Compendium aromatariorum,* ed. LEO ZIMMERMANN. Leipzig 1929 p. 92. A different means of detecting adulteration of *balsamus* occurs in Rufinus, *Herbal.* ed. LYNN THORNDIKE. Chicago 1949 p. 51.

23) The rationale for using gums and other aromata in a plague-drink is concisely expressed in a plague-tract attributed to GENTILE: Et quia cor maxime per

odorativa confortatur, odoribus huiusmodi potioni utantur ... SUDHOFF, "Pestschriften ..." Sudhoffs Archiv 5 (1911) p. 339 lines 96—98. ALBERTUS MAGNUS, ed. cit. p. 404 explains their use in terms of the four elements.

24) Cf. the careful directions for the preparation of illuminators' materials in DANIEL V. THOMPSON, The *De Clarea* of the so-called "Anonymous Bernensis". Techn. Stud. Fine Arts 1 (1932), 8—19, 69—81.

25) For the use of gums in collyria, cf. PIERO GIACOSA, Un ricettario del secolo XI esistente nell' Archivio Capitolare d'Ivrea. Atti d. R. Accad. Torino (ii) 37 (1886), 643—663.

26) Cf. JERRY STANNARD, Joachim Camerarius' Contributions to Pharmacy and Medicine: Observations on his *De theriacis et mithridateis commentariolus* (in press).

27) MAURICE JUSSELIN, L'embaumement d'un roi de France au XIVe siècle. France Médicale 54 (1907), 297—298.

28) PIERRE RUELLE, *L'Ornement des Dames (Ornatus mulierum). Texte anglonormand du XIIIe siècle.* Bruxelles 1967.

29) C. KÜLZ und E. KÜLZ-TROSSE, *Das Breslauer Arzneibuch.* Dresden 1908 p. 42. Cf. the recommendation to chew mastyke for "euyll in pe mouthe", MARGARET S. OGDEN, *Liber de diversis medicinis.* London 1938 p. 2.

30) JULIUS JÖRIMANN, *Frühmittelalterliche Rezeptarien.* Zürich 1925 p. 11 sect. vi.

31) Mastix is used with anise in preparing chrysomelones (apricots, *Prunus Armeniaca L.*) for the table, BAPTISTA MASSA, *De Fructibus vescendis.* Venice [pr. of AUSONIUS, ca. 1472 fide Klebs 672. 1] p. 17v; campfora is recorded by LUIS FARAUDO DE SAINT GERMAIN, El Libre de Sent Sovi. Recetario de cocina catalana medieval. Bol. d. R. Acad. Buenas Letras, Barcelona 24 (1951/52) p. 65. In general, cf. JERRY STANNARD, The botanico-medical Background of Baptista Fiera's *Coena de herbarum virtutibus* (in) *Civilità dell' Umanesimo.* Atti del VI, VII, VIII Conregno del Centro di Studi Umanistici. Firenze, 1972 pp. 327—344.

32) HANS VERMEER, Technisch-naturwissenschaftliche Rezepte aus einer Harburger Handschrift. Sudhoffs Archiv 45 (1961) p. 119 nr. 31.

33) VERMEER, op. cit. p. 120 nr. 37.

34) RUFINUS, op. cit. p. 47 s. v. *attramentum.* Two recipes are found in WILHELM WACKERNAGEL, Kochbuch von Maister Hannsen des von Wirtenberg koch. Ztschr. f. deutsches Altertum 9 (1853) p. 369 nr. vi; 370 nr. IX.

35) RUFINUS *op. cit.* p. 150: gummi ... ponitur in incausto, quia facit ipsum tenacem.

36) ADALBERT JEITTELES, Färbemittel und andere Recepte. Germania 29 (1884) p. 338; HANS WISWE, Mittelalterliche Rezepte zur Färberei sowie zur Herstellung von Farben und Fleckenwasser. Niederdeutsches Jahrbuch 81 (1958), p. 58 nr. 18.

37) ALBERTUS MAGNUS, *ed. cit.* p. 345.

38) HELMUTH BOSSERT und WILLY STORCK, edd. *Das mittelalterliche Hausbuch.* Leipzig 1912 fol. XXI line 3.

39) BOSSERT u. STORCK, *op. cit.* fol. XXI lines 18—25.

40) BOSSERT u. STORCK, *op. cit.* fol. XX lines 13—14.

41) Cf. ANTON BIRLINGER, Bairische Besegnungen. Germania 24 (1879), p. 74 nr. 4 for weiss weyroch in a charm and the scriptural associations implied by the names "Caspar, Balthasar, Melchior." In a similar manner, *pilulae pestilenciales,* which normally contained gums, were enhanced by their association with Maria Magdalena, SUDHOFF, "Pestschriften" Sudhoffs Archiv 9 (1916), 131 lines 541—545.

42) Cf. GERHARD EIS, Fasciculus myrrhae. Leuvense Bijdragen 49 (1960), 90—96.

Acknowledgement: This publication was supported in part by N. I. H. GRANT LM 02054 from the National Library of Medicine. I am especially grateful to Mrs. MARILYN CLARK and Mrs. MARIANNE GRIFFIN, Reference Department, University of Kansas Libraries, for their efficiency and cooperation in obtaining some of the material necessary for this study.

XVIII

A Fifteenth-Century Botanical Glossary

(Huntington Library MS HM 64) *

Huntington Library MS HM 64 is a collection in 197 folios of some seventeen tracts of varying size and content. It was written in England apparently about 1475. This provisional date is based on the relative chronology of a series of related glossaries, several of which will be discussed below. That this date applies to the codex as a whole may be inferred from the fact that it is written throughout in the same hand and that the pagination in a contemporary, if not the same, hand is continuous.

What little is known of the later history of the codex will be found in W. J. Wilson's study of alchemical manuscripts.[1] It may be, owing to the heterogeneous nature of the material in the codex, that it has not been examined as carefully as other English writings of a similar nature and date. The contents were described by Wilson, though he did not examine it personally. His interest in the alchemical portions precluded a detailed analysis of all of the tracts. Consequently, his descriptions of the non-alchemical portions are brief and, in at least one place, of questionable accuracy. In addition to the usual extracts and anonymous pieces, there is a plague tract, several astronomical and alchemical tracts, and three glossaries. These latter occupy folios 125ʳ–135ʳ, 176ᵛ–183ᵛ, and 184ʳ–190ᵛ. It is the last of these three glossaries, described by Wilson as "A dictionary of herbs in Latin," which I shall discuss here.

These fourteen leaves contain a glossary of plants, minerals, and animal products used as drugs plus a few miscellaneous items denoting pharmaceutical equipment and other matters of interest to the physician and apothe-

* This paper was delivered at the annual meeting of the History of Science Society, Bloomington, Indiana, 6 April 1963.

† Rutgers – The State University. Research on the Huntington manuscript was made possible by a summer research position, School of Medicine, University of California at Los Angeles. To Professor C. D. O'Malley, who made this possible, and to Dr. Herbert C. Schulz, Curator of Manuscripts, Henry E. Huntington Library, I am grateful for many suggestions. The present paper was completed through the assistance of USPHS RG 9361.

[1] W. J. Wilson, "Catalogue of Latin and Vernacular Alchemical Manuscripts in the United States and Canada," *Osiris*, 1939, *6*: 1–836. The present manuscript is described on pp. 408–418.

Reprinted from *Isis: Journal of the History of Science Society*, 1964, Volume 55, No 3, by kind permission of the University of Chicago Press.

cary. The form of the glossary is basically quite simple. The entry, in Latin with an initial capital, stands at the left margin of either of the two columns constituting a leaf. It is followed by a short description whose aim sometimes is that of identifying the item in question by supplying one or more synonyms. In other cases, it appears that the purpose is to describe the medicinal value of the drug. The description contains one or more of the following pieces of information: the physiological action of the drug, expressed in terms either of its properties or " faculties " or in terms of its temperament or " grade "; some physical characteristics of the drug; and occasionally a remark on where the drug was obtained. The description, when present, may be followed by one or more synonyms, also in Latin. Their identity with the item in question is usually indicated by a simple *idem*, but it may also be graphically represented by lines running from each of the synonyms, all of which converge upon the entry. The Latin synonyms may be followed by one, two, or rarely three vernacular synonyms or these may follow directly the entry without any intervening Latin synonyms. The vernacular synonyms are in Anglo-Norman and Middle English and constitute the only non-Latin words in the glossary. Many of the Latin terms are transliterations, often in corrupt form, of names known from Greek medical texts; one Arabic term is likewise almost disfigured beyond recognition.[2] The entire gloss may run from a single synonym to some thirty words, but the average length is about ten words.

This simple plan of the typical gloss is susceptible of many modifications. The flexibility of this arrangement is increased by the absence of a strict alphabetical order within the main divisions of the alphabet. The letters *G*, *H*, *Q*, and *T* through *Z* are lacking; *K* is included under *C* (though one vernacular synonym has a clearly written lower case *K*) and *I-J* are run together, though initial *I* is orthographically distinguishable from *J*. Among the many puzzles created by the glossary — for example, duplication of entries, absence of some of the terms always found in similar glossaries, missing vernacular synonyms — none is more challenging than the cross-references, in both the margins and in the text, to names which do not appear. Some of these puzzles may be partially solved by a comparison with other glossaries whose relation to the present one will be discussed below.

Of the 309 entries, the majority, some 278, refer to plants, while twelve denote minerals, and the remainder form a miscellaneous group. Because of the preponderance of plant names, my remarks will be confined to the botanical content of the glossary.

Before we turn to a more detailed analysis of the plant names and an identification of some of the plants, the sources of the glossary must be considered. This is necessary for two reasons. Many of the plant names are corrupt forms of names known from classical Greek and Latin authors. This, plus the fact that several of the glosses are truncated and almost unintelligible

[2] The history of several of these names — *bissara* (184 va), *alabice* (186 ra) from Arab. *basal*, and *piganum* (189 va) — has been examined by Jerry Stannard, " The Plant Called Moly," *Osiris*, 1962, *14*: 254–307.

as they stand, requires that if one is to understand them, their sources must first be identified in order to use them to reconstitute a faulty text and to restore the lacunae. But there is another reason quite apart from these paleographical considerations, and that concerns the plants themselves. For, as I shall mention below, one of the sources for our glossary was written with reference to a Mediterranean flora. Accordingly, when one encounters a strange or unfamiliar Latinized plant name, glossed with an equally unknown Anglo-Norman or Middle English synonym, one must first identify both the original name, probably Greek, *and* the Mediterranean plant which it denoted. This must precede the identification of all plants appearing in a botanical glossary which refers to plants not indigenous to the locality in which it was written. Historians of botany usually recognize that in glossaries such as ours there is sometimes an attempt to adjust the Latin names to a local flora by assigning vernacular synonyms.[3] An example of this sort of adjustment will be mentioned below. It goes without saying that frequently the Latin name and the vernacular name are not synonyms at all, but represent two different plants.

The identification of the plants named or described in the glossary is made more difficult by the lack of agreement in medieval botany on what we would call generic names and specific epithets. This directly concerns the present glossary for many of the entries represent generic terms which in turn include two, sometimes three, specific names. The latter are usually distinguished by affixing to the generic name some qualifying adjective such as *major* and *minor* or *alba* and *nigra*. These distinctions do not, in most cases, accord with modern taxonomy. But there is just enough of a correlation that groupings under generic terms cannot be ignored completely.

There is, finally, and this will bring us to the subject of sources, a third difficulty which applies particularly in the case of identifying plants recorded in early English writings. For English botany, like English culture in general, is a fusion of two originally quite distinct traditions — the Germanic or Anglo-Saxon and the later Norman or Latin traditions. Both left their mark on the formation of English plant names as well as in their medicinal uses.

There are many reasons for believing that there was a native Anglo-Saxon botanical tradition of an early date. Its extent is impossible to determine, however, for lack of satisfactory literary material. What has survived is fragmentary or, in the case of Anglo-Saxon translations of Dioscorides, Pseudo-Apuleius, and other Graeco-Roman medical texts, contaminated with continental materials.[4] This is especially true as regards plant names, some Anglo-Saxon names being really nothing but transliterations of Latin names

[3] Cf. E. Landgraf, " Ein frühmittelalterlicher Botanicus," *Kyklos*, 1928, *1*: 114–146. Landgraf points out that the anonymous compiler of MS St. Gall 217 (saec. IX) has omitted those chapters in Pseudo-Apuleius describing plants which were unknown in Central Europe and has supplied an equal number of medicinal plants which occurred locally.

[4] Cf. J. F. Payne, *English Medicine in the Anglo-Saxon Times* (Oxford: Clarendon Press, 1904).

which, as often as not, were corruptions of Greek names.[5] Nonetheless, in order even to speak of Anglo-Saxon botany, we must resort to the available material such as it is. Our only concern with this material at present is to note that some of the plants described in our glossary are very likely identical to plants mentioned in Anglo-Saxon leech books and known to have been used in English folk medicine right up to the fifteenth century. There is an abundance of medical recipes written in the margins and endpapers of books of all kinds. These, and the equally numerous commonplace books, contain material of immense value when seeking to trace the early English use of plants for specific medicinal purposes. Since this material usually is independent of the formal glossaries, and often shows little evidence of continental nomenclature, valuable light is shed on vernacular synonyms, especially so when they are correlated with anything amounting to a description of the plant.

Although there are no Anglo-Saxon words in our glossary, some of the Middle English words are variants of older plant names or cognate to them etymologically. The significance of a native English botanical tradition will become apparent when we deal with some glosses which cannot be traced to the two primary sources.

The first of the two identifiable sources for our glossary is the well-known *Circa instans* written, probably at Salerno about 1150, by Matthaeus Platearius (d. 1161). This antidotarium, based on the Latin Dioscorides, treats in 273 chapters 229 medicinal plants, the remainder being drugs of mineral and animal origin.[6] It is not surprising, therefore, that over three-fourths of the plants mentioned in our glossary are also found in this work. This, in itself, does not prove that it was a direct source. But when one finds entire glosses agreeing *verbatim* with the corresponding passages in the *Circa instans,* the likelihood of coincidence or literary convention is considerably decreased.

On the basis of these parallels, I think it safe to say that the *Circa instans* was one of the sources for our glossary. This does not mean, however, that the compiler of our glossary had a copy of it in front of him. Although manuscript copies of it are known in England at this date and an incunable edition of 1497 is recorded, it is rather more probable that the compiler was following an epitome or another glossary which contained extracts.[7] My

5 Examples are provided by the series μίνθη, Lat. *menta,* AS *mintan* or πέπερι, Lat. *piper,* AS *pipor.* Both AS names occur in the "Lacnunga," J. H. G. Grattan and Charles Singer (eds.), *Anglo-Saxon Magic and Medicine* (London: Oxford University Press, 1952), p. 116. In such a series as πεντάφυλλον (*lit.* five-leaf), Lat. *pentafilon,* AS *fifleafan,* an attempt has been made to translate, rather than to transliterate, the original. For other examples, cf. Johannes Hoops, *Über die altenglischen Pflanzennamen,* Dissertation, Albert-Ludwigs-Universität, Freiburg im Breisgau, 1889.

6 Præpositus Nicolaus, *Dispensarium . . . ad aromatarios nuper diligentissime recognitum. Item cum pluribus additionibus. . . . Platearius vulgo Circa instans nūcupatus de simplici medicina . . .* (Lyons: J. Crespin for S. de Gabiano, 1537), folio. Our text occupies folios 70v–96r.

7 Ludwig Hain, *Repertorium bibliographicum . . .* (2 vols., bound as 4) (Stuttgart: Cotta, 1826–1838), No. 14695 = A. C. Klebs ("Incunabula Scientifica et Medica," *Osiris,* 1938, *4*: 1–359), No. 911.2. Several fifteenth-century manuscripts are in the Wellcome Historical Medical Library, one of which (No. 626) is of French origin and dated *c.* 1480–1500, cf. S. A. J. Moorat (ed.), *Catalogue of Western Manu-*

reasons for this rest on the fact that in several cases a gloss in the Huntington glossary contains but one of the several sentences constituting the corresponding passage in the *Circa instans*. This, plus the equally revealing fact that some passages in our glossary are incomplete sentences, the whole of which is again found in the *Circa instans*, makes direct copying improbable.

Up to the time of the so-called scholar-naturalists, botanists like Otto Brunfels, Leonhard Fuchs, and Hieronymus Bock, there were no botanical writings exempt from Salernitan influence. The Salernitan tradition was quickly and widely diffused; for example, the herbal of Henrik Harpestreng, Bishop of Roskilde in Denmark (d. 1244) shows direct borrowing from Salernitan medical botany.[8] Another channel by means of which Salernitan botany was diffused concerns the fountainhead of all medical glossaries, the anonymous *Alphita*. According to Salvatore de Renzi, this glossary containing over 1,200 entries was written in the late fourteenth century.[9] We know something of its influence abroad, for it was adapted to local, for example, English, use about the middle of the fifteenth century. Actually, there were several English adaptations, but only two concern us here. These glossaries, while abstracting from the *Alphita*, also added both vernacular synonyms to the main entries and introduced some new entries, usually in Latin but occasionally in Anglo-Norman. The longer of the two English glossaries, also called *Alphita*, is based on two manuscripts, B. M. Sloane 284 and Bodleian Library, Selden B. 35.[10] (These will be referred to henceforth simply as the Sloane and Selden glossaries.) The Selden glossary has been dated about 1465 and the Sloane is probably slightly older. As in the case of our glossary, both of these are copies of a now lost or unknown prototype. This has an important bearing on our glossary for we can sometimes explain an unfamiliar word or a suspect passage by referring to these slightly earlier glossaries. Nothing can be inferred about sources when words are correctly spelled, but the likelihood that the same bizarre spelling found in some glosses would occur independently to two scribes is difficult to entertain. There are, of course, more substantial reasons for believing that these two glossaries, and another even earlier English glossary, the *Synonyma Bartholomei*, form the second of the sources for our glossary.[11] My evidence here is twofold. The first point, alluded to above, is that some strange spellings, otherwise unknown, are found in both the English glossaries and in ours. This, coupled with what in some cases is an identical gloss, points to a second source. The second reason concerns the vernacular synonyms. When these are added to an otherwise identical or nearly identical gloss, it is doubly difficult to resist the conclusion that the Huntington glossary is an independent copy of the ancestor common to the Selden and

scripts on Medicine and Science in the Wellcome Historical Medical Library, Vol. I (London: Wellcome Historical Medical Library, 1962), pp. 482–483.

8 Henrik Harpestraeng, *Liber herbarum*, ed. Poul Hauberg (Copenhagen: Carl Kretzschmer, 1936).

9 Salvatore de Renzi, *Collectio Salernitana* (5 vols.) (Naples: Sebezio, 1852–1859), Vol. III (1854), pp. 271–322.

10 J. L. G. Mowat, *Anecdota Oxoniensia*, Vol. I, Pt. ii. *Alphita* . . . (Oxford: Clarendon Press, 1887).

11 For the *Synonyma Bartholomei*, cf. J. L. G. Mowat, *Anecdota Oxoniensia*, Vol. I, Pt. i (Oxford: Clarendon Press, 1882).

Sloane glossaries, which has, up to this time, not been identified. That our glossary was not copied directly from either of the English glossaries is conclusively proven by observing that some glosses, with vernacular synonyms, are not to be found in either the Selden or Sloane glossaries. The fact also that the Selden glossary breaks off in the middle of the letter *S*, as does our glossary, is further, but only presumptive, evidence of a common ancestor.

Since there are some glosses with vernacular synonyms which cannot be traced to either of the aforementioned glossaries, it is possible that a third source was utilized. It is only conjecture, of course, but this postulated source may not have been a glossary at all. The presence of several observations concerning plants — their habit of growth and their habitat — and several references to the apothecary's name for a plant, especially when it differs from the standard glossarial name, suggest that this conjectural third source was a herbal. Later, I hope to be more positive on this matter, but this must await an examination of over a score of English manuscripts which I began this year.

Turning now from the form of the glossary to its contents, some of the plants mentioned in the glossary will be examined and their identification attempted. Glossaries, especially when they are representatives of a long tradition as is ours, seldom contain anything very original. And, save for some hitherto unrecorded vernacular names, there is little here that cannot be traced to one of the aforementioned sources. But the manner in which the material has been compiled and arranged, plus some stray bits here and there, gives the glossary, an otherwise lifeless list of names, a certain individuality of its own. Thus, we find that *castoreum*, the dried preputial follicles of the European beaver, is glossed as *agnus castus*, the name of a well-known tree (*Vitex agnus castus* L.). The common name of this tree, chaste tree, derives from its reputed power of making women continent. *Agnus*, the ordinary Latin term for lamb, was sometimes thought to be derived from ἁγνός, "pure." I cannot explain why *agnus castus* appears as a synonym for *castoreum* nor do I know of any precedent for this use.[12] That the animal drug was intended is clearly indicated by the remainer of the gloss which begins, *testiculus cuiusdam animalis* (185 rb). The juxtaposition of two items which have such widely differing associations for us, and for some early reader of the manuscript who made an asterisk in the margin opposite this entry, forms a pleasant interlude in a list of names.

There is no reason to suppose that the compiler had any special interest in plants beyond his task of recording what others had said, and from the preceding illustration his knowledge of plants was seemingly not profound. And yet, in speaking of *malua domestica*, probably our common hollyhock (*Althea rosea* L.), he states that they are found *in ortis*, "in gardens" (188 rb). More will be said about this passage below. Here, it suffices

[12] Although later in time, Bacon was probably following an earlier source when he observed that *castoreum*, musk, and the seeds of rue and *agnus castus* have the same properties and are useful for the same disorders. *Sylva sylvarum*, Section 966, in *The Works of Francis Bacon* . . . , ed. Basil Montagu (3 vols.) (Philadelphia: A. Hart, 1852), Vol. II, p. 133.

to point out that the above passage, plus the occasional remarks about apothecaries, and one reference to *magistros*, which I take to mean "physicians," leaves little doubt that our glossary was intended for the very practical purpose of identifying medicinal plants and describing their virtues. There is no trace, one may note, of the magic, charms, and superstitions common enough in contemporary English medical writings.

Whether or not the glosses are based on empirical data gathered at a remote place and time, there are several reasonably accurate descriptions of plants. These descriptions and the synonyms associated with them permit, in most cases, the identification, at least on a generic level, of the plant in question. Except in rare instances, the descriptions are based on gross morphological characteristics such as the leaves or the blossoms. The almost infinite variety in the shape, color, and texture of leaves has resulted in their assuming an important, if erroneous, role in pre-Linnaean taxonomies.

Attention to leaves as a distinguishing characteristic for the identity of the plant is evident in our glossary. A good example of the graphic abilities of early plant description is provided by the gloss on *cimbalaria. Id est pennywort*, the gloss begins, and then the plant is described as having "the stem in the middle of the leaf. The leaf is round like a denarius" (185 va). One might expect, in virtue of the vernacular synonym, "like a penny," yet the tradition of the glossary dictated that Latin must remain, despite the easy adaptation to the vernacular. We should all be grateful that this little plant (*Umbilicus rupestris* Salisb.) has come to be known, even today, as pennywort rather than "denarius wort."

Another example is interesting for several reasons in addition to indicating the limitations of botanical description inherent in a glossary. The plant called *cruciata* is first connected with two synonyms by means of an *idem*. Then follows the description, "It has an erect stem. The leaves are arranged step wise <and> in all directions. Above it has yellow flowers in the form of a cross. In French it is called croyse" (186 ra). First, as to the identity of the plant: the three references to a cross immediately suggest either that the floral structure is 4-merous, petals and sepals being arranged quadrangularly, or that the leaves are arranged in a similar fashion. Actually both are the case, for there is little difficulty in identifying the plant as *Galium cruciata* (L.) Scop., known in John Gerarde's day as in ours as crosswort, and to Henry Lyte (1529–1607) as golden crossewort, in reference to its yellow flowers.[13] But interest also attaches to this gloss for quite a different reason. We have noted above that, although our glossary frequently follows the Sloane and Selden glossaries, there are good reasons for believing that another English source was also used.

Of the synonyms listed for *cruciata* in our glossary, one of them, *spargula*

13 "It is called *Cruciata*, and *Cruciatis*, of the placing of the leaves in manner of a Crosse: in English, Crosse-woort," John Gerarde, *The Herball or Generall Historie of Plantes, very much enlarged and amended by Thomas Johnson* (London: A. Islip, J. Norton and R. Whitaker, 1633), pp. 1123–1124. Lyte also mentions that in French it is called *Croysée* and "in base Almaigne, Crusette," Rembert Dodoens, *A Niewe Herball or Histoire of plantes . . . nowe first translated . . . by Henry Lyte* (London: Gerard Dewes, 1578), p. 541.

maior, is also found in the Selden glossary. But that entry reads as follows: " Spergula maior, herba cruciata idem. * * * gradatim et stipitem erectum. gallice * * * " These words cannot satisfactorily be put into English for the reason that the phrase *gradatim et stipitem erectum* is a phrase containing an adverb and a noun plus its modifier in the accusative case. There is no verb, and the plant name following *gallice* – that is, " in French " – is lacking. In other words, there are two lacunae in the Selden glossary while the plant is not entered in the Sloane glossary. Thus, we have here certain evidence that neither one of these two glossaries was the source for the entry. The fact that our gloss supplies the words which have dropped out of the Selden glossary proves that ours is independent of the latter and is thus a copy of the original which was also followed by the writer of the Selden glossary.

This may be the place to mention, before we proceed to the next description, that this particular entry provides one of the clues for the conjectural third source. There are four Middle English manuscripts, all from the fifteenth century, though one may be as early as 1400, which contain variations of the passage in question.[14] And these manuscripts, be it noted, are not glossaries but herbals. An example will be given below from one of these manuscripts in order to show how closely they resemble the Latin glossaries and may thereby be used to supplement our readings.

In addition to the shape of the leaves or their arrangement on the stem, attention was sometimes directed to the divisions of the leaf. Sometimes these were simple leaves with pronounced indentations while in other cases they were true compound leaves, each of the so-called parts actually representing what we would call a leaflet. I chose an example of the former class because it also brings out another interesting matter concerning the transmission of botanical knowledge via the glossary.

To the rustic or to those who looked for edification from a world endowed with moral reminders, the shape of the leaves often suggested animal, bird, or human forms. From these resemblances, usually quite fanciful, moral allegories were often developed. As an example of human form, *palma christi* provides a specimen of how a plant name may determine anthropomorphic descriptions. Christ's hand, as *palma christi* may be rendered, "⟨is⟩ like archangelica, but it has larger leaves and much more divided as if [quasi] they were five fingers. It has a square stem, a little on the dark side. In English it is called moderworte " (189 rb). Despite theological overtones, the descriptive part of the gloss is really rather good. *Leonurus cardiaca* L., as a member of the mint family, possesses the typical square stem. The stem often is a reddish violet and the leaves are digitately five lobed and five nerved at the base – what a botanist would call a palmate-shaped leaf. The deep indentations obviously have become the fingers. Finally this plant today is known as *motherwort*, a name which readily

[14] Cf. Gosta Brodin, *Agnus Castus* (Upsala: Lundeqvist, 1950). Pages 85–102 contain a description of the manuscripts on which the corresponding passage, s. v. *Herba cruriatica* (sic!) (p. 161), is based.

derives from the Middle English *moderworte* by the simple substitution of an unvoiced " th " for the voiced " th " common in early English. It is, perhaps, amusing to note that an eminent botanist, ignoring the straightforward description in favor of something more romantic, transferred the digitate leaf to a palmate root and come out with castor bean! [15]

Equally interesting is the fact that the word translated above as " divided " (*fissa* in Latin) has been misread by the copyist as *spissa.* Mistakes between the old form of upright *S* and lower case *f* are of course easily made, but *spissa,* a good botanical adjective which appears elsewhere in our glossary, meaning " dense," " compact," " thick," etc., makes no sense here.[16] There is good authority for my emendation of *spissa* to *fissa* in this gloss. The Selden glossary, which reads almost word for word with ours, has *fissa.* Moreover, one of the Middle English herbals mentioned above has the following passage: " an herbe that me cleputh mannes moderwort or palme de dieu or stedefast. this herbe hat leues lich to blyndenetele. bot thes buth more grettur and more endented and he hath a stalk that is foursquare. . . ." [17]

The replacement of *fissa* or *endented* by the meaningless *spissa* is clearly due to scribal carelessness. Although this is a comparatively harmless example, there are other cases whereby misreading the copy in front of him the scribe has created new names and new plants. The frequency with which errors of this sort abound in our glossary illustrates the dangers of relying upon a single glossary for botanical information and a single manuscript for purposes of editing such material. But the presence of these errors, on the other hand, enhances the value of our glossary in precisely those places where it has preserved the better or more difficult reading.

Unusual markings on the leaves, such as spots or blotches of one sort or another, are often mentioned in medieval botanical writings owing to the widespread belief in the doctrine of signatures. According to this belief, a plant indicates by its mere external appearance the disease or portion of the body for which it was intended and for which it was, for that reason, of special therapeutic value.[18] There is no trace of this doctrine in our glossary except in the sense of, perhaps unknowingly, preserving the names and descriptions of these plants previously associated with this belief.

Such a plant was *saturion,* one of several closely related species of the genus *Orchis,* the most common of British orchids. This plant, so runs the gloss: " has leaves with small blotches and nodes [lit. *testiculos*] on the root. Its virtue is that of drawing from distant parts " (190 va). Our gloss stops abruptly at this point. If one consults the *Circa instans,* almost certainly the source for this passage, it will be found that there is one further clause, to wit, " hence it aids coitus." [19] I leave it to others to determine

15 G. Henslow, *Medical Works of the Fourteenth Century* (London: Chapman and Hall, 1899), p. 233. As an example of the transformation of the palmate root into the likeness of a human hand, cf. Tullia Gasparrini Leporace, *Un Inedito Erbario Farmaceutico Medioevale* (Florence: Olschki, 1952), p. 51.

16 For some of the medieval uses of *spissus* as a botanical term, cf. T. A. Sprague, " Botanical Terms in Albertus Magnus," *Kew Bulletin,* 1933, *9*: 440–459 (Ref. p. 456).

17 Brodin, *op. cit.*, p. 197.

18 Cf. Henri Leclerc, " La Médecine des signatures magiques," *Janus,* 1918, *23*: 5–28.

19 The phrase, *virtutem habet atrahendi a remotis partibus, unde et coitum adiuvat* occurs

whether the omission was due to carelessness or to prudery.

A peculiarity of the leaf finally is recorded in a gloss which is of special interest in that it preserves an unrecorded plant name. The gloss concerns the plant called *centrum foramina* (185 va), literally " a hundred openings " or " a hundred passages." Although the gloss contains no description of the plant, its identity is assured by the four synonyms. Two of them, *ypericon* and *herba perforata* are so common in medieval writings that there is no difficulty is assigning the name *centrum foramina* to *Hypericum perforatum* L. This plant today is commonly known as St. John's Wort, thus agreeing with another of the synonyms, *herba johannis.* The leaves of St. John's Wort have been described recently by an English botanist as follows: " when held up to the light they show a number of small transparent dots." [20] These dots are, evidently, the hundred openings alluded to in its very name.

Neither leaf structure nor the color of the flower alone is a satisfactory means of identification. Both, however, are important as field marks and, as we have noted in connection with *palma christi,* they must not be ignored if a satisfactory identification is to be attempted. Since the color of the blossom will be noted in several instances discussed below, I shall not take up the subject separately at this time. Instead, I shall turn to other forms of describing plants found in our glossary and note how these also aid in the identification.

The plant called *sparagus* is described with commendable brevity, " it has a spiny fruit " (190 rb). Despite the superficial resemblance between this word and *asparagus,* they designate two quite unrelated plants. That which is referred to here is *Ulex Europaeus* L., commonly known as gorse or furze. Its spiny foilage is much better known that the property to which it probably owes its name *sparagus.* On a hot summer day, a mature seed pod may burst spontaneously. This is accompanied by an audible crack or pop and the seed is ejected, perhaps one-half of a meter. It is to this characteristic that its etymology may be traced back to the Greek σπαργάω, " to burst," and traces of the old etymology may be found in writers who have observed this phenomenon.[21]

The use of analogy in plant description is as old as Theophrastus and is a particularly common device in nonbotanical writings when mention of a plant is required, but where science can be, momentarily at least, ignored or subordinated to poetry. We have already noticed one example of this technique, that in which the round leaves of *cimbalaria* were likened to a

in the 1497 edition on the authority of Lynn Thorndike, *The Herbal of Rufinus* (Chicago: University of Chicago Press, 1949), p. 281. I have not been able to see this edition. A similar meaning, but not the exact wording, occurs in the 1537 edition, fol. 94r s.v. satirion: *virtutē potētissimā h⟨abet⟩ augmētādi sperma, e⟨t⟩ purgat extrema, et ad incitandū libidinē . . . ⟨et⟩ promovet libidinem.* In the *Alphita,* ed. Mowat, p. 158, s.v. *satirion* has a still different reading, but a similar sense: *radix potata venereos actus provocat* and glosses with the AS *stondenegousse.*

20 John Hutchinson, *British Wild Flowers* (2 vols.) (Harmondsworth: Penguin Books, 1955), Vol. I. p. 183. For the other names, and some beliefs connected with the plant, cf. Heinrich Marzell, " Das Johanniskraut," *Natur,* 1918–1919, *10*: 138–140.

21 Cf. Henslow, *op. cit.*, p. 257; Hutchinson, *op. cit.*, Vol. I, p. 47.

denarius. A closely related method is to describe a plant on the basis of another plant which, presumably, requires no description. From the several instances of this, we choose the description of *asarabaccara*. The following description provides a perfect illustration of how, if the plants to which a reference is made are themselves known, the plant in question may be identified readily: " It is a mountain plant whose leaves are angular like the leaves of ivy and whose flowers are like those of henbane " (184 ra). A further sentence about the odor of the root does not concern us, for on the basis of the above, the plant can be identified as hazelwort (*Asarum Europaeum* L.). It may be worth noting that often in late Latin writings and in some medieval glossaries, a second, quite distinct plant, is denoted by *asarabaccara*, viz., water-avens (*Geum rivale* L.).[22] That the present entry refers to hazelwort and not water-avens depends precisely on its likeness to ivy and henbane. Our plant, it is said, has flowers like henbane.[23] Since henbane is *Hyoscyamus niger* L., well known because of its poisonous seeds, it can easily be observed that it has a bell-shaped blossom. And so too, do both of the candidates for *asarabaccara*. But when one turns to the second portion of the description, leaves like ivy, we can quickly eliminate water-avens, whose leaves bear no resemblance to ivy. The leaves of hazelwort, on the other hand, are so like ivy that even John Ray observed, " folia habet rotunda . . . ut Hederae." [24]

Sometimes, in addition to a morphological description, the gloss contains information regarding the plant's habitat. *Nenufer*, undoubtedly the white water lily (*Nymphaea alba* L.) still quite common in pools and backwaters over much of England, is described with terse simplicity: " it has broad leaves found in aquatic places . . . its white flowers are used in medicine . . ." (188 vb). Another water plant, by the name of *lentigo*, is described as follows: " it swims in the water, food for ducks. In English it is called *enedegras* " (187 vb). Were we to know nothing more of this plant than what is preserved in this gloss, it would still not be difficult to identify it as a species of *Lemna*. *L. minor* L. or duckweed, the most common species of this genus, is nothing more than an adaptation of *enedegras* <OF *anet* <Lat *anas*.[25] As to the plant swimming on the water, this is not as irresponsible a guess as may at first sight appear. Duckweed, which usually accounts for the green scum covering quiet ponds, is a true aquatic plant. It lacks roots,

22 Both *Geum rivale* L. and *G. urbanum* L. were denoted by *avancia* or *avencia*. Cf. Hermann Fischer, *Mittelalterliche Pflanzenkunde* (Munich: Akademie-Verlag, 1929), p. 270. The word *avens* is derived from the Anglo-Norman *avense* which is listed as one of the synonyms for *avencia* in Mowat, *Alphita*, p. 17.

23 The similarity between these two plants was noted by Ibn Baithar (d. 1248) who probably saw neither one; cf. J. von Sontheimer (trans.), *Grosse Zusammenstellung über die Krafte der bekannten einfachen Heil- und Nachrungsmittel von . . . Ebn Baithar. Aus dem Arabischen* . . . (2 vols.) (Stuttgart: Hallberger, 1840–1842), Vol. I, p. 31, s. v. *asārūn*.

24 John Ray, *Methodus plantarum* (London: H. Faithorne and J. Kersey, 1682), p. 67.

25 The word *endmete* (*lit.* duckmeat) conceals the same etymology and refers to the same, but mistaken, belief that ducks eat the plant. Actually, they strain it through their bills in search of small aquatic animals which live among the often thick growth. For *endmete* and other English vernacular names, cf. James Britten and Robert Holland, *A Dictionary of English Plant-Names* (London: Trübner, 1878), s. v. *Lemna*.

distinct stems and leaves, and, as a microscopic examination will reveal, it possesses a reduced and primitive flower. This insignificant plant with its primitive flower occupies an important, not to say unique, place in the history of science, for when Anton van Leeuwenhoek placed some of the green scum from his favorite canal under his lens, it was the first of the higher plants *to be so diagnosed* due to a microscopic examination.[26]

One further gloss of ecological interest concerns *crassula minor*, one of several species of the genus *Sedum*, a small succulent often used as a ground cover. Due to its fondness for a rocky habitat, several species are known collectively by the common name, stonecrop. This term is probably connected with *stonehore*, the synonym listed in our glossary for *crassula minor*. The brief description, " it grows on a rock," *super lapidem crescit* (185 vb), is not nearly so interesting as the fact that in the Selden glossary the preposition *super* has dropped out. And without it, even to ears attuned to vulgar Latin, the grammar is intolerable. Since our glossary has preserved the preposition, I feel confident in reasserting its independence from its better-known relative.

Early in this paper, it was stated that glossaries, by the addition of vernacular synonyms, provide evidence that an attempt was made to adjust the Latin names, originally referring to a Mediterranean flora, to an indigenous English flora. *Eviscus*, a generic term in our glossary, is followed by two synonyms, *altea* and *bismalua*, related to the former by an *idem*. The gloss continues: " It is called wymauve in French and seehocke in English. It has narrower leaves than the other malua and its grows higher. It grows next to the sea " (186 vb). Except for the vernacular synonyms, this passage resembles the *Circa instans* and the Selden glossary.[27] Unlike the latter, however, which lists St. Cuthbertscole as the English synonym, our glossary has *seehocke*. The plant in question is *Althea officinalis* L., better known as marshmallow. By distinguishing the present plant from " the other malua," mentioned above as hollyhock and glossed separately s. v. *malua domestica*, our glossary has apparently tried to adjust the Latin names to a local flora. Marshmallow is indigenous to England and flourishes in maritime marshes. Hollyhock, on the other hand, which was said above to grow in gardens, is an import from the Mediterranean basin where it is indigenous.

Another distinction, hinted at in this gloss, is that between a genus and a species. *Eviscus*, first of all, represents our generic term *Althea*. This

[26] Hooke, of course, had priority in examining the microscopic structure of plants. But in the *Micrographia*, his investigations were either on cryptogams, e. g., the blight on rose leaves, or on plants whose taxonomical status was already known, e. g., the seed of an oat. Robert Hooke, *Micrographia* . . . , facsimile edition (New York: Dover, 1961), pp. 121–156. The taxonomic status of crytogamous plants was not advanced by Hooke because of his reliance on the old doctrine of zoophytes " which are of a middle nature " (sc. between plants and animals) (*ibid.*, p. 124). On the other hand, Leeuwenhoek begins by examining the truth of what " the common people say " that duckweed " is generated in the ground underneath." Upon examination, " I always found that one of these plants is produced by another, as with trees and other vegetables." Letter 147, translated by Clifford Dobell, *Antony van Leeuwenhoek and His ' Little Animals '* (New York: Russell and Russell, 1958), p. 276.

[27] *Circa instans*, fol. 86v, s. v. *malua*; Mowat, *Alphita*, p. 110, s. v. *malua*.

genus covers two species known to the glossator: "the other malua," or hollyhock, which botanists have designated *Althea rosea* L. and, as a second species, that which botanists have termed *A. officinalis* L., or marshmallow, and which may be called variously *altea* or *bismalua*. The relation between these two plants is obvious when one examines the seed capsules and flowers, though in other respects they do not bear a particularly close resemblance to one another.

The possibility of error always exists in glossaries, especially where they are copies of copies. Most of what appear to be errors in our glossary can probably be explained as a result of careless copying. Nevertheless, there remains one domain where the glossator's knowledge was demonstrably weak. This concerns plant geography, and it is especially evident when he reports the localities whence the drugs came. Following classical precedent, the unknown East is compressed into a vague geographical entity known as India. Thus, we find *nux muscata* or *nutmyggis* (*Myristica fragrans* Houtt.) described as "fruit of a tree in India" (188 vb), or *mirra* (from *Balsamondendron* spp.) described as "gum of a tree in India" (188 rb), or *costus* (*Saussurea lappa* Clarke) "root of a tree in India" (185 vb). About all that can be said for these descriptions is that nutmegs are, botanically speaking, the seeds of a fruit whose pericarp supplies mace, that the myrrh of commerce is a gummy exudation, though from North Africa and the Levant, and that *costus* is a root, though hardly from a tree.

Pharamacognosy long ago abandoned the belief that by examining the etymology of the name of a drug, something of value might be learned concerning its medicinal uses. This custom was almost universal in ancient times and was continued far into the Middle Ages. Thus, Macer Floridus (Odo of Meung), writing in the twelfth century, fancifully derived the name "malva" from the phrase *molliat alvum*, "it mollifies the stomach." [28] There are no such vivid examples as this in our glossary, yet the belief continues. Compare, for example, the following gloss on *staphisagria*: "it is also called 'head purge' because it purges the head, and herba pedicularis . . . its seeds are especially effective" (190 va). This plant, known in Shakespeare's day as stavesacre, is *Delphinium staphisagria* L., a closer relative of our common larkspur than to the hybrid delphiniums of the hothouses.[29] A plant known as *herba pedicularis*, literally lousy herb and later actually called lousewort, would certainly, on no other evidence than this, be considered useful in the days when keeping lice down to a tolerable minimum was a daily chore.[30]

If we believe that an etymology adds little evidence for a drug's specific

28 Macer Floridus, *De viribus herbarum*, ed. Ludwig Choulant (Leipzig: Voss, 1832), verse 1962; *Dixerunt malvam veteres quod molliat alvum*.

29 "Staves-acre! that's good to kill vermin. Then, belike, if I serve you, I shall be lousy," Christopher Marlowe, *The Tragical History of Doctor Faustus*, Act I, Scene 4, ed. Havelock Ellis (New York: Hill and Wang, 1956). Cf. also Dodoens, *op. cit.*, pp. 371–372.

30 Cf. Gerarde, *op. cit.*, p. 495. Britten and Holland, *op. cit.*, p. 315, s. v. Lousewort, also mention a Lycebane. In the first century A. D., the name *herba pedicularis* was explained *quod pediculos nectat*, Scribonius Largus, *Compositiones VIII*, ed. Georg Helmreich (Leipzig: Teubner, 1887), p. 8.

nature, so too do we feel that a line of verse, howsoever charming or skillfully wrought, will tell us little regarding the nature of a plant. But to the medieval mind, nourished by authority and comforted with verses cited by heart, there was nothing unusual in the following description given for *camomila* (186 ra): "*fetet amarusca, redolet similis camomila* (amarusca stinks and camomille too)."[31] While it is perfectly true that *amarusca* (*Anthemis cotula* L.) has an unpleasant odor, and that the closely related camomille (*A. camomilla* L.) has an only slightly less pronounced odor, these are not sufficient grounds for distinguishing between the two, let alone of identifying them among a host of other plants with equally pungent foliage.

Up to this point our discussion has centered on the higher plants in contradistinction to the so-called lower plants or cryptogams, which include ferns, mosses, liverworts, and the like. The word cryptogam, literally "a hidden marriage," refers to the absence of reproductive organs of the floral type considered to be the norm in early anthological theories. Theophrastus queried this matter, and many systems were, in later times, devised in a more or less desperate attempt to account for what happened to be a botanical anomaly.[32] We must not suppose that our glossator was troubled about such taxonomical niceties since, in the five references to cryptogams, they are treated as if they were no different than other plants. *Polypodium*, or as it is still called in some quarters, polypody (*Polypodium vulgare* L.), is glossed as follows: "it is also called oaken fern. It has red spots under the leaves. That is best which grows upon oaks. In English it is called polypode" (189 vb). The accuracy of this description is improved only by our technical terminology, in which case the modern botanist would translate as "golden brown sori on the undersides of the fronds."

There are, finally, two further types of botanical descriptions which refer to the medicinal properties of the plants. The first concerns the milky sap, common to many genera and even to some families. The presence of milky sap is useful even today as a clue to the field identification of a plant suspected of being a member of a very large and diverse family — the Euphorbiaceae or spurge family. Oddly, there is no reference to the milky sap of the one certain member of this family mentioned in our glossary. Rather, in describing *catapucia* (185 rb) reference is made to its seeds. These, as we know from many sources, were taken internally in the form of a pill as a drastic purge. In fact, the Latin *catapucia* is no other than a corrupt form of *καταπότια*, "pill." But the milky sap of other plants is often mentioned, including the dandelion or *dens leonis* (*Taraxacum officinale* Web.) (186 va), which was probably not as common then as it is today.

31 The same verse occurs twice in Mowat's *Alphita*, p. 8, s. v. *amaracus* and p. 28, s. v. *camamilla*. I have not traced the origin of the verse for its does not occur in the *Circa instans* or in the Salernitan *Alphita*. Macer Floridus, *ed. cit.*, vv. 551–555 has a similar sentiment and, as Choulant notes *ad loc.*, in one manuscript the present verse occurs following v. 555.

32 E. g., Ernst Hallier, *Die pflanzlichen Parasiten des menschlichen Körpers* (Leipzig: Engelmann, 1866), believed that fungi developed from bacteria and that both were degenerate forms of higher plant life transformed by differing environments and nutritive media.

Another plant, *celidonia*, whose name has become modified as celandine (*Chelidonium maius* L.), is a member of the poppy family whose most famous lactiferous member is the opium poppy. A recent manual has described the thick fleshy rootstalk of the celandine as " full of yellow juice," a description no more or no less exact than that of our glossary, " it emits a yellow sap when broken " (185 rb).[33]

Almost all portions of the plant have been used, at one time or another, for medicinal purposes. Sometimes it makes little difference whether the leaves or the fruit are used. But there are cases where chemical tests reveal the presence of an active principle in only one portion of a plant. Of *coriandrum*, our coriander (*Coriandrum sativum* L.), the glossary states, " the seed is used " (185 vb) probably in reference to the fact that the foliage when crushed emits an offensive odor while the seed is quite free from any odor. A more striking case, and one that is fully justified by clinical observation, concerns the drug known then, as it was up to 1937 when it was still listed in the U. S. Pharmacopoea, as *nux vomica*. (The drug as it is known in commerce is the dried and cleaned seed of the *Strychnos nux vomica* L.) It must be carefully freed from the pulp, for the latter contains strychnine, often in sufficient concentration to cause death. The seeds, however, are free from strychnine and are useful as a purge and as a nervine. The bark, on the other hand, is exceedingly bitter, though it possesses neither the strychnine of the pulp nor the purgative powers of the seed.[34] As if with reference to these matters, our glossary states, " the interior is used, not the cortex " (188 vb). I think there is little doubt that this refers to the seed while " cortex " refers not to the bark but to the pulp around the seed — a possible though not common meaning of " cortex." Of the few early samples of the seeds which I have personally examined, fragments of pulp can be detected adhering to the seed. With even lower standards of purity in the fifteenth century, the quantity of pulp may have been sufficiently large to produce toxic results. If so, our glossary does well to warn the prospective user and to throw out a broad hint to the apothecary concerning the way to package his wares.

33 Hutchinson, *op. cit.*, Vol. I, p. 256.

34 F. A. Flückiger and D. Hanbury, *Pharmacographia* (London: Macmillan, 1874), pp. 384–387.

INDEX

For Product Safety Concerns and Information please contact our EU representative GPSR@taylorandfrancis.com Taylor & Francis Verlag GmbH, Kaufingerstraße 24, 80331 München, Germany

Batch number: 10725917

Printed by Printforce, the Netherlands